Advances in Tropical Crop Protection

Mui-Yun Wong
Editor

Advances in Tropical Crop Protection

Springer

Editor
Mui-Yun Wong
Department of Plant Protection
Faculty of Agriculture
Universiti Putra Malaysia
Serdang, Selangor, Malaysia

Institute of Plantation Studies
Universiti Putra Malaysia
Serdang, Selangor, Malaysia

ISBN 978-3-031-59267-6 ISBN 978-3-031-59268-3 (eBook)
https://doi.org/10.1007/978-3-031-59268-3

This Springer imprint is published by the registered company Springer Nature Switzerland AG
The registered company address is: Gewerbestrasse 11, 6330 Cham, Switzerland

Preface

The United Nations (UN) has proclaimed the year 2020 as the International Year of Plant Health with a slogan "Protecting Plants, Protecting Life". This shows the utmost importance of healthy plants to humans and planet Earth. This proclamation is expected to increase awareness about the importance of healthy plants in achieving the Sustainable Development Goals (SDGs) where they have a direct impact on five SDGs: SDG1 end hunger, SDG2 reduce poverty, SDG8 boost economic development, SDG13 mitigate climate change, and SDG15 protect the environment. Besides, healthy plants ensure food security for the growing global populations which is expected to reach 9 billion in 2050.

Modern crop production systems utilize intensification model involving intensive use of farm input such as water, fertilizers, and pesticides. A major component of crop production is pest and disease management. FAO estimates that up to 40% of food crops are lost due to plant pests and diseases annually. To reduce yield losses caused by pests and diseases, synthetic pesticides are commonly used, often injudiciously, due to their immediate and indiscriminate effects, relatively low cost, and easy availability in the market. Global use of pesticide went up 62% between 2000 and 2021 (FAO, 2023). Extensive use of synthetic pesticides causes hazard on living organisms as well as pollution to underground water, degradation of good soil, reduction in biodiversity, and pest resistance—issues of grave concern among many stakeholders.

While pesticide use is necessary to increase yield to feed a growing population, it is considered the last option in the integrated pest management approach. Non-chemical strategies for crop protection that are safe for living organisms and the environment are available. Conservation of natural enemies and high species diversity in the agroecosystems often reduce the risk of pests and diseases in general. This book entitled *Advances in Tropical Crop Protection* focuses on innovative nature-based solutions for sustainable crop protection in plantation and food crops.

There are 16 chapters in this book and organized into three parts. First part describes research advances related to tropical insect pest management including bagworm and red palm weevil bio-based management; sustainable management of thrips infesting vegetables; biology and potential application of baculoviruses as biopesticides; nature-based substances for enhanced biopesticide efficacy against insect pests; as well as biology and integrated pest management (IPM) of pink pineapple mealybug. Second part covers advances in plant disease management in the tropics including description of a newly emerging rubber disease; descriptions of common diseases in pineapple and their management; microbiome of banana and its manipulation for management of Fusarium wilt disease; utilization of plant-based product in post-harvest disease management of fruits; updates on infection and control strategies of basal stem rot disease of oil palm in Malaysia; revolution of the next-generation sequencing and its application in phytobacterial diseases; and lastly utilization of plant and microbial diversity in disease management. Third part embarks on research advances related to management of other tropical pests including vertebrate pest management; sustainable rat management model in rice field; threats and management of invasive apple snails; and development of bioherbicides for sustainable weed management.

With stand-alone chapters written in an accessible language, the book is recommended for research scientists in academia and industry, readers particularly interested in crop protection, and those working within the broader area of plant biosecurity.

To keep plants healthy and feed the increasing world population, nature-based crop protection for sustainable agriculture is the way forward. The solutions that nature has developed provide us the inspiration for further innovations in crop protection.

Serdang, Malaysia Mui-Yun Wong

Acknowledgements

The editor would like to sincerely thank all contributing authors for accepting this invitation and making this book a reality. Their continuous cooperation and commitment in this project are much appreciated. I am also thankful to Associate Professor Dr. Norida Mazlan, the Head of Department of Plant Protection, Faculty of Agriculture, Universiti Putra Malaysia, for her support in the publication of this book. Furthermore, I am extremely grateful to Dr. Go Wen Ze for her professional and technical assistance in the process of preparing the book manuscript. Last but not least, I appreciate the opportunity given by Springer Publisher and its team for their support and valuable advice in the production of this book.

Contents

Advances in Tropical Insect Pest Management

Bagworm and Red Palm Weevil—Major Pest and Potential Threat to the Malaysian Oil Palm Industry

Muhamad Haziq Hadif Zulkefli, Syari Jamian, Nur Azura Adam, Sumaiyah Abdullah, Siti Izera Ismail, and Mohamed Mazmira Mohd Masri

Abstract *Elaeis gunineensis* is the golden crop of Malaysia that contributes to the economic export of the country and the planters' livelihood. Despite control measures to overcome the issue, bagworm remains the most reported infesting oil palm cultivation in the last four decades. Furthermore, the coconut areas in oil palm plantations documented an invasive alien species infestation, the *Rhynchophorus ferrugineus* or red palm weevil (RPW). The widespread devastation the pests caused on palm cultivation worldwide has served as warning to the industry, primarily in Malaysia, where palm oil is the main commodity. The present study highlights the current report on the biology, control, and technological advancements in bagworm management. A recent report on the RPW concerning its biology, detection, the scenario in oil palm plantations, and control methods investigated in past studies was also discussed. This study could offer new insights on the pest and its potential threats to oil palm cultivation in Malaysia.

Keywords Oil palm · Bagworm · Red palm weevil

M. H. H. Zulkefli · S. Jamian (✉) · N. A. Adam · S. Abdullah · S. I. Ismail
Department of Plant Protection, Faculty of Agriculture, Universiti Putra Malaysia, 43400 Serdang, Selangor, Malaysia
e-mail: syari@upm.edu.my

S. Jamian · S. I. Ismail
Laboratory of Climate-Smart Food Crop Production, Institute of Tropical Agriculture and Food Security (ITAFoS), Universiti Putra Malaysia, Serdang Malaysia, 43400 Serdang, Selangor, Malaysia

M. M. Mohd Masri
Malaysian Palm Oil Board, 6 Persiaran Institusi, Bandar Baru Bangi, 43000 Kajang, Selangor, Malaysia

M. Wong (ed.), *Advances in Tropical Crop Protection*,
https://doi.org/10.1007/978-3-031-59268-3_1

1 Introduction

Oil palm is a crucial commodity in Southeast Asian countries, especially Malaysia and Indonesia. In 2021, the Malaysian Palm Oil Board (MPOB) estimated the area of palm oil cultivation in Malaysia at approximately 5.8 million hectares (ha) (MPOB, 2021), comprising almost 90% mature, while the remaining percentage was immature palms. The cultivated land was segregated between plantations, around 70%, and smallholders (MPOB, 2021).

Parveez et al. (2021) discovered that reduced cultivation area, low fresh fruit bunch (FFB) yield, and decreased palm oil and palm oil-based product export in 2020 did not impact palm oil export value. Conversely, an approximate increase of 9% (RM 73.25 million) was documented compared to the past year. In 2020, palm oil products were the fourth largest export, accounting for 4.7% of Malaysian exports (Malaysia External Trade Development Corporation, 2021).

In oil palm cultivation, pests threaten the potential FFB yield. Bagworm is the most significant pest plaguing the industry (Moslim et al., 2019). The pest feeds on defoliating fronds. In severe cases, the fronds appear burned. At a 10–50% infestation rate, the insect has impacted oil palm yield by 43% (Basri & Kevan, 1995).

Bagworms were first documented in the late 1950s. The pest is commonly found in Peninsular Malaysia (Wood & Kamarudin, 2019) in three species: *Metisa plana* (*M. plana*), *Pteroma pendula* (*P. pendula*), and *Mahasena corbetti* (*M. corbetti*) (Mazmira et al., 2011; Moslim et al., 2019). Bagworm management is crucial to ensure healthy and productive oil palms as the life span of the plants could exceed 25 years. Meanwhile, the potential threat of *Rhynchophorus ferrugineus* or the red palm weevil (RPW) has become a new concern in the Malaysian oil palm industry besides bagworm. The pest has been reported to be plaguing coconuts, a subsidiary commodity industry of the country. Globally, RPW has also been established as a significant pest of the palm species (Dembilio & Jaques, 2015). In 2007, the Malaysian Department of Agriculture (DOA) reported RPW issues in all coconut plantations in Terengganu, Malaysia (Wahizatul et al., 2013b). The symptoms of a severe RPW attack are collapsed crowns and instant death of coconut palms (Wahizatul et al., 2013b). As coconut shares the similar Aceraceae family as oil palm (Perera, 2014), RPW threatens the Malaysian oil palm industry, which could devastate the golden crop and planters' livelihood.

This study summarises current reports on bagworm infestation and management methods and discussed integrated pest management (IPM) for bagworms in the field. Moreover, findings on coconut infestation and oil palm plantation surveillance at Malaysian climates were included. Moreover, the study provides a brief review of the current scenario on the pests and possible management approaches to control the pests.

2 The Primary Malaysian Oil Palm Pest

2.1 Occurrence, Lifecycle, and Damage Caused by Bagworms in the Malaysian Climate

Bagworm is the major pest attacking the oil palm industry in Malaysia, which could result in devastating impacts due to ineffective management. *M. plana* Walker, *P. pendula* Joannis, and *M. corbetti* Tams (Lepidoptera: Psychidae) are the three common bagworm species documented in the country. Wood and Kamarudin (2019) reported that the bagworm outbreak in Malaysia began in late 1950 but remained under control from the mid-1960s to 1990.

A survey between 2000 and 2005 recorded 16% of the surveyed area, reported bagworm outbreaks, mainly by *M. plana* and *P. pendula* and a few with *M. corbetti* (Norman & Basri, 2007). Additionally, Moslim et al. (2019) documented a significant bagworm attack in 2013 in the Perak and Johor states of Peninsular Malaysia, while a smallholder reported that 24, 385.9 ha was infested with bagworm in 2014s (Kamarudin et al., 2015).

The life cycle of a bagworm varies according to species. Currently, the bagworm of the *M. plana* species is the most widely reported and documented with the most complete life cycle (see Fig. 1.1) (Basri & Kevan, 1995; Kok et al., 2011). The life cycle of bagworms from egg to adult is 90–103 days. Nonetheless, *P. pendula* could live almost three months (Radaha, 1977; Syed, 1978), while *M. corbetti* has been recorded with a 101–124 day life cycle (unpublished MPOB data; Ramlah et al., 2018).

Wood and Kamarudin (2019) suggested that systemic pesticide applications led to infestation surges considering its impact on the biological control of the pest. Accordingly, the bagworm life cycle has been utilised in the field as an indication for pesticide applications. Pesticides commonly focus on the eating behaviours of larvae as the chemicals could not penetrate bagworm bagging. Consequently, bagworm management is recommended during the early larvae stages when its population exceeds the economic threshold level (ETL) (Moslim et al., 2019) of 5–10 larvae per frond for *M. plana* and *P. pendula* and 1–5 larvae per frond for *M. corbetti* (Chung, 1998; Chung et al., 1995; Syed & Speldewinde, 1974).

Matured palms of over eight years are susceptible to bagworm defoliations (Ramlah et al., 2018). Young bagworm larvae scrape the epidermis layers of the oil palms, whereas the older larvae strip the epidermis and create holes in the oil palm leaflets (Moslim et al., 2019; Ramlah et al., 2018). The symptoms of bagworm damage initiate on the older fronds and continue to the younger fronds (Moslim et al., 2019). Severe damage is indicated by burn-like or brownish areas (Moslim et al., 2019).

Wood et al. (1973) reported that 50% defoliated fronds results in 43% yield loss. In another study, Basri (1993) revealed 44% of yield loss from 10 to 13% leaflet bagworm defoliation. Both studies demonstrated that bagworm defoliation under 20% could impact FFB production to almost 50%, indicating significant effects by

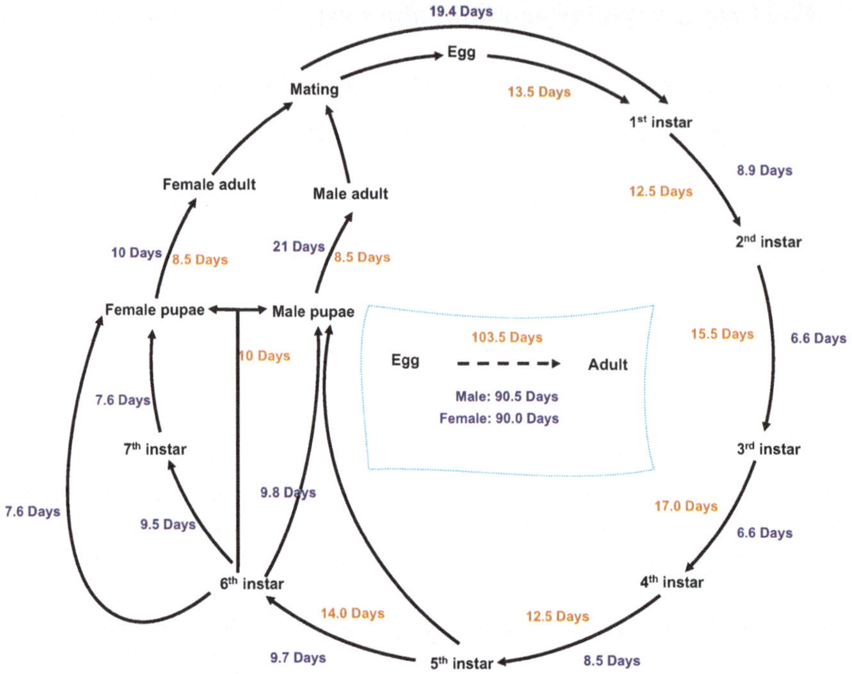

Fig. 1.1 Complete life cycle of *Metisa plana* summary indicated by day for each stage. *Note* The colour of the day denotes reference citation; purple: Basri and Kevan (1995) and orange: Kok et al. (2011)

the pest could have on oil palm cultivation, primarily when not maintained below the ETL. Nonetheless, notable and recent advancements have been reported on bagworm management through integrated pest management (IPM).

2.2 *Latest Technology Advancements and Research on Bagworms in Malaysia*

Artificial intelligence for bagworm census tool is available and among them is the Oto-BaC™ an advanced artificial intelligence (AI) detector manufactured to detect live and dead bagworm larvae of the *M. plana* species during field censuses compared to the manual censuses conducted at the moment (Ahmad et al., 2021). The data evaluated during trials indicated that the detector could detect live bagworm larvae and pupae at 77–88% accuracy, while 75–79% accuracy was recorded for dead larvae and pupae. Conversely, manual census documented 100% accuracy.

The Oto-BaC™ detector documented a shorter census period per frond, approximately 14 min/frond, compared to the manual census, which recorded approximately

17 min/frond (Ahmad et al., 2021). In a similar study by Ahmad et al. (2021), a 30 cm distance and a closed system enabled 100% detection of bagworms with the deep learning algorithm, while the colour processing documented under 60% of detection.

Several factors still require consideration to improve the accuracy of the detector. Ahmad et al. (2021) stated that low detection resulted from bagworm sample size, light, field surface area, and the technology of the detector. Nevertheless, the detector has initiated advancements in bagworm detection in Malaysia, which could aid field censuses.

Live pheromone trapping has been employed to manage adult moths. Several studies utilised live female adult *M. plana* bagworms in pheromone trapping to manage bagworm regeneration in oil palm cultivations (Ahmad et al., 2017; Kamarudin & Arshad, 2006; Kamarudin et al., 2010). Kamarudin and Arshad (2006) employed three trapping models to evaluate the most suitable approach for trapping *M. plana* bagworms. Therefore, the vane trap employing four female bagworm adults as pheromone sources was the best design that captured the highest number of adult male catches per trap. Furthermore, the model reduced subsequent generation population and offspring.

Kamarudin et al. (2010) revealed no significant difference in the number of moths trapped when sticky vane traps on 2- and 4-m wooden poles were employed in areas with taller oil palms. Nevertheless, the report documented diminished second-generation larvae, fewer baggings with eggs, and fewer larvae per frond. The study also recorded minimal frond damage and heavier bunches of the trapping areas (Kamarudin et al., 2010).

Pheromone trapping has been proven effective in reducing bagworm populations to manageable levels. Nevertheless, the approach necessitates correct and timely assessments of female adult receptivity and a prior stage census of the infestation in the treated areas to prevent recurrent larvae in the next generation (Ahmad et al., 2017). According to Kamarudin and Arshad (2006), female *M. plana* adults are receptive to mating for nine days. Consequently, perfect timing is critical to ensure more populations are suppressed from being mated by the male moths to achieve new generation reduction. In another accordance, the incorporation of *Bacillus thuringiensis* (Bt), aerial spraying into pheromone trapping areas resulted in a 94% decrement in the bagworm population, which was higher compared to non-incorporated sections (Ahmad et al., 2017).

Biopesticides offer potential applications in oil palm plantations to manage bagworm populations and reduce the synthetic pesticide impacts on beneficial insects. For example, cypermethrin application has reportedly led to high mortality in pollinating weevil *Elaeidobius kamerunicus* (Najib et al., 2009, 2012; Norhayu & Nurnisa, 2020; Yusdayati & Hamid, 2015).

Three consecutive treatment cycles with Bt or Ecobag-1 (EC) through aerial spray significantly reduced bagworm larvae per frond of the oil palms cultivated in Perak (Kamarudin et al., 2017). The first, second, and third applications diminished the larvae population by 89.9, 92.9, and 88.2%. Moreover, the final assessment on the third cycle of the larvae per frond was, on average, 6.1 lower than the proposed ETL of 10 larvae per frond. Damaged (burned or brownish appearance) oil palm fronds also exhibited improved health (green) after two years of Bt application.

Masri et al. (2021) applied EC via drones to manage bagworms in multiple stages. Two treatment rounds of the biopesticide suppressed the larvae population to 9.3 larvae per frond. Furthermore, no pupae were documented on the final assessment post the second-round application. The first treatment cycle considerably reduced the number of larvae by 96%, while 91.5% was documented in the second-round application. Significant differences were also observed between the control plot, which did not receive the Bt treatment, and the treated areas after two years. The palms sprayed with EC demonstrated improved conditions, from a burned-like appearance to lush green (Masri et al., 2021).

Mazuan et al. (2021) employed different pesticides with mist blowers to control bagworm populations. The report recorded absence of notable variations between the EC and synthetic pesticides, flubendiamide, and cypermethrin, employed. All pesticides in the report recorded the ability to suppress the *M. plana* population up to 45 days post-application. The study revealed the potential of biopesticide to control pest populations to manageable levels below ETL in oil palm plantations.

Predators and parasitoids are natural organisms employed in oil palm cultivations to manage pest populations. According to Cheong et al. (2010), bagworm mortalities associated with predators, parasitoids, and fungal infections were 37%, 35.9%, and 27.2%, respectively. *Colmelestes picticeps, Callimerus arcufer* (Cheong et al., 2010), and *Sycanus dichotomus* (Ramlah et al., 2018) are some examples of bagworm predators. *Colmelestes picticeps* resulted in 60–100% bagworm mortality, while *C. arcufer* led to 23–40% of deaths. Nevertheless, the mortality rates of both predators differed based on the bagworm type (Cheong et al., 2010). Furthermore, a higher number of *C. picticeps* than *S. dichotomus* was reported during bagworm outbreaks (Jamian et al., 2017).

Based on a taxonomy report, 18 parasitoid species are correlated to bagworms (Kamarudin et al., 1996). According to Cheong et al. (2010), the primary parasitoids identified affecting bagworms were *Pediobius imbrues*, *Dolichogenidae matesae*, *P. elasmi*, and *Aulosaphes psychidivorus*, which respectively documented approximately 70, 10, 9, and under 8.5% parasitising the bagworm samples assessed. In another study, the largest number of parasitoids emerging from bagworms were *Cotesia metesae* (51%), *Brachymeria carinata* (21%), *Buysmania oxymora* (13%), and *Goryphus bunoh*, *Pediobius* sp., and *Eupelmus cotoxanthae* (under 9%) (Halim et al., 2017).

Jamian et al. (2017) reported that the abundance of *S. dichotomus* and *C. picticeps* predators increased with enhanced beneficial plant richness and ground vegetation coverage and shorter oil palm stands. Beneficial plants and weeds are crucial for parasitoid and predator sustainability and in one study by Kamarudin and Wahid (2010) documented that parasitoids *Cassia cobanensis* and *Asystasia gangetica* preferred areas with beneficial plants and weeds over grounds covered with other plants, such as weeds and ferns. The beneficial plant *C. cobanensis* exhibited a higher insect predator and parasitoid diversity than covered ground sections in oil palm plantations (Kamarudin & Arshad, 2016).

The lower number of parasitoids in covered grounds was believed to be due to the scarcity and unavailability of food sources. *C. picticeps* and *C. arcufer* documented in the area compared to absolutely no *S. dichotomus* captured on the grounds indicated the dependency of *S. dichotomus* on beneficial plants (Kamarudin & Arshad, 2016). Beneficial plants are one of the natural supports of predator and parasitoid populations in oil palm cultivated areas.

3 Potential Threats to the Malaysian Oil Palm Industry

3.1 The RPW Biology, History, and Scenario in Malaysia

The RPW, or *Rhynchophorus ferruginius* Olivier (Coleoptera: Curculionidae), is a potential threat to Malaysian oil palm cultivation. The pest was reported as the world's most destructive pest (Abbas, 2010), with numerous reports of infestations in the palmae family. In Malaysia, RPW was detected in *Cocos nucifera*, a coconut species cultivated in the country (Azmi et al., 2017; Wahizatul et al., 2013b).

The DOA reported the first RPW report in Malaysia was in 2007, infesting coconut plantations in Terengganu (Wahizatul et al., 2013b). The severity of the spread and damage was documented in 58 locations within seven Terengganu districts, including areas approaching the borders of Kelantan and Pahang. Nevertheless, the infestation increased to 858 locations in 2011 (DOA, 2011), and by 2016, it had spread to Kedah, Kelantan, Terengganu, Perlis, and Pulau Pinang (DOA, 2016). The observations proved the rapidness of RPW infestation (DOA, 2016).

Wahizatul et al., (2013b) hypothesised that RPW was introduced to Malaysia through dates and date palm tree importation from the Middle East. The pest was reported to attack four common coconut cultivars in Malaysia: MATAG, Malayan tall, MAWA, and Pandan (Wahizatul et al., 2013b). The shoots, trunks, and root systems of coconut trees were reported as RPW entry points (Wahizatul et al., 2013b). The RPW attacks straight to the cabbage of coconut palm shoots and drill or utilises holes made by *Oryctes rhinoceros* beetles on the trunks. Adult RPW also digs the soil around the coconut roots before tunnelling into the root systems (Wahizatul et al., 2013b).

Severe RPW damage could lead to crowns falling and instant death to coconut palms (Wahizatul et al., 2013b), thus demonstrating its grave threat to palm cultivation. In oil palm plantations, several studies conducted censuses to determine the presence of the pest. Samplings conducted via trapping in several oil palm cultivation areas, including FELDA, FELCRA, and TDM, documented 40% of RPW among the weevil accumulated (Idris et al., 2014). The study also recorded RPW at the centre of the plot. Although no damage symptoms were documented, the pest still concerns the oil palm industry as it is the primary source of income for the planters in Malaysia. Consequently, utilising advanced sensing technologies, such as acoustic detection (Khudri et al., 2021), requires consideration to confirm the presence of feeding larvae.

3.2 The RPW—A Potential Threat to Oil Palm

Studies have conducted through censuses in oil palm plantations in order to observe the presence and population of RPW. The surveys employed baiting with RPWlure and Ferrolure and acoustic detection monitoring devices. Based on the baiting approach, 40% of the RPW captured in several oil palm cultivation locations (FELDA, FELCRA, and TDM) was of the *R. ferrugenius* species (Idris et al., 2014). Moreover, the pest was documented at the centre of the plots. However, no damage symptoms on the oil palm were observed (Idris et al., 2014).

In another study, DOA collected the larvae, pupa, and adult RPW from the apical shoots of two dying oil palm trees at a smallholder's plantation in Tanah Merah, Kelantan (Khudri et al., 2021). The MPOB randomly installed IoTree smart sensors on 38 trees in the study site. Five suspension noise movements were recorded in the palms after a month of installation, and on the second month of observation only one tree remained infested. A further acoustic detection with an AED 2010 sensor implanted on the stem of the infested palm yielded a negative for RPW. Further inspection recorded three male RPW were found on the frond of a suspected infested palm post chopping of the palm. The findings demonstrated the effectiveness of the IoTree sensor in detecting RPW movement. Based on the results, acoustic detection possessed promising results for early RPW detections. Nonetheless, RPW could still bore the tree without being detected by the sensor. The findings demonstrated the possibility of RPW presence in oil palm trees without resulting in significant observable damages. Consequently, regular monitoring is necessary to ensure that the weevil does not shift host, which could damage the plantation.

A food consumption study by Zulkifli et al. (2018) documented that coconut and oil palm cabbages and sago stems were the preferred nourishment for RPW. Nevertheless, when comparing development periods, oil palm cabbages induced a faster growth at one month and nine days and only six instar stages before the RPW larvae entered the pupal stage. Meanwhile, RPW larvae that consumed coconut cabbages required almost three months, while sago stems recorded almost four months of growth. The RPW larvae that fed on both food sources recorded seven instar stages before pupation (Zulkifli et al., 2018). The data demonstrated that RPW larvae required double the time to develop when consuming coconut cabbages and sago stems than oil palm cabbages. The study also documented a critical finding that associated proteins with RPW larvae RPW diet; the trypsin (cationic trypsin) and aminopeptidase enzymes. It was also noted that protein is essential for the digestive and defence systems of the insect. Accordingly, trypsin and aminopeptidase could be utilised to create biological control agents for RPW and disrupt its feeding behaviour. The report revealed the possibility of RPW larvae surviving and developing faster on oil palm cabbages. Consequently, successful RPW infestation could significantly damage oil palm plantations.

Harith-Fadzilah et al. (2020) artificially infected oil palm seedlings with RPW larvae in an investigation. The larvae required approximately nine weeks to impose serious damage to two-year-old oil palm seedlings, which caused the fronds to collapse. The study divided the level of damage into four stages. The first stage included no visual damage, while the seedlings classified in the second stage exhibited smelly saps, dark ooze, and the presence of sawdust on their trunks. The third stage included seedlings with signs of frond collapse. In the final stage, the trees collapsed and demonstrated brown colouration on their fronds. Despite the severe damage the larvae could impose on seedlings, an in-depth study is still necessary to assess its impact on matured palms.

Photosynthesis rate could be another damage severity indicator of RPW-infested palms, but to set it as a standard, thorough investigations are still required (Harith-Fadzilah et al., 2020). Harith-Fadzilah et al. (2020) suggested that RPW larvae feeding on oil palm seedlings reduced the photosynthesis rate of the plants in the third week of infestation, during which no signs of damage were observed. Nevertheless, the seedlings sustained severe damage by the end of the study. No significant variation in the total chlorophyll, stomatal conductance, and seedling growth was reported between the wounded and control palms. The report suggested a more immediate impact that RPW larvae had on younger oil palm seedlings without visible damage until the infestation reached the severe stage.

The effect of invasive alien species (IAS) on the native plant population could lead to local extinction through reduced host plant population and plant death (Kenis et al., 2009; Pyšek et al., 2020). This phenomenon could indirectly affect local plant communities (Kenis et al., 2009). Invasive insects reportedly displace local insect populations. For example, another *Bemisia tabaci* biotype had displaced the local species by interfering with the mating of the local species (Kenis et al., 2009). Similarly, RPW resulted in the faster spread of IAS within local coconut palm hosts in Malaysia, which was documented in several districts in the country (DOA, 2011, 2016). The IAS has determinant effects on locally cultivated plants and insect populations competing in the same niche. Consequently, monitoring and preventing RPW spread is crucial to curbing invasive organisms.

3.3 Other Studies Related to RPW in Malaysia

Identification is vital to differentiate between the *Rhynchophorus vulneratus* and *R. ferrugineus* species. *R. vulneratus* is native to Malaysia and is a pest of coconut and sago palms. On the other hand, *R. ferrugineus*, or RPW, is an organism of concern. Wahizatul et al., (2013b) documented over 30 size, colour, and pronotal marking variations of adult RPW that attacked coconut plantations in Terengganu, Malaysia. Nevertheless, early morphological evaluations could not distinguish the RPW species from native sago and coconut pest, *R. vulneratus* (Azmi et al., 2017).

In a morphometric identification between RPW males and females, the significantly different characteristics in adult weevils were the rostrum length and the shape of the front tibia and pygidium (Wahizatul et al., 2012). The varied black spot pronotal markings commonly observed on the insect could not be relied on an identification mark. Consequently, other features (molecular analysis) should be considered.

A molecular analysis was conducted to assess genetic variations of *R. ferrugeneus* and the out-group *R. vulneratus* sampled from six districts in Terengganu, Malaysia (Chong et al., 2015). The study, which employed DNA amplification with mitochondrial cytochrome c subunit 1 (CO1) marker, successfully differentiated the species. The results also demonstrated no nucleotide base variations between the samples obtained from the different districts in Terengganu, indicating low genetic diversity. All the evaluated specimens also recorded a single haplotype, which was correlated to El-Mergawy et al.'s (2011) H8 haplotype. The observation suggested that the samples collected in Malaysia originated from the Mediterranean region.

In another report, a combination of cytochrome oxidase subunit I (COI) and cytochrome B (Cytb) data successfully distinguished *R. ferrugineus* from *R. vulneratus*. The time divergence assessment performed during the study recorded an earlier divergence of *R. vulneratus* than *R. ferrugineus* (Aman-Zuki et al., 2021). The findings demonstrated the possibility of *Rhynchophorus* species determination via several molecular analysis methods.

Several reports on pest infestation determination recorded effective RPW luring with pheromone traps. In a five-year study (2007–2012), RPWlure was employed in India and Ferrolure in Costa Rica to trap RPW (Idris et al., 2014). The RPWlure documented 80% of success, while the Ferrolure recorded 96%. Furthermore, 60% of *R. vulneratus* and 40% of *R. ferrugineus* were trapped in oil palm sampling areas. No damage symptoms were observed on the trees infested with the pest.

RPW trapping study conducted in coconut cultivation areas utilising pheromone (ferrugineol) and pineapple baits recorded the highest RPW trapped, followed up by other pheromone and bait combinations (Azmi et al., 2014). Despite the high numbers of RPW caught, pheromone and pineapple or sugarcane bait was costly and required twice weekly change compared to when employing sago palm bait. Daily rainfall also negatively affected the number of RPW trapped. Azmi et al. (2017) suggested that pheromone trapping was effective for RPW. Nonetheless, the approach was labour-intensive, expensive, could induce resistance, and risk environmental harm.

Haris-Hussain et al. (2020) reported that the highest attractant for adult RPW was pineapple, while the lowest was oil palm. The laboratory experiment utilised several baits, either synthetic or natural. The highest responses were documented by female RPW at almost 65% compared to male RPW. In another kairomone-releasing food bait assessment performed in a Terengganu field, Muhammad Firdaus et al. (2020) documented that 12 adult RPW per trap were caught with pineapple bait, which was the highest. Oil palm petiole trap recorded high adult RPW trapped at almost 11 per trap. The oil palm petiole caught more RPW than other baits employed, sugarcane, coconut, and control, demonstrating that RPW preferred oil palm petiole (Firdaus et al., 2020). The study also reported female RPW as the most caught for all baits at almost 76%. The experiment provided new information regarding the possibility of RPW preference for oil palm, hence requiring further monitoring.

Entomopathogenic fungi (EPF) have been studied as a potential RPW management technique in Malaysia. A preliminary investigation by Wahizatul et al., (2013a) isolated 43 fungi from adult RPW, including *Acremonium* sp., *Aspergillus* sp., *Cladosporium* sp., *Curvularia lunata, Geotrichum*sp., *Penicillium citrinum, Trichoderma* sp., and several unidentified species. *Aspergillus flavus* recorded the highest occurrence, approximately 26%, and most of the fungi detected were saprophytic. Potential pathogenic insect fungi were *Aspergillus, Curvularia,* and *Penicillium,* with 15 other fungi requiring further identification for potential EPF application (Wahizatul et al., 2013a).

Lin et al. (2017) reported the isolation of seven strains of fungus, which were named MetGra-1 to 7. The study employed *Tenebrio molitor* as soil bait and was conducted in Terengganu, the epicentre of RPW occurrence. The isolated strains were of the *Metarhizum anisopliae* species. The fungi were also evaluated against adult RPW for pathogenicity. MetGra-4 and MetGra-7 documented pathogenicity against the weevil, resulting in mortality within four and five days post-treatment, respectively. Furthermore, at 19 and 20 days, the total deaths of the inoculated RPW caused by MetGra-4 and MetGra-7, while the other strains recorded 0–60% mortality.

The MetGra-4 was assessed with molecular analysis, identifying the fungus as *M. anisopliae* (Ishak et al., 2020). The fungus spores were mass-produced to enable pathogenicity evaluations against adult RPW. Resultantly, 85% spore viability led to 92% total RPW infection with lethal time (LT_{50}) to kill 50% of the RPW population at almost nine days, while the lowest spore viability, 39%, documented an LT_{50} of 21 days. The MetGra-4 strain also exhibited virulent and infective properties on RPW 21 days after treatment. The report demonstrated the possibility of controlling the RPW population in Malaysia through EPF, where the approach caused high mortality of the weevil. Nonetheless, further research is necessary to mass produce the fungicide for effective delivery to the target organism, especially the RPW larvae commonly deposited inside palm stems.

4 Conclusions

The current bagworm management methods, which are equipped with the latest biopesticides and integrated with machinery, could curb outbreaks to a minimum amplitude below the threshold level. Nevertheless, Malaysian planters are concerned of the emergence of invasive alien RPW species currently affecting the subsidiary coconut industry. The pest could lead to a devastating impact to the oil palm plantations in the country if not monitored. Preventive measures and detection and management steps are necessary before host shift occurrence. Biopesticides may eliminate the damaging pest, and further studies would contribute to the successful pest management of the coconut industry and offer safety to other commodities in Malaysia.

Acknowledgements The authors would like to express their sincere gratitude to Dr Ramle Moslim and Mr Saharul Abillah Mohamad for their guidance, comments, and support throughout this study.

References

Abbas, M. S. T. (2010). IPM of the red palm weevil, *Rhynchophorus ferrugineus*. In A. Ciancio & K. G. Mukerji (Eds.), *Integrated management of arthropod pests and insect borne diseases. Integrated management of plant pests and diseases* (Vol. 5, pp. 209–233). Springer. https://doi.org/10.1007/978-90-481-8606-8

Ahmad, M. N., Kamarudin, N., Ahmad, S. N., Arshad, O., Masri, M. M. M., Moslim, R., & Kushairi, A. (2017). Efficacy of pheromone trapping and aerial spraying of *Bacillus thuringiensis* (Bt) for controlling bagworm, *Metisa plana* walker (Lepidoptera: Psychidae) in Yong Peng, Johor, Malaysia. *Journal of Oil Palm Research, 29*(1), 55–65. https://doi.org/10.21894/jopr.2017.2901.06

Ahmad, M. N., Shariff, A. R. M., Aris, I., Halin, I. A., & Moslim, R. (2021). Oto-BaCTM: An automated artificial intelligence (AI) detector and counter for bagworm (Lepidoptera: Psychidae) Census. *Advances in Agricultural and Food Research Journal, 2*(1), 1–16. https://doi.org/10.36877/aafrj.a0000218

Aman-Zuki, A., Ghazali, S. Z., Badrulisham, A. S., Hazmi, I. R., Wahida, O. N., & Yaakop, S. (2021). Proof on the divergence times of two sympatric species, *Rhynchophorus ferrugineus* and *R. vulneratus* (Coleoptera: Curculionidae) by molecular clock analysis. *Journal of the Entomological Research Society, 23*(1), 11–26. https://doi.org/10.51963/jers.v23i1.1851

Azmi, W. A., Chong, J. L., Ahamad Zakeri, H., Yusuf, N., Omar, W. B., Yong, K. W., Zulkefli, A. N., & Hussain, M. H (2017). The red palm weevil, *Rhynchophorus ferrugineus*: Current issues and challenges in Malaysia. *Oil Palm Bulletin, 74*(May), 17–24.

Azmi, W. A., Daud, S. N., Hussain, M. H., Wai, Y. K., Chik, Z., & Sajap, A. S. (2014). Field trapping of adult red palm weevil *Rhynchophorus ferrugineus* Olivier (Coleoptera: Curcilionidae) with food baits and synthetic pheromone lure in a coconut plantation. *Philippine Agricultural Scientist, 97*(4)

Basri, M. W. (1993). *Life history, ecology and economic impact of the bagworm,* Metisa plana *Walker (Lepidoptera: Psychidae) on the oil palm,* Elaeis guineensis *Jacquin (Palmae) in Malaysia.* University of Guelph.

Basri, W. M., & Kevan, P. (1995). Life history and feeding behaviour of the oil palm bagworm, *Metisa plana* Walker (Lepidoptera: Psychidae). *Elaeis, 7*(1), 18–34.

Cheong, Y. L., Sajap, A. S., Hafidzi, M. N., Omar, D., & Abood, F. (2010). Outbreak of bagworm and their natural enemiesin an oil palm, *Elaeis guineensus*, plantation at Hutan Melintang, Perak, Malaysia. *Journal of Entomology, 7*(3), 141–151.

Chong, J. L., H'ng, T. M., Azmi, W. A., & Amansuria, N. H. (2015, September). Genetic variation and invasion history of the invasive red palm weevil (*Rhynchophorus ferrugineus* (Olivier)) in Terengganu. *International Journal of Agriculture, Forestry and Plantation, 1*, 34–43.

Chung, G. F. (1998). Strategies and methods for the management of leaf-eating caterpillars of oil palm. *The Planter, 74*(871), 531–558.

Chung, G. F., Wahid, M. B., & Ariffin, D. (1995). *Recent development in plant protection of the Malaysian oil palm industry—1990 to 1995.* Proceedings of the 1995 PORIM National Oil Palm Conference—Technologises in Palntations—The Way Forward, pp. 105–124.

Dembilio, Ó., & Jaques, J. A. (2015). Biology and management of red palm weevil. In W. Wakil, J. R. Faleiro, & T. A. Miller (Eds.), *Sustainable pest management in date palm: Current status and emerging challenges. Sustainability in plant and crop protection* (pp. 13–36). Springer. https://doi.org/10.1007/978-3-319-24397-9_2

DOA (Department of Agriculture). (2011). *Report on current status of attack of the Red Palm Weevil,* Rhynchophorus ferrugineus *in Terengganu.* Biosecurity Division, Department of Agriculture, Malaysia.
DOA (Department of Agriculture). (2016). *Report on current status of attack of the Red Palm Weevil,* Rhynchophorus ferrugineus *in Terengganu.* Biosecurity Division, Department of Agriculture, Government Press, Malaysia.
El-Mergawy, R. A. A. M., Faure, N., Nasr, M. I., Arman, A. F., Rochat, D., & Silvain, J. F. (2011). Mitochondrial genetic variation and invasion history of red palm weevil, *Rhynchophorus ferrugineus* (Coleoptera: Curculionidae), in middle-east and Mediterranean basin. *International Journal of Agriculture and Biology, 13*(5), 631–637.
Halim, M., Muhaimin, A. M. D., Zulaikha, S. A. S., Atikah, A. R. N., Masri, M. M. M., & Yaakop, S. (2017). Evaluation of infestation in parasitoids on *Metisa plana* Walker (lepidoptera: Psychidae) in three oil palm plantations in peninsular Malaysia. *Serangga, 22*(2), 135–149.
Haris-Hussain, M., Kamarudin, N., & Azmi, W. A. (2020). Efficacy of baits for red palm weevil (RPW), *Rhynchophorus ferrugineus* Olivier under constant laboratory condition. *Journal of Oil Palm Research, 32*(2). https://doi.org/10.21894/jopr.2020.0026
Harith-Fadzilah, N., Haris-Hussain, M., Abd Ghani, I., Zakaria, A., Amit, S., Zainal, Z., Azmi, W. A., Jalinas, J., & Hassan, M. (2020). Physical and physiological monitoring on red palm weevil-infested oil palms. *Insects, 11*(407), 1–12. https://doi.org/10.3390/insects11070407
Idris, A. B., Mokhtaruddin, H., Zazali, C., Wahida, O. N., Yaakop, S., & Hazmi, I. R. (2014). The potential of red palm weevil infesting and destroying oil palm industry in Malaysia. *The Planter, 90*(1058), 329–335.
Ishak, I., Ng, L. C., Haris-Hussain, M., Jalinas, J., Idris, A. B., Azlina, Z., Samsudin, A., & Wahizatul, A. A. (2020). Pathogenicity of an indigenous strain of the entomopathogenic fungus *Metarhizium anisopliae* (Hypocreales: Clavicipitaceae) (MET-GRA4 Strain) as a potential biological control agent against the red palm weevil (Coleoptera: Dryophthoridae). *Journal of Economic Entomology, 113*(1), 43–49. https://doi.org/10.1093/jee/toz233
Jamian, S., Norhisham, A., Ghazali, A., Zakaria, A., & Azhar, B. (2017). Impacts of 2 species of predatory Reduviidae on bagworms in oil palm plantations. *Insect Science, 24*(2), 285–294. https://doi.org/10.1111/1744-7917.12309
Kamarudin, N., Ahmad, S. N., Arshad, O., & Wahid, M. B. (2010). Pheromone mass trapping of bagworm moths, *Metisa plana* Walker (Lepidoptera: Psychidae), for its control in mature oil palms in Perak, Malaysia. *Journal of Asia-Pacific Entomology, 13*(2), 101–106. https://doi.org/10.1016/j.aspen.2009.11.003
Kamarudin, N., & Arshad, O. (2006). Potentials of using the pheromone trap for monitoring and controlling the bagworm, *Metisa plana* Wlk (Lepidoptera: Psychidae) on young oil palm in a Smallholder Plantation. *Journal of Asia-Pacific Entomology, 9*(3), 281–285. https://doi.org/10.1016/S1226-8615(08)60304-0
Kamarudin, N., & Arshad, O. (2016). Diversity and activity of insect natural enemies of the bagworm (Lepidoptera: Psychidae) within an oil palm plantation in Perak, Malaysia. *Journal of Oil Palm Research, 28*(3), 296–307. https://doi.org/10.21894/jopr.2803.06
Kamarudin, N., Ali, S. R. A., Masri, M. M. M., Ahmad, M. N., Manan, C. A. H. C., & Kamarudin, N. (2015, October). *Controlling* Metisa plana *walker (Lepidoptera: Psychidae) outbreaks at an Estate in Slim River, Perak Using* Bacillus thuringiensis. Proceedings of the PIPOC 2015 International Palm Oil Congress—Oil Palm: Powering the World, Sustaining the Future: Module 1–Agriculture, Biotechnology and Sustainability Conference, 40, pp. 111–116.
Kamarudin, N., Ali, S. R. A., Masri, M. M. M., Ahmad, M. N., Manan, C. A. H. C., & Kamarudin, N. (2017). Controlling *Metisa plana* Walker (Lepidoptera: Psychidae) outbreak using *Bacillus thuringiensis* at an oil palm plantation in Slim River, Perak, Malaysia. *Journal of Oil Palm Research, 29*(1), 47–54. https://doi.org/10.21894/jopr.2017.2901.05
Kamarudin, N., & Wahid, M. B. (2010). Interactions of the bagworm *Pteroma pendula* (Lepidoptera: Psycidae) and its natural enemies in an oil palm plantation in Perak. *Journal of Oil Palm Research, 22.*

Kamarudin, N., Walker, A. K., Wahid, M. B., LaSalle, J., & Polaszek, A. (1996). Hymenopterous parasitoids associated with the bagworms *Metisa plana* and *Mahasena corbetti* (Lepidoptera: Psychidae) on oil palms in Peninsular Malaysia. *Bulletin of Entomological Research, 86*(4), 423–439. https://doi.org/10.1017/s000748530003501x

Kenis, M., Auger-Rozenberg, M.-A., Roques, A., Timms, L., Pere, C., Cock, M. J. W., Settele, J., Augustin, S., & Lopez-Vaamonde, C. (2009). Ecological effects of invasive alien insects. *Biological Invasions, 11*, 21–45. https://doi.org/10.1007/s10530-008-9318-y

Khudri, N. A. F. R. S., Mohd Masri, M. M., Maidin, M. S. T., Kamarudin, N., Hussain, M. H., Abd Ghani, I., & Jalinas, J. (2021). Preliminary evaluation of acoustic sensors for early detection of red palm weevil, *Rhynchophorus ferrugineus* incidence on oil palm and coconut in Malaysia. *International Journal of Tropical Insect Science*. https://doi.org/10.1007/s42690-021-00499-3

Kok, C. C., Eng, O. K., Razak, A. R., & Arshad, A. M. (2011). Microstructure and life cycle of *Metisa plana* Walker (Lepidoptera: Psychidae). *Journal of Sustainability Science and Management, 6*(1), 51–59.

Lin, G. L. E., Mohd Salim, J., Ahmad, M. F., & Azmi, W. A. (2017). Entomopathogenic fungi isolated from the soil of Terengganu, Malaysia as potential bio-pesticides against the red palm weevil, *Rhynchophorus ferrugineus*. *Journal of Sustainability Science and Management, 12*(2), 71–79.

MATRADE. (2021). *Top 10 major import products, 2020*. https://www.matrade.gov.my/en/malaysian-exporters/services-for-exporters/trade-market-information/trade-statistics/28-malaysian-exporters/trade-statistics/5082-top-10-major-export-products-2020

Mazmira, M. M. M., Ramlah, A. A. S., Najib, M. A., Norman, K., Kushairi, A., & Basri, M. W. (2011). Pest management of bagworm in Southern Perak by aerial spraying with *Bacillus thuringiensis*. *Oil Palm Bulletin, 63*, 24–33.

Mazuan, S., Mohamed, S., Ishak, I., Omar, D., & Asib, N. (2021). Optimization of motorized backpack mistblower for efficient application of insecticides against the bagworm, *Metisa plana* Walker. *Pakistan Journal of Agricultural Research, 34*(2), 479–485. https://doi.org/10.17582/journal.pjar/2021/34.2.479.485

Masri, M. M. M., Kamarudin, N., Napiah, N. R. M. A., & Keni, M. F. (2021). Effectiveness of *Bacillus thuringiensis* aerial spraying against the bagworm, *Metisa plana* Walker (Lepidoptera: Psychidae) outbreak in oil palm using drone. *Journal of Oil Palm Research*. https://doi.org/10.21894/jopr.2021.0045

Moslim, R., Kamarudin, N., Ahmad Ali, S. R., & Masijan, Z. (2019). Kumbang pendebungaan, perosak sawit dan kawalannya. In A. Kushairi, G. K. A. Parveez, & N. Kamarudin (Eds.), *Perusahaan sawit di Malaysia: Satu panduan* (4th ed., pp. 279–318). Lembaga Minyak Sawit Malaysia (MPOB).

MPOB. (2021). *Oil palm planted area 2020*. https://bepi.mpob.gov.my/images/area/2020/Area_summary.pdf

Firdaus, M. H. M., Chuah, T. S., & Wahizatul, A. A. (2020). Synergistic effect of synthetic pheromone and kairomone-releasing food baits in mass trapping system of red palm weevil, *Rhynchophorus ferrugineus*. *IOP Conference Series: Earth and Environmental Science, 494*(1), 1–7. https://doi.org/10.1088/1755-1315/494/1/012015

Najib, A. M., Ramlah, A. A. S., Mazmira, M. M. M., & Basri, W. M. (2009). Effect of *Bacillus thuringiensis*, Terakil-1, and Teracon-1 against oil palm pollinator, *Elaeidobius kamerunicus* and beneficial insects associated with *Cassia cobanensis*. *Journal of Oil Palm Research, 21*, 667–674.

Najib, A.M., Ramlah, A. A. S., Mazmira, M. M. M., & Basri, W. M. (2012). Effect of Bt products, Lepcon-1, Bafog-1 (S) and Ecobac-1 (Ec), against the oil palm pollinating weevil, *Elaeidobius kamerunicus*. *Journal of Oil Palm Research, 24*, 1440–1445. https://doi.org/10.13140/2.1.2374.6249

Norhayu, A., & Nurnisa, N. M. (2020). Effect of six insecticides on oil palm pollinating weevil, *Elaeidobius kamerunicus* (Coleoptera: Curculionidae). *Serangga, 25*(2), 1–9.

Norman, K., & Basri, M. W. (2007). Status of common oil palm insect pests in relation to technology adoption. *The Planter, 83*(975), 371–385.

Parveez, G. K. A., Tarmizi, A. H. A., Sundram, S., Soh, K. L., Ong-Abdullah, M., Poo Palam, K. D., Mohamed Salleh, K., Mohd Ishak, S., & Idris, Z. (2021). Oil palm economic performance in Malaysia and R&D progress in 2020. *Journal of Oil Palm Research, 33*(2), 181–214. https://doi.org/10.21894/jopr.2021.0026

Perera, S. A. C. N. (2014). Oil palm and coconut. In J. K. A. Pratap (Ed.), *Alien gene transfer in crop plants* (Vol. 2, pp. 231–252). Springer. https://doi.org/10.1007/978-1-4614-9572-7

Pyšek, P., Hulme, P. E., Simberloff, D., Bacher, S., Blackburn, T. M., Carlton, J. T., Dawson, W., Essl, F., Foxcroft, L. C., Genovesi, P., Jeschke, J. M., Kühn, I., Liebhold, A. M., Mandrak, N. E., Meyerson, L. A., Pauchard, A., Pergl, J., Roy, H. E., Seebens, H., … Richardson, D. M. (2020). Scientists' warning on invasive alien species. *Biological Reviews, 95*(6), 1511–1534. https://doi.org/10.1111/brv.12627

Radaha, K. (1977). Larval biology of *Crematopsyche pendula* Joannis. *The Planter, 53*, 381–394.

Ramlah, A. A. S., Norman, K., Basri, W. M., & Kushairi, A. (2018). *Sustainable oil palm insect pest management in Malaysia.* Malaysian Palm Oil Board (MPOB).

Syed, R. A. (1978). Bionomics of the three important species of bagworms on oil palm. *Malaysian Journal of Agriculture, 51*(4), 392–398.

Syed, R. A., & Speldewinde, H. V. (1974). Pest detection and census on oil palms. *The Planter, 49*, 488–490.

Wahizatul, A. A., Abdul Rahman, A. R., Chong, J. L., & Wong, A. S. Y. (2012). Scanning electron microscopy of the red palm weevil (RPW), *Rhynchophorus ferrugineus* Olivier (Coleoptera: Curculionidae): A new invasive pest of coconut palms in terengganu. *Malaysian Journal of Microscopy, 8*(1), 148–152.

Wahizatul, A. A., Ngadin, A., Ng, L. C., & Pong, K. K. (2013a). Identification and characterization of fungi associated with red palm weevil, *Rhynchophorus ferrugineus*: A microscopy study. *Malaysian Journal of Microscopy, 9*(1), 127–132.

Wahizatul, A. A., Zazali, C., Abdul Rahman, A. R., & Izzah, A. G. N. (2013b). A new invasive coconut pest in Malaysia: the red palm weevil (Curculionidae: *Rhynchophorus ferrugineus*). *The Planter, 89*(1043), 97–110.

Wood, B. J., Corley, R. H. V., & Goh, K. H. (1973). Studies on the effect of pest damage on oil palm yield. In R. L. Wastie & D. A. Earp (Eds.), *Advances in oil palm cultivation* (pp. 360–374). Incorporated Society of Planters.

Wood, B. J., & Kamarudin, N. (2019). Bagworm (Lepidoptera: Psychidae) infestation in the centennial of the Malaysian oil palm industry—A review of causes and control. *Journal of Oil Palm Research, 31*(3), 364–380. https://doi.org/10.21894/jopr.2019.0032

Yusdayati, R., & Hamid, N. H. (2015). Effect of several insecticide against oil palm pollinator's weevil, *Elaeidobius kamerunicus* (Coleoptera : Curculionidae). *Serangga, 20*(2), 27–35.

Zulkifli, A. N., Ahamad Zakeri, H., & Azmi, W. A. (2018). Food consumption, developmental time, and protein profile of the digestive system of the red palm weevil, *Rhynchophorus ferrugineus* (Coleptera: Dryophthoridae) larvae reared on three different diets. *Journal of Insect Science, 18*(5), 1–7. https://doi.org/10.1093/jisesa/iey093

Biology, Damage Caused, and Management of Thrips (Thysanoptera: Thripidae) Infesting Vegetable Crops in Malaysia

Syari Jamian, Siti Izera Ismail, Norsazilawati Saad, Johari Jalinas, Sumaiyah Abdullah, and Ibrahim Sani

Abstract Thrips (Thysanoptera: Thripidae) are regarded as one of the most pernicious insect pests that afflict vegetable crops across the globe. They cause damage to plant tissues during the pre-flowering to flowering stages, with the most harm occurring to pollen and flowers. The insect undergoes a gradual type of metamorphosis with six developmental stages: an egg, two active larval stages (L1 and L2), an active and non-feeding pupal stage (pre-pupa and pupa), and an adult stage (male and female). Both nymphs and adults can cause direct damage to vegetable crops through feeding and oviposition and indirect damage through the transmission of tospoviruses. Vegetables that are damaged due to infestation of thrips or viral diseases can become unmarketable, leading to financial losses to the producers. Control of thrips is mainly reliant on synthetic chemical pesticides, although alternative measures such as cultural, biological, and botanical controls have been reported to be promising and safer.

Keywords Biology · Malaysia · Management · Thrips · Tospovirus · Vegetable

---

S. Jamian (✉) · S. I. Ismail · N. Saad (✉) · S. Abdullah · I. Sani
Department of Plant Protection, Faculty of Agriculture, Universiti Putra Malaysia, 43400 Serdang, Selangor, Malaysia
e-mail: syari@upm.edu.my

N. Saad
e-mail: norsazilawati@upm.edu.my

S. Jamian · S. I. Ismail
Laboratory of Climate-Smart Food Crop Production, Institute of Tropical Agriculture and Food Security (ITAFoS), Universiti Putra Malaysia, 43400 Serdang, Selangor, Malaysia

J. Jalinas
Laboratory of Applied Entomology, Centre for Insect Systematics, Department of Biological Sciences and Biotechnology, Faculty of Science and Technology, Universiti Kebangsaan Malaysia, 43600 Bangi, Selangor, Malaysia

I. Sani
Department of Biology, Faculty of Natural and Applied Sciences, Umaru Musa Yar'adua University, Katsina P.M.B 2218, Katsina State, Nigeria

M. Wong (ed.), *Advances in Tropical Crop Protection*,
https://doi.org/10.1007/978-3-031-59268-3_2

1 Introduction

Vegetables account for 15% of daily food intake in Malaysia and provide job opportunities for farmers (Jipanin et al., 2001). The most commonly grown vegetables in Malaysia are cabbage, lettuce, tomato, leaf mustard, long bean, cucumber, water convolvulus, spinach, chili, and okra (Mohamed & Rokiah, 2006). Many vegetable crops suffer from insect pests that can cause major yield losses. These pests include caterpillars, beetles, aphids, whiteflies, mites, and thrips. Thrips (Thysanoptera: Thripidae), identified as a particularly problematic pest that targets vegetable crops, can be especially devastating (Tan et al., 2016). Thrips are tiny insects, ranging in size from 0.1 to 15 mm, and can be found worldwide. They have two pairs of wings that are fringed with cilia and have bristles on the veins. Their bodies are slender and elongated, and they come in a range of colors from light to dark, including some intermediate morphs (Mound, 2005; Palomo et al., 2015).

Numerous studies have identified *Frankliniella occidentalis* (Per-gande), *Thrips palmi* (Karny), and *Thrips tabaci* (Lindeman) as the primary pests in cucumber, pepper, tomato, onion, and zucchini. Thrips pose a significant threat to plants worldwide, as they not only cause direct damage during feeding and oviposition but also transmit plant viruses such as tomato spotted wilt virus (TSWV) and the impatiens necrotic spot virus (INSV). The consequences of their feeding activity can result in deformities, discoloration, and reduced yield, which can cause considerable economic losses (Ng & Zaimi, 2018). Unfortunately, the quarantine risks associated with these thrips species have been shown to have a negative impact on the global vegetable trade (Kumar et al., 2013; Palomo et al., 2015).

While thrips can cause substantial harm to vegetable crops, there appears to be a dearth of information regarding their presence in commonly cultivated vegetables in Malaysia. The available research on thrips and their impact on vegetable crops is scarce and largely restricted to their identification and distribution within the country. Nonetheless, the presence of *T. palmi, Thrips parvispinus, Megalurothrips usitatus, Thrips hawaiiensis,* and *T. parvispinus* has been reported on certain vegetable fields (Tan et al., 2016). A comprehensive survey conducted in 2011 recorded a total of 32 species of Thripidae, with a significant number of them belonging to the genus thrips (Azidah, 2011). Meanwhile, a separate study conducted in Peninsular Malaysia also recorded 23 species of the same genus. These findings suggest a diverse population of Thripidae in the region and provide valuable insights into the distribution and abundance of these species (Mound & Azidah, 2009).

The primary method for controlling thrips is through the use of chemical pesticides (Merel et al., 2018). However, alternative approaches such as cultural techniques, natural enemies, botanicals, and fungal pathogens are also available (Chinniah et al., 2016; Cote & Day, 2015). Our objective for this paper is to provide a comprehensive review of past studies on the biology, economic importance, and management of thrips that affect vegetables specifically in Malaysia.

2 Taxonomy, Host Range, and Distribution

In 1744, De Geer designated the name *Physapus* for thrips. However, in 1758, Linnaeus renamed them Thrips, as documented by Hamaseh (2011). The order Thysanoptera has been divided into two suborders: Terebrantia and Tubulifera. The order encompasses nine families, with the largest being Phlaeothripidae followed by Thripidae (Iftikhar et al., 2016). Family Thripidae is divided into 4 subfamilies, of which special attention has been given to the members of the subfamily Thripidae because they are the most economically important pest species (Ng & Zaimi, 2018). The Thripidae family encompasses a staggering 20,000 recognized species and can be found throughout the globe wherever vegetation thrives (Sartiami & Laurence, 2013). With around 6,200 species in existence, thrips are ubiquitous insects that inhabit forests, grasslands, bushes, leaves, and flowers and are distributed across both tropical and temperate zones (Iftikhar et al., 2016; Ng & Zaimi, 2018).

In Malaysia, the data on thrips distribution are not much. However, (Mound & Azidah, 2009) provided a key to the 23 species of the genus Thrips recorded from Peninsular Malaysia. Meanwhile, there are about 92 known species of Thripidae family recorded based on the checklists on thrips in Peninsular Malaysia, and this includes 17, 6, and 63 species from subfamilies Panchaetothripinae, Sericothripinae, and Thripinae, respectively (Azidah, 2011; Mound & Azidah, 2009; Ng & Zaimi, 2018; Ng et al., 2014). Thrips species in the genus, *Frankliniella* which is believed to be an invasive thrips species, have been found to be abundant in highland areas (Tan et al., 2016).

There is limited data available on the distribution of Thrips in Malaysia. However, 23 species of the Thrips genus from Peninsular Malaysia have been documented (Mound & Azidah, 2009). According to checklists of thrips in the region, there are approximately 92 known species in the Thripidae family, including 17, 6, and 63 species from the subfamilies Panchaetothripinae, Sericothripinae, and Thripinae, respectively (Azidah, 2011; Mound & Azidah, 2009; Ng & Zaimi, 2018; Ng et al., 2014). Notably, the invasive thrips species Frankliniella appears to be thriving in highland areas (Tan et al., 2016).

The thrips have a wide range of plant hosts, frequently feeding on them, which can deform and stunt the plants (Riley et al., 2018). Western flower thrips target various vegetable crops including cucumber, onion, pepper, potato, lettuce, and tomato (Capinera, 2001). These species of thrips are important for pollination of the plants they feed on, as they feed on flower tissues and pollen (Eliyahu et al. 2015; Suetsugu et al., 2019).

3 Biology of Thrips

Generally, members of the Thripidae family have a characteristic life cycle that consists of an egg, 2 active feeding larval instars, 2 relatively immobile non-feeding pupal instars, and adult stages (Eliyahu et al., 2015; Mound, 2018; Reitz, 2009). One of the major factors affecting the rate at which thrips develop is temperature. The number of degree days required for most of the genera of *Frankliniella* to complete one generation (egg to adult) is demonstrated to be 234 (lower threshold 10.5 °C) and 253 (6.5 °C), respectively (Riley et al., 2018).

Thrips are reported to be haplodiploid; that is, males have half the number of chromosomes (haploid number) and the other half (the diploid number) found in females (the diploid number). Males thrips are usually smaller than the females and are produced from unfertilized eggs (Parthenogenesis) (Funderburk, 2009).

3.1 Eggs and Larvae

Female thrips lay bean-shaped eggs inside leaves or floral parts of plant tissue (McLeod, 2008). The eggs take about 3 days to hatch, after which two nymphs emerge and start feeding on the plant (Cote & Day, 2015). These nymphs go through two larval stages, L1 and L2, which can last anywhere from 2 to 13 days, depending on the species (Cook et al., 2011). For instance, onion thrips (*T. tabaci*) take about 1.95 ± 1.42 and 4.12 ± 0.92 days to complete their first and second larval instar stages, respectively (Pourian et al., 2009). Western flower thrips (*F. occidentalis*), on the other hand, take 2 to 3 days to complete their first instar stage, and their second larval instar can last from 3 to 4 days (Gill et al., 2015).

3.2 Pupa and Prepupa

In the larval feeding stages, the insect transitions to a mobile prepupal stage, where feeding ceases. Following this, the insect descends to the soil and enters the pupal stage after approximately a week, protected within a cocoon (Cook et al., 2011; Eliyahu et al., 2015). The development of both prepupal and pupal stages in onion and western flower thrips typically occurs within 1 to 2 days (Pourian et al., 2009; Zhang et al., 2007). Once the pupation process is complete, the adults emerge and begin to feed (Cote & Day, 2015).

3.3 Adult

Various species of thrips have distinct life spans as adults. Thrips can develop from an egg to an adult in less than 19 days when conditions are favorable (Sani & Umar, 2017). Meanwhile, the western flower thrips can complete their entire egg-to-adult life cycle in approximately 10 days at 80° F (Cote & Day, 2015). In a study on the life history of the western flower thrips by Zhang et al. (2007), *F. occidentalis*, on excised leaves of five vegetables—cucumber, cabbage, bean, capsicum, and tomato leaf—developmental times from egg to adult were found to be 9.22 ± 0.13, 10.19 ± 0.08, 10.42 ± 0.06, 12.15 ± 0.07, and 12.91 ± 0.04 days, respectively. Research conducted by Gill et al. (2015) found that onion thrips require around 20 days to develop from an egg to an adult. Adult thrips typically live for no more than 23 days. Meanwhile, adult *F. occidentalis* females have a lifespan of 4–5 weeks at a temperature of 30 °C and lay an average of 50 eggs (Riley et al., 2018). In contrast, adult *T. tabaci* females' longevity ranges from 16 to 42 days on garlic and 28 to 30 days on onion (Gill et al., 2015).

4 Damages and Losses Causes by Thrips on Vegetable Crops Feeding Damage

Thrips are phytophagous insects that cause direct damage to plants during feeding and oviposition. The damage is due to the mechanical action of their mouthparts (Tommasini & Stefano, 1995). Some species of thrips cause damage to a single ornamental and vegetable plant part of either flower buds, leaf buds, leaves, fruit, or flowers, while others attack two or more of these plant parts (Bruce et al., 1995).

Adult *F. occidentalis* have been reported to inhabit the flowers of tomato, pepper, and eggplant, where they feed on pollen and flower tissues (Funderburk, 2009). Of all the vegetable crops attacked by thrips, tomato is the most seriously injured crop through oviposition (Capinera, 2001). Female insects laying their eggs in plant flowers can result in unsightly dimples and holes on the leaves that may persist until the fruit reaches maturity (Funderburk, 2009). Conversely, a particular insect species known as *T. tabaci* exhibits a preference for feeding and oviposition on onion leaves, ultimately leading to lowered yields. While they generally do not inflict significant economic damage on potatoes, these insects have also been known to target sweet potatoes and mustard (Gill et al., 2015).

Nymphs and adult thrips use their piercing-sucking mouthparts to infest vegetable plants by feeding on plant tissues, which consequently interrupts and decreases the leaf's photosynthetic capability (Palomo et al., 2015). Flower thrips primarily feed on plant pollens and flower tissues (Riley et al., 2018). Infested leaves and flowers can become distorted, malformed, and silvery due to the loss of plant fluids (Cloyd, 2010). Improper control measures of *T. tabaci* can cause yield loss ranging from 6 to 63% in onion crops. Severe infestation can lead to the complete destruction of young plants (Zereabruk et al., 2019).

Despite being a prevalent pest in Europe and the USA, thrips have not received much attention as a pest problem in Malaysian vegetable crops. However, recent reports indicate that *T. palmi* and *T. parvispinus* may be feeding on the flowers of eggplants, chilies, and bell peppers in Cameron Highlands, Malaysia (Tan et al., 2016). Zarifah and Azida (2018) recognized three genera of thrips; *Megalurothrips usitatus*, *T. hawaiiensis*, and *T. parvispinus* affecting some leguminous plants. The chili thrips, also known *as Scirtothrips dorsalis* Hood, belonging to the genus *Scirtothrips*, have been found in Cameron Highland. This particular species is known to attack a wide range of crops, including chili, avocadoes, and mangos (Ng et al., 2014). Recently, Jamian et al. (2019) have reported that thrips infestations have been observed on lettuce grown in commercial hydroponic farms in Perak, Malaysia.

5 Thrips as Vector

Tomato spotted wilt virus (TSWV) and *Impatiens necrotic spot virus* (INSV) are two species of tospovirus that are frequently vectored by thrips and are known to cause substantial damage to vegetable crops. Western flower thrips are known to be the most efficient vector of both 2 virus species (Reitz, 2009). The virus is commonly acquired by larvae in their first and second instars and later transmitted to vegetable plants by adult insects (Cloyd, 2010). Over the past three decades, thrips and viruses have become major pest problems for crops like tomato, pepper, and peanut in northern Florida (Funderburk, 2009). Reports suggest that the tomato spotted wilt virus (TSWV) alone can cause a loss of over $1 billion in annual yield (Reitz, 2009).

Symptoms of tospovirus infection in plants include necrotic lesions, wilting, yellowing, leaf distortion, and ring spots, although in some cases, infected plants may harbor viruses without showing any symptoms (Cote & Day, 2015). However, the expression of these symptoms can vary depending on the plant's growing conditions and environment. Fruits and vegetables exhibiting viral symptoms are deemed unsuitable for human consumption, thereby decreasing market value and causing producers to incur financial losses (Szostek et al., 2017). Additionally, Cloyd (2010) highlights that wounds caused by thrips, which occur during feeding or oviposition, can pave the way for other plant pathogens, including fungi, to invade the plant.

6 Management of Thrips

Integrated Pest Management (IPM) has earned global recognition as a pest control approach. Its primary objective is to limit the ecological and health consequences of chemical pesticides. The management of thrips pests via IPM typically consists of four strategies: cultural and mechanical measures, host plant resistance, chemical control, and biological control (Loomans, 2013).

6.1 Cultural and Physical Management

There are several methods to manage thrips through cultural control tactics. One proven technique is to use fine mesh screening that can prevent thrips from entering greenhouses. However, it is crucial to ensure that vent openings are appropriately sized to provide adequate ventilation. Recent research has revealed that reflective materials, such as aluminum foil, placed around vents can also repel thrips (Cote & Day, 2015).

According to a study conducted by Zereabruk et al. (2019), planting onions with cabbage and carrots can significantly decrease the number of thrips present and the harm they cause. This method also encourages the growth of natural predators of thrips. An effective way to culturally control thrips is through the use of ultraviolet-reflective mulch. Research has shown that this can prevent the spread of *F. occidentalis* adults, thereby reducing the initial transmission of the tomato spotted wilt virus (Funderburk, 2009). Additionally, weed control has been found to be useful in minimizing the thrips population, as many weeds can harbor thrips and act as virus reservoirs (Cote & Day, 2015; Riley et al., 2018).

Other cultural management practices by smallholder farmers include crop rotation and the use of resistant or tolerant cultivars (Sibanda et al., 2000). Jipanin et al. (2001) revealed that a significant number of Malaysian farmers are reluctant to employ pest control measures. The research found that only a small proportion of farmers in Malaysia have attempted to address pest-related issues through the adoption of recommended methods such as organic farming, crop rotation, resistant varieties and net-house planting methods. This lack of action may be due to several factors, including a lack of awareness regarding effective pest management practices, inadequate access to relevant information, or resistance to changing conventional farming methods. Further research is required to identify the underlying reasons for this phenomenon and to develop strategies that encourage Malaysian farmers to adopt effective pest control measures.

6.2 Biological Management

Biological control of pests involves the conservation of natural enemies, the use of parasitoids (specific natural enemies), predators (non-specific natural enemies), the sterile male method, microorganisms, and sex attractants (Singh & Satyanarayana, 2009).

Natural enemies of thrips include predatory thrips, predatory mites (such as *Amblyseius* spp.), anthocorid bugs or minute pirate bugs (*Orius* spp.), ground beetles, lacewings, hoverflies, and spiders (Sani & Umar, 2017). The immature and adult bugs *O. insidiosus* (Say) and *O. laevigatus* (Fieber) (Hemiptera: Anthocoridae) are the well-known commercial thrips predators that feed on the foliar stage (adult and immature stages) of western flower thrips (Gholami & Amin, 2016). Microorganisms, especially fungal pathogens, can also be used as biological control agents to manage thrips infestations. Various studies have shown the effectiveness of fungal pathogens in controlling different species of thrips. Kivett et al. (2016) conducted a field study and found that three different entomopathogenic fungi (*Beauveria bassiana, Isaria fumosoroseus,* and *Metarhizium anisopliae*) caused significant mortality among western flower thrips larvae. These three fungi have also been found to be effective in controlling thrips in onion and chili crops under laboratory conditions (Arthurs et al., 2013; Thungrabeab et al., 2006).

6.3 Botanical Control

Botanical pesticides with multiple toxic effects on insect biological activity are also one of the major components of IPM used to reduce synthetic pesticides (Singh & Satyanarayana, 2009). These biological activities of botanical insecticides on insects include behavior and feeding deterrence effects, fumigant toxicity, knockdown activity, and lethal toxicity (Gholami & Amin, 2016).

In Malaysia, powdered dried roots of jerangau (*Acorus calamus*) were first used as botanical pesticides against termites in rubber (Sivapragasam, 2009). Since then, researchers from different parts of Malaysia have recorded many successes in using botanicals to control different insect pests attacking crops. The botanical pesticides garlic, pyrethrum, neem, nicotine, rotenone, ryania, and sabadilla have been suggested for thrips control (Sani & Umar, 2017). However, only neem tree *Azadirachta indica*, garlic *Allium cepa*, and *Citronella* spp are currently being used as sources of registered botanical pesticides in Malaysia (Sivapragasam, 2009).

6.4 Chemicals Management

The utilization of agrochemicals as pesticides by Malaysian farmers is prevalent and represents the primary method for producing unblemished vegetables and increasing farm productivity (Jipanin et al., 2001). Pesticides from the organochlorine, organophosphate, and pyrethroid groups are commonly used by farmers to general vegetable pests (Jeyanthi & Kombairaju, 2005). The application of these chemicals at 5-day intervals, as recommended, can greatly reduce thrips infestations in crop (Gill et al., 2012). Chemical control measures should only be applied when the pest population is above the economic threshold, while targeting the vulnerable immature stages (Tey & Cheong, 2013). However, the indiscriminate use of these chemicals has led to numerous problems, such as accidental poisoning of humans and detrimental effects on the environment and non-target organisms (Jipanin et al., 2001). Thus, the IPM principle does not prohibit the application of chemical pesticides but rather should be the last option to be used prudently by the farmers if other alternatives are not effective against pest infestation on the crops (Halimatunsadiah et al., 2016).

7 Perspectives

This review aims to contribute to the existing literature on thrips by providing comprehensive information on the insect's biology, distribution, host range, damage causes, and management practices in Malaysia and other regions. While there has been a significant amount of literature published on thrips infesting vegetables, there has been a lack of information on the economic losses caused by thrips in Malaysian vegetable crops. It is important to note that thrips can transmit plant viruses in addition to directly damaging crops. Therefore, it is crucial to implement precautionary and effective control measures to prevent outbreaks of these insect pests and the diseases they can cause on crops.

References

Arthurs, S. P., Luis, F. A., & Pasco, B. A. (2013). Evaluation of entomopathogenic fungi against Chilli Thrips, *Scirtothrips dorsalis* evaluation of entomopathogenic fungi against Chilli Thrips, *Scirtothrips dorsalis*. *Journal of Insect Science, 13*(31), 1–16. https://doi.org/10.1673/031.013.3101

Azidah, A. A. (2011). Thripidae (Thysanoptera) species collected from common plants and crops in Peninsular Malaysia. *Scientific Research and Essays, 6*(24), 5107–5113. https://doi.org/10.5897/SRE10.1094

Bruce, L. P., Margaret, S., & Trevor, L. (1995). Thrips biology and management. In M. Skinner & T. L. B. L. Parker (Eds.), *Thrips biology and management*. Springer Science + Business Media. https://doi.org/10.1007/978-1-4899-1409-5.

Capinera, J. L. (2001). Order Thysanoptera—Thrips. In *Handbook of vegetable pests* (pp. 535–550). Elsevier. https://doi.org/10.1016/b978-012158861-8/50013-2

Chinniah, C., Ravikumar, A., Kalyanasundaram, M., & Parthiban, P. (2016). Management of sucking pests, by integration of organic sources of amendments and foliar application of entomopathogenic fungi on chilli. *Journal of Biopest, 9*(1), 34–40.

Cloyd, A. (2010). Western flower Thrips management on greenhouse-grown crops. Kansas State University. https://www.bookstore.ksre.ksu.edu/pubs/mf2922.pdf

Cook, D. A., Herbert, D., Akin, S., & Reed, J. (2011). Biology, crop injury, and management of Thrips (Thysanoptera: Thripidae) infesting cotton seedlings in the United States, pp. 1–9.

Cote, K. W., & Day, E. R. (2015). *Thrips*. Virgenia Cooperative Extension Publishing, Virgenia State University. https://www.pubs.ext.vt.edu/444/444-281/444-281.html

Eliyahu, D., McCall, A. C., Lauck, M., Trakhtenbrot, A., & Bronstein, J. L. (2015). Minute pollinators: The role of thrips (Thysanoptera) as pollinators of pointleaf manzanita, Arctostaphylos pungens (Ericaceae). *Journal of pollination ecology, 16*, 64–71.

Funderburk J. (2009). Management of the western flower Thrips (Thysanoptera: Thripidae) in fruiting vegetables. *Florida Entomologist, 92*(1), 1–6. https://doi.org/10.1653/024.092.0101

Gholami, Z., & Amin, S. (2016). Management strategies for western flower thrips in vegetable greenhouses in Iran: a review. *Plant Protection Science, 52*(2), 87–98. https://doi.org/10.17221/2/2015-PPS

Gill, S., Ethel, D., Michael, R., & John, D. (2012). Thrips management in greenhouses. University of Maryland, pp. 1–6. http://extension.umd.edu

Gill, H. K., Harsh, G., Arshdeep, K. G., Jennifer, L. G., & Brian, A. N. (2015). Onion thrips (Thysanoptera: Thripidae) biology, ecology, and management in onion production systems, pp. 1–9. https://doi.org/10.1093/jipm/pmv006

Halimatunsadiah, A. B., Mazlan, N., Omar, Z., & Kamarulzaman, N. H. (2016). Application of pesticide in pest management: The case of lowland vegetable growers. *International Food Research Journal, 23*(1), 85–94.

Hamaseh, A. (2011). *Biology and population dynamics of mango flower thrips (Thysanoptera) and effectiveness of neem* (Azadirachta indica) *oil for their control*. Universiti Sains Malaysia. http://eprints.usm.my/43286/1/HAMASEHALIAKBARPOUR.pdf

Iftikhar, R., Inaam, U., Stan, D., & Muhammad, A. (2016). Deciphering thysanoptera: A comprehensive study on the distribution and diversity of thrips fauna in Pakistan. *Pakistan Journal of Zoology, 48*(5), 1233–1240.

Jamian, S., Ismail, S. I., Hata, E. M., Saad, N., & Sani, I. (2019, October 18–20). *Case study: First report of thrips infestation and gray mold disease on hydroponically grown lettuce in Perak Malaysia*. ISSAAS International Congress and General Meeting 2019. Universiti Putra Malaysia, Serdang, Selangor, pp. 99.

Jeyanthi, H., & Kombairaju, S. (2005). Pesticide use in vegetable crops: Frequency, intensity and determinant factors. *Agricultural Economics Research Review, 18*(2), 209–221.

Jipanin, J., Alinah, A. R., Jackson, R., & Jaimi, P. K. (2001). Management of pesticide use on vegetable production: Role of department of agriculture sabah. *6th SITE Research Seminar* (pp. 1–21). Sabah.

Kivett, J. M., Cloyd, R. A., & Bello, N. M. (2016). Evaluation of entomopathogenic fungi against the western flower thrips (Thysanoptera: Thripidae) under laboratory conditions. *Journal of Entomological Science, 51*(4), 274–91. https://doi.org/10.18474/jes16-07.1

Kumar, V., Garima, K., Cindy, L., Dakshina, R., & Lance, S. (2013). An overview of Chilli Thrips, Scirtothrips Dorsalis (Thysanoptera: Thripidae) biology, distribution and management. In *Weed and pest control—Conventional and new challenges* (pp. 54–77). https://doi.org/10.5772/55045

Loomans, A. J. M. (2013). *Parasitoids as biological control agents of thrips pests parasitoids as biological control agents of thrips pests*. Wageningen University.

McLeod, P. (2008). *Identification, biology and management of insects attacking commercially produced vegetables*. Arkansas Paul McLeod Department of Entomology University of Arkansas Fayetteville, Arkansas. http://citeseerx.ist.psu.edu/viewdoc/download?doi=10.1.1.616.241&rep=rep1&type

Merel, S., Ahmed, A. E. H., Petra, B., Marcel, D., Rocio, E. B., Gang, C., Haring, M. A., et al. (2018). Thrips advisor: Exploiting thrips-induced defences to combat pests on crops. *Journal of Experimental Botany, 69*(8), 1837–1848. https://doi.org/10.1093/jxb/ery060

Mohamed, M. S., & Rokiah, M. Y. (2006). Tropical fruits and vegetables in Malaysia: Production and impact on health. In *Fruits and vegetables for health workshop* (pp. 1–5). https://doi.org/10.1073/pnas.95.19.11435

Mound, L. A. (2005). THYSANOPTERA: Diversity and interactions. *Annual Review of Entomology, 50*(1), 247–69. https://doi.org/10.1146/annurev.ento.49.061802.123318

Mound, L. A. (2018). Biodiversity of Thysanoptera. In R. G. Foottit & P. H. Adler (Eds.), *Insect biodiversity* (1st ed., Vol. II). John Wiley & Sons Ltd. https://doi.org/10.1002/9781118945582.ch18

Mound, L. A., & Azidah, A. A. (2009). Species of the genus thrips (Thysanoptera) from Peninsular Malaysia, with a checklist of recorded thripidae. *Zootaxa, 2023*, 55–68. https://doi.org/10.11646/zootaxa.2023.1.4

Ng, Y. F., Mound, L. A., & Azidah, A. A. (2014). The genus scirtothrips (Thysanoptera: Thripidae) in Malaysia, with four new species and comments on biltothrips, a related genus. *Zootaxa, 3856*(2), 253–266. https://doi.org/10.11646/zootaxa.3856.2.6

Ng, Y. F., & Zaimi, J. S. (2018). The economically important thrips from Malaysia, with a key to species (Thysanoptera, Thripinae). *ZooKeys, 810*, 113–126. https://doi.org/10.3897/zookeys.810.28457

Palomo, L. A. T., Nestor, B. M., Roberto, J. N., Jesus, R. N., Obdulia, S. L., Hussein, S. A., & Jorge, V. G. (2015). Population fluctuations of thrips (Thysanoptera) and their relationship to the phenology of vegetable crops in the Central Region of Mexico. *Florida Entomologist, 98*(2), 430–438. https://doi.org/10.1653/024.098.0206

Pourian, H., Majid, M., Marzieh, A., & Szilvia, O. (2009). Study on biology of onion thrips, thrips Tabaci Lindeman (Thysanoptera: Thripidae) on cucumber (Var. Sultan). *Laboratory Conditions, 49*(4), 4

Singh, T. V. K., & Satyanarayana, J. (2009). Insect outbreaks and their management. In R. Peshin, & A. K. Dhawan (Eds.), *Integrated pest management: Innovation-development process*. Springer Science+Business Media B.V. https://doi.org/10.1007/978-1-4020-8992-3.

Reitz, S. R. (2009). Biology and ecology of the western flower thrips (Thysanoptera: Thripidae): The making of a pest. *Florida Entomologist, 92*(1), 7–13. https://doi.org/10.1653/024.092.0102

Riley, D., Alton, S. Jr., Rajagopalbab, S., George, K., Greg, F., John, S., & Steve, O. (2018). Thrips: Biology, ecology, and management. In *Sustainable management of arthropod pests of tomato* (pp. 49–71). Elsevier Inc. https://doi.org/10.1016/B978-0-12-802441-6.00003-6.

Sani, I., & Umar, K. M. (2017). Biology and management of legume flower thrips (*Megalurothrips sjostedti*) (Thysanoptera: Thripidae), a major insect pest of cowpea: A review. *Annals of Experimental Biology, 5*(1), 14–17.

Sartiami, D., & Laurence, A. M. (2013). Identification of the terebrantian thrips (Insecta, Thysanoptera) associated with cultivated plants in Java, Indonesia. *ZooKeys, 306*, 1–21. https://doi.org/10.3897/zookeys.306.5455

Sibanda, T., Dobson, H. M., Cooper, J. F., Manyangarirwa, W., & Chiimba, W. (2000). Pest management challenges for smallholder vegetable farmers in Zimbabwe. *Crop Protection, 19*(2000), 807–815.

Sivapragasam, A. (2009). *Biopesticides from Malaysian Flora—resources for sustainable pest*. Proceedings of the National Seminar on New Crops and Bio-Resources, pp. 125–132.

Suetsugu, K., Shintaro, T., Masayoshi, K. H., & Tadaaki, T. (2019). Thrips as a supplementary pollinator in an orchid with granular Pollinia: Is this mutualism? *Ecology, 100*(2). https://doi.org/10.1002/ecy.2535

Szostek, S. A., Priscila, R., Jasmin, S., Scott, A., & Rayapati, A. N. (2017). Western flower thrips can transmit tomato spotted wilt virus from virus-infected tomato fruits. *Plant Health Progress, 18*(1), 1–6. https://doi.org/10.1094/PHP-RS-16-0057

Tan, J. L., Ooi, P. A. C., & Khoo, G. (2016). Thrips on eggplant, chilli and bell pepper in Cameron Highlands, Malaysia. *Serangga, 21*(1), 71–85.

Tey, C. C., & Cheong, Y. L. (2013). *Challenges in integrated pest management (IPM)*. 10th NATSEM 2013-Confronting Management Challenges in The Oil Palm Industry, pp. 117–127. Sarawak, Malaysia.

Thungrabeab, M., Peter, B., & Cetin, S. (2006). Possibilities for biocontrol of the onion thrips tabaci Lindeman (Thys., Thripidae) using different entomopathogenic fungi from Thailand. *Mitteilungen der Deutschen fur Allgemeine und Angewandte Entomologie, 15*, 299–304.

Tommasini, M. G., & Stefano, M. (1995). Biological control of thrips pests. *Wageningen Agricultural University Papers, 95*(1).

Zarifah, Z., & Azida, A. A. (2018). Diversity and population of thrips species on legumes with special reference to *Megalurothrips usitatus* diversity and population of thrips species on legumes with special reference to *Megalurothrips usitatus, 47*(3), 433–439. https://doi.org/10.17576/jsm-2018-4703-02

Zereabruk, G., Mulatu, W., & Gashawbeza, A. (2019). Management of onion thrips [*Thrips tabaci* Lind (Thysanoptera: Thripidae)] on onion using eco-friendly cultural practices and varieties of onion in Central Zone of Tigray, Ethiopia. *Journal of Agriculture and Ecology Research International, 18*(2), 1–10. https://doi.org/10.9734/jaeri/2019/v18i230053

Zhang, Z., Wu, Q., Li, X., Zhang, Y., Xu, B., & Zhu, G. (2007). Life history of western flower thrips, *Frankliniella occidentalis* (Thysan., Thripae), on five different vegetable leaves. *Journal of Applied Entomology, 131*(5), 347–354. https://doi.org/10.1111/j.1439-0418.2007.01186.x

Baculoviruses: Discovery, Interaction, and Application

Wei Hong Lau and Erise Anggraini

Abstract Baculoviruses are a family of viruses that primarily infect insects. They have a rich history that spans several decades since their discovery. The discovery of these viruses marked the beginning of extensive research into their unique characteristics, interactions, and potential applications in pest control. Baculoviruses are host-specific, especially those in the orders Lepidoptera, Hymenoptera, and Diptera. This specificity, evolved over time in conjunction with host insects, makes them precise biocontrol agents. Baculoviruses are named for the distinctive bacilliform or rod-shaped appearance of their occlusion bodies that protect the viral particles and contribute to their environmental stability. With a unique mode of action, baculoviruses offer a promising approach to counteracting insecticide resistance in pest populations. Their practical application in agriculture adheres to integrated pest management principles, providing a safe, targeted solution that spares beneficial insects. The natural degradation of baculoviruses minimizes environmental impact, a crucial factor in sustainable pest management. Baculoviruses stand at the forefront of environmentally friendly pest control, reflecting a harmonious integration of scientific discovery and practical application in modern agriculture.

Keywords Nucleopolyhedroviruses · Granuloviruses · Biological control · Mode of action · Application

---

W. H. Lau (✉)
Department of Plant Protection, Faculty of Agriculture, Universiti Putra Malaysia, Serdang, Malaysia
e-mail: lauweih@upm.edu.my

E. Anggraini
Study Program of Agroecotechnology & Study Program of Plant Protection, Faculty of Agriculture, Universitas Sriwijaya, Palembang, Indonesia
e-mail: erise.anggraini@unsri.ac.id

M. Wong (ed.), *Advances in Tropical Crop Protection*,
https://doi.org/10.1007/978-3-031-59268-3_3

1 Introduction

Baculoviruses, belonging to the Baculoviridae family, are large DNA viruses that infect insects (Wang & Hu, 2019). The family Baculoviridae is categorized into four genera through phylogenetic analyses: Alphabaculoviruses and Betabaculoviruses (infecting lepidopterans); Gammabaculoviruses (infecting hymenopterans); and Deltabaculoviruses (infecting dipterans) (Harrison et al., 2018). A comprehensive compilation of over 600 baculoviruses has been identified, originating from Lepidoptera (butterflies and moths), Hymenoptera (sawflies), and Diptera (mosquitoes) (Herniou et al., 2011).

Baculoviruses are characterized by the presence of rod-shaped nucleocapsids that are encased in a proteinaceous matrix. Granuloviruses (GVs) contain a solitary enveloped nucleocapsid per occlusion body, whereas nucleopolyhedroviruses (NPVs) comprise an assortment of nucleocapsids that extend throughout the polyhedral occlusion body (OB) matrix and are either single-enveloped or multiply enveloped. Baculovirus genomes vary in length from eighty to one hundred eighty thousand base pairs and replicate within the nuclei of infected cells (Harrison & Hoover, 2012). Like other large DNA viruses, baculoviuses contain an abundance of accessory genes that regulate cellular processes including apoptosis and the cell cycle (Clem & Passarelli, 2013).

Baculoviruses are exceptionally specific to their hosts and are friendly to the environment. They infect a negligible number of insect species, demonstrating their extreme host specificity. The distinctive association between the virus and its insect host is underscored in the virus's nomenclature, thereby placing further emphasis on the significance of host context in the study of viruses (D'Amico & Slavicek, 2012). Baculovirus nomenclature generally incorporates a reference to the insect host from which they were initially isolated, often combining the name of the host species with a descriptor associated with the virus (Shu et al., 2020). For example, AcMNPV is a baculovirus that was initially isolated from the alfalfa looper moth (*Autographa californica*). "*Autographa californica*" refers to the host species, "multiple" describes the polyhedral inclusion bodies that contain multiple virions per occlusion body, and " nucleopolyhedrovirus" is the genus of baculoviruses. The naming system serves to facilitate the identification of the insect host and furnish details regarding the properties of the virus (Cotmore et al., 2014).

2 Discovery

The origins of baculoviruses are intertwined with the ancient silk industry or sericulture in China, dating back over 5000 years. The cultivation of silkworms extended its influence across Asia, reaching Japan through Korea circa 300 CE and making its way to Europe around 550 CE (Millward, 2013; Reid, 1994). By the twelfth century, the silk industry had firmly established itself in Italy and Spain, and by the 1500 s, it had expanded to include France, England, and Mexico (Schoeser, 2007). Both China and Japan have maintained a strong cultural connection to silk production (Li, 1981). Although silk production no longer holds the prominence it once did, the Japanese royal family continues to engage in it symbolically (Steinhaus, 1975).

The diffusion of sericulture from Asia to Europe had introduced several diseases that had a notable impact on silk-producing regions. Among these, two prominent diseases are grasserie (also known as jaundice disease or Nuclear Polyhedrosis Virus—NPV) and flacherie (attributed to non-occluded viruses and bacteria) (Capinera, 2008). Grasserie, or jaundice disease, affects the larval stage of silkworms. It is caused by a virus that replicates in the nuclei of infected cells, leading to their enlargement and discoloration (Babu et al., 2009). Infected larvae display sluggish behavior, and their color may turn yellowish-white (Federici, 1997). This disease has the potential to rapidly spread within silkworm colonies, resulting in elevated mortality rates and a reduction in silk production (Chopade et al., 2021). Flacherie represents another affliction for silkworms, caused by non-occluded viruses and bacteria (Rahul et al., 2019). Symptoms of this disease can vary, but infected larvae typically exhibit lethargy, cease feeding, and eventually succumb to the illness (Maciel-Vergara & Ros, 2017). The consequences of flacherie include significant losses in both silk production and quality (Kalagi et al., 2022).

In the sixteenth century, Marcus Hieronymus Vida, an Italian poet and cleric, penned "De bombyce", a Latin poem. Within this poetic work, Vida artfully delves into various facets of silkworm life, including their metamorphosis, cocoon formation, and silk production. He described the rupture and discharge of fluid from afflicted silkworm larvae, shedding light on the darker aspects of silkworm cultivation (Boucias & Pendland, 1998; Steinhaus, 1975). While lacking a comprehensive grasp of the microbiological intricacies involved, Vida's keen observations offer valuable insights into the challenges faced by silk breeders during that era.

In 1679, Maria Merian, in her published work on the metamorphosis of caterpillars, made notable references to jaundice (Gelbsucht). She attributed the occurrence of jaundice to thunderstorms. Fast forward to 1856, when Italian scientist Agostino Maestri conducted the first scientific examination of the agent responsible for jaundice in silkworm larvae, marking a pivotal moment for the silk industry. Utilizing a compound microscope, Maestri identified and described polyhedral bodies or occlusions present in the fluids and tissues of diseased silkworms. These polyhedral bodies, later confirmed as occlusion bodies, represented a significant leap in understanding the intricate relationship between pathogens and diseases.

In the late nineteenth century, J. Bolle, the director of the Austrian Imperial Sericultural Experiment Station in Gorz, recorded the discovery of proteinaceous crystals in silkworms. These crystals demonstrated resistance to various solvents but susceptibility to alkaline solutions. Interestingly, the virus maintained its infectious properties even after the dissolution of the polyhedra in alkali (Bergold, 1947). Electron microscopic studies unveiled rod-shaped virions in *Bombyx mori* and *Lymantria dispar* polyhedral viruses. These virions were observed within a central chromatic mass, forming bundles of rods encircled by a membrane (Bergold, 1947). The polyhedral bodies were formed through the deposition of a material encircling these bundles

(Hughes, 1953). Subsequent observations indicated that healthy silkworms, when fed with these polyhedral bodies, displayed signs of jaundice (Benz, 1986).

In 1909, wilt disease, characterized by a distinct shape of polyhedral granules compared to those in jaundiced insects, was identified in nun moth larvae (*Lymantria monacha*) that had been fed with polyhedral bodies (Wahl, 1909). By 1913, Glaser and Chapman proposed the idea that the "wilt disease" might be caused by a filterable virus with the ability to pass through a diatomaceous Berkefeld Grade "N" filter (Glaser & Chapman, 1913). Contrary to earlier beliefs, Glaser and Chapman (1916) asserted that the polyhedral bodies themselves were not the causal agent of the wilt disease (Glaser & Chapman, 1916).

In the late 1930s, French scientist André Paillot discovered a novel viral disease in the cabbage butterfly *Pieris brassicae*, characterized by the production of a significant quantity of granules, contrasting with NPVs (Paillot, 1926). Paillot and his colleagues confirmed the presence of infectious entities associated with these granules in high-speed pellets through the utilization of high-speed centrifugation. This virus, named pseudo-grasserie by Paillot, proliferated in the adipose tissue and led to the white coloration of infected larvae. In the early 1930s, Paillot documented a similar ailment in the cutworm *Euxoa segetum*, causing nuclear hypertrophy in infected fat body cells, followed by the formation of granules. In the 1930s, Trager observed the formation of polyhedral granules and virus multiplication in actively dividing cells after infecting cultured cells with baculoviruses (Trager, 1935). The presence of pseudo-grasserie was also reported beyond France, specifically in the larvae of the variegated cutworm *Peridroma margaritosa* in 1947 (Steinhaus, 1947). Notably, granulosis viruses have only been identified in lepidopteran hosts to date. These viruses exhibit high specificity, infecting either a single species or species within the same genus, making them among the most host-specific insect viruses.

In general, three distinct types of insect diseases have been identified: nuclear polyhedrosis, cytoplasmic polyhedrosis, and granulosis (Rohrmann, 2019). Nuclear polyhedrosis is a type of viral infection in insects caused by viruses known as nucleopolyhedroviruses (NPVs). These viruses belong to the family Baculoviridae and are characterized by their ability to form polyhedral inclusion bodies within the cell nuclei of infected insects. Cytoplasmic polyhedrosis, in the other hand, refers to a type of viral infection in insects caused by viruses known as cytoplasmic polyhedrosis viruses (CPVs) (Zhang et al., 2020). Unlike the rod-shaped NPVs containing DNA, CPVs exhibit icosahedral capsids and belong to the Reoviridae virus family under the genus Cypovirus, possessing segmented, double-stranded RNA genomes (Hill et al., 1999). They are characterized by their ability to form crystalline structures called polyhedra within the cytoplasm of infected cells (Zhou et al., 2014). Granulosis is a viral disease affecting insect larvae caused by granuloviruses. These viruses form granular inclusion bodies in infected cells (Paillot, 1926). These granules are distinct from the polyhedra seen in NPVs and CPVs. Each type of insect disease is associated with specific viruses and has unique characteristics in terms of the formation and location of inclusion bodies within the infected cells.

3 Characteristics of Nucleopolyhedrovirus

NPVs, classified as large double-stranded DNA viruses, exhibit rod-shaped nucleocapsids enveloped individually (SNPV type) or in groups (MNPV type) within a membrane-like envelope, surrounded by polyhedral protein crystals (polyhedra) (Rohrmann, 2014). These polyhedra play a crucial role in protecting and preserving the viral particles, aiding in the transmission of the virus to other susceptible hosts. From a phylogenetic perspective, NPVs are the most prevalent and widely distributed baculoviruses. Their presence has been reported in over 400 insect species spanning seven insect orders, with Lepidoptera being the most predominant hosts (Murphy et al., 1995). Although less frequently reported, NPVs have been associated with Hymenoptera, Diptera, Coleoptera, Thysanura, and Trichoptera (Lauzon et al., 2004). Some NPVs even affect shrimp (class Crustacea, order Decapoda) (Sivakumar et al., 2020).

Remarkably, virus replication in all these arthropod groups is confined to the midgut epithelium, except for lepidopteran-targeting viruses and a specific cranefly disease affecting *Tipula paludosa* (Federici, 1997). NPVs exclusively invade the midgut epithelial nuclei of mosquito, sawfly, and caddisfly larvae, leading to the formation of occlusion bodies (Federici, 1997; Miller, 1997; Slack & Arif, 2006). Following the establishment of a transient infection in the midgut of lepidopterans lacking occlusion bodies, the virus proceeds to infect other tissues (Clem & Passarelli, 2013).

4 General Characteristics of Granulovirus

Granulosis viruses (GVs) generate virions biochemically and structurally akin to those of NPVs, but these virions are individually occluded within small granules referred to as occlusion bodies (Consigli et al., 1983). These granules have an average diameter of 120 to 300 nm and a length ranging from 300 to 500 nm, with the length varying based on the size of the occluded virion, which can differ among various viruses (Federici, 1997; Krell, 2008).

GVs have been exclusively identified in members of the order Lepidoptera, with approximately 100 lepidopteran species currently recognized as susceptible to this virus type (Granoff & Webster, 1999). The reproductive strategy of almost all known GVs parallels that of lepidopteran NPVs (Federici, 1999). The virus initiates a transient infection in the midgut epithelium, and depending on the GV type, it subsequently invades either the fat body or both the fat body and other major tissues. In these invaded tissues, the virus undergoes replication initially in the nucleus and later in the cytoplasm after the rupture of the nuclear envelope (Miller, 1997). The GV associated with the Western grape leaf skeletonizer, *Harrisina brillians*, deviates from this typical strategy, replicating and producing occlusion bodies solely in the midgut epithelium, akin to mosquito and sawfly NPVs (Federici & Stern, 1990).

Nonetheless, the cytopathology of this tissue is comparable to that of the fat body and other GV-invaded tissues (Tanada & Hess, 1984).

5 Mode of Action

During the life cycle of a baculovirus, two phenotypic virions are produced at different chronological stages: budded virions (BV) and occlusion-derived virions (ODV). BV is produced during the initial phase of infection, playing a role in secondary infection. It emerges from the cytoplasmic membrane as a single nucleocapsid per envelope. On the other hand, ODV is formed in the final phase of infection and is enveloped in a crystalline protein matrix known as occlusion bodies (OB). Despite sharing genome similarities, these two virion types differ in their origins, envelope compositions, and functions in the viral life cycle (Haase et al., 2015). The primary proteins constituting the crystalline matrix in OBs are called polyhedron in NPVs and granule in GVs. These proteins exhibit remarkable stability outside of their host environment. Consequently, OBs can endure various environmental conditions and maintain their infectious state indefinitely.

The primary infection initiates when susceptible lepidopteran larvae consume OBs *per os*. The infection takes place in the gastrointestinal tract of the insect, which is anatomically divided into three regions: foregut, midgut, and hindgut (Saxena et al., 2018). The midgut's alkaline pH triggers the degradation of OBs, releasing ODVs upon ingestion. The primary infection occurs in the midgut epithelial cells. ODVs pass through the protective peritrophic membrane (PM), composed of chitin and glycoprotein, shielding cells from abrasion, cuts, or insect pathogens (Terra, 2001). Some baculoviruses encode a metalloproteinase enzyme with divalent cations as an integral part of their structure, aiding in the penetration of the PM. Enhancin, a metalloproteinase, is thought to break down mucin, a component of the PM, allowing the virus access the epithelial cell surface (Wang & Hu, 2019). The enhancin of GV may enhance the infectivity of certain NPVs (Slavicek & Popham, 2005).

Following PM traversal, ODVs attach to ODV-specific receptors and fuse with the brush border of microvilli on columnar epithelial cells, releasing nucleocapsids (see Fig. 3.1). Baculovirus replicates in the nucleus, where nucleocapsids traverse the cytoplasm toward the cell's basal membrane (Harrison & Hoover, 2012). Subsequently, early genes undergo transcription and translation, leading to the translocation of specific proteins into the nucleus for nucleocapsids and viral particle assembly. These nucleocapsids emerge from the cell's basal membrane as BV, enveloped with GP64, a protein recognizing the target cell membrane and facilitating BV attachment to the cell membrane (Hefferon et al., 1999). In the systemic infection phase, BVs disperse within the hemolymph to infect other cells (Becnel & White, 2007; Rohrmann, 2019), bypassing the midgut basal lamina to initiate secondary infections in larval tissues (Clem & Passarelli, 2013). BVs target diverse cells, including tracheal cells, hemolymph cells, body fat cells, and epidermis cells, entering through

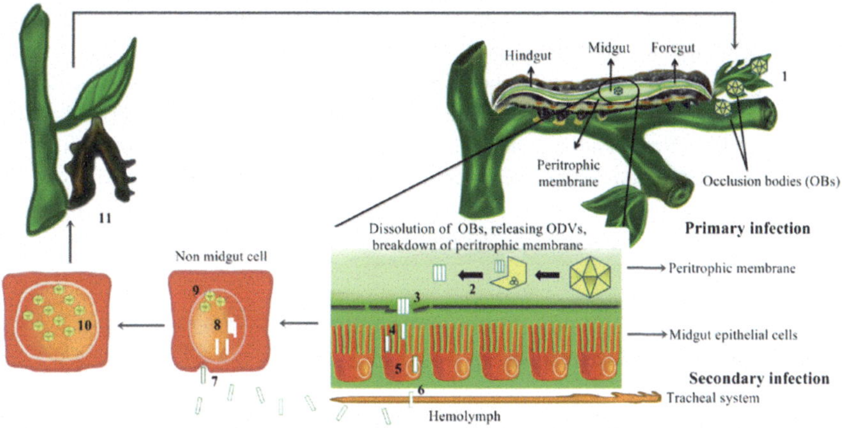

Fig. 3.1 Nucleopolyhedrovirus multiplication and transmission. (1) During the primary infection, insects consume OBs. (2) OBs are dissolved in the insect's midgut, and (3) ODVs are released after passing through the peritrophic membrane and (4) fusing with the microvilli of midgut epithelial cells to release nucleocapsids. (5) Nucleocapsids enter the nucleus and release viral DNA, allowing replication to begin. Progenies of nucleocapsids form in the nucleus, bud through the nuclear membrane, and develop into budded viruses (BV). (6) These BVs disperse in the hemolymph and infect other tissues during secondary infection. (7) Endocytosis allows BVs to enter cells and multiply in the nucleus. (8) Newly formed nucleocapsids can either form BVs to infect other cells or (9) be enveloped to form OB-occluded ODVs. (10) Fully formed OBs are seen in the nucleus during the late stage of infection. (11) Upon death, the larvae cling to the host plant's uppermost leaves, and the integument ruptures, releasing the OBs into the environment. Thc infection cycle will restart if other insects consume the contaminated leaves (Anggraini et al., 2022)

receptor-mediated endocytosis and replicating in the nucleus (D'Amico & Slavicek, 2012; Khorramnejad et al., 2021).

During the late stage of infection, nucleocapsids undergo encapsulation to form ODVs, which are subsequently enveloped in the crystalline protein matrix to become OBs. These OBs accumulate within the nucleus (Sajap et al., 2000). The accumulation of viral proteins and virions in the nucleus induces swelling, ultimately leading to host cell lysis, tissue liquefaction, and the weakening and rupture of the cuticle (Monteiro et al., 2012). The liquefaction contaminates the foliage, perpetuating the viral transmission cycle to other insect hosts (Elderd, 2013; Harrison & Hoover, 2012), potentially establishing a self-reinforcing positive-feedback loop (Elderd, 2013). In certain cases, only the midgut and fat body of infected insects exhibited symptoms, while other tissues remain unaffected (Senthil-Nathan, 2015). Some insects that ingest viral OBs undergo sublethal infection, exhibiting no discernible signs of symptoms or mortality (Williams et al., 2017).

Throughout the infection process, there are noticeable changes in the color of larvae integument, accompanied by a reduction in larvae development, feeding, and mobility (Harrison & Hoover, 2012). For example, *Spodoptera litura* larvae infected with *Spodoptera litura* nucleopolyhedrovirus exhibited polyhedrosis symptoms. The integument becomes flaccid and pink, leading to a cessation of feeding after three days post-inoculation (Javier-Hila & Caoili, 2020; Nazli-Huda et al., 2012). After five days post-inoculation, the dorsal, lateral, and ventral sides of infected larvae exhibit a pinkish hue compared to healthy larvae. After seven days post-inoculation with a single dose of 2.4×10^7 OBs, the larvae rupture, releasing white body fluid containing OBs (Nazli-Huda et al., 2012). Prior to succumbing to the infection, the infected larvae exhibit a distinctive behavior of migrating to the upper portions of the plant, hanging themselves from pseudopods, a behavior associated with baculovirus-induced tree-top disease (Goulson, 1997).

6 Efficacy of Baculovirus

The efficacy of baculovirus is subject to variation based on factors such as the target pest, environmental conditions, and the specific strain employed. Generally, a single baculovirus strain is employed to address a particular insect pest. The co-evolution of distinct baculovirus strains with specific insect species renders them highly host-specific. This specificity proves beneficial as it minimizes the risk of harm to non-target organisms, thus contributing to the preservation of ecological balance. Table 3.1 provides a summary of some studies evaluating the pathogenicity of a single baculovirus against various insect pests.

7 Interaction of Baculovirus

7.1 Mixed Infection

Different strains or species of baculovirus exhibit different levels of infectivity and virulence (Cory & Myers, 2003). Combining these strains may enhance the efficacy of insect species control by increasing the likelihood of successful infection or suppression. Certain strains may excel in infecting specific life stages or preferences of the target species compared to others. Employing multiple baculoviruses broadens the coverage, as they target diverse aspects of the insect's biology. Dependence on a single baculovirus strain can result in the evolution of insect resistance. Relying on a single baculovirus strain may lead to the development of insect resistance over time. The application of multiple baculoviruses can complicate the emergence of resistance in insects, as they would need to adapt to multiple strains or species simultaneously (Wang & Hu, 2019). Baculoviruses may vary in their sensitivity to environmental factors such as temperature, humidity, and sunlight (Anggraini et al., 2022). Utilizing a combination of baculoviruses provides an alternative approach to increase the likelihood that one or more will remain active and effective under diverse environmental conditions.

Table 3.1 Efficacy of baculoviruses

Species	Host range	LC_{50}/LD_{50}	References
Autographa californica MNPV (*Ac*MNPV)	*Pseudoplusia includens* *Trichoplusia ni*	• LD_{50}: 40.8 OBs/larva • LD_{50}: 3.1 OBs for larvae reared on artificial diet • LD_{50}: 16.6 OBs for larvae reared on cabbage • LD_{50}: 61.6 OBs for larvae reared on potato	Chikhalya et al. (2013) Chen et al. (2018)
Autographa californica MNPV (AcMNPV)	2nd instar *Trichoplusia ni* larvae	• AcMNPV-HR3, LD_{50}: 6.1 OBs/larva • AcMNPV-FV#11, LD_{50}: 10.9 OBs/larva • AcMNPV- FV#13, LD_{50}: 3.4 OBs/larva	Erlandson et al. (2007)
Arctornis submarginata GV	*A. submarginata*	• LC_{50}: 4.46×10^4 OBs/ml	Mukhopadhyay and De (2009)
Helicoverpa armigera single nucleopolyhedrovirus	*H. armigera* and *S. exigua*	•LC_{50}: 0.7×10^5 OBs/ml for *H. armigera* • LC_{50}: 1.9×10^7 OBs/ml for *S. exigua*	Gençer (2023)
Helicoverpa armigera NPV (HearNPV-BR1, HearNPV-BR2, HearNPV-BR3)	*Helicoverpa armigera*	• HearNPV-BR1, LD_{50}: 8.0×10^4 OBs/ml • HearNPV-BR2, LD_{50}: 4.1×10^4 OBs/ml • HearNPV-BR2, LD_{50}: 3.5×10^5 OBs/ml	da Costa et al. (2019)
Mamestra brassicae NPV-K1	3rd instar *M. brassicae* larvae	• LC_{50}: 3.7×10^4 OBs/larva at 20° C • LC_{50}: 9.9×10^2 OBs/larva at 25° C • LC_{50}: 3.8×10^2 OBs/larva at 30° C	Choi et al. (2009)
Spodoptera frugiperda NPV (SfNPV- BR4)	*Helicoverpa armigera*	• LD_{50}: 5.0×10^5 OBs/ml	da Costa et al. (2019)

(continued)

Table 3.1 (continued)

Species	Host range	LC_{50}/LD_{50}	References
Spodoptera littoralis NPV (isolate Cab2, P2 and P3)	*S. littoralis*	• Cab2, LC_{50}: 1.59 × 10^4 OBs/ml • P3, LC_{50}: 1.23 × 10^6 OBs/ml • P2, LC_{50}: 2.7 × 10^4 OBs/ml	Yasser et al. (2022)
Spodoptera littoralis GV	*Spodoptera littoralis*	• LD_{50}: 3.74 × 10^4 OBs/1st larva • LD_{50}: 6.86 × 10^5 OBs/3rd larva • LD_{50}: 1.1 × 10^7 OBs/5th larva	Hatem et al. (2011)
Spodoptera litura NPV	*Spodoptera litura*	• LD_{50}: 438 OBs/mL • LC_{50}: 0.42 OBs/ mm^2 cabbage • LC_{50}: 0.61 OBs/ mm^2 cotton • LC_{50}: 0.75 OBs/ mm^2 potato • LC_{50}: 0.93 OBs/ mm^2 groundnut • LC_{50}: 1.28 OBs/ mm^2 rose	Sanjaya et al. (2010) Ravishankar and Venkatesha (2010)
Trichoplusia ni SNPV (TnSNPV)	2nd instar *Trichoplusia ni* larvae	• TnSNPV-RJ, LD_{50}: 5.6 OBs/larva • TnSNPV FV#5, LD_{50}: 5.4 OBs/larva • TnSNPV FV#26, LD_{50}: 13.6 OBs/ larva • TnSNPV- FV#34), LD_{50}: 2.1 OBs/larva	Erlandson et al. (2007)

Mixed infections may result in diverse outcomes, and the interactions among distinct baculovirus strains within the same host can be complex. Distinct baculovirus strains coexisting within the same host may engage in competition for resources. Beperet et al. (2021) reported that when a susceptible insect undergoes coinfection by two distinct alphabaculovirus species, it leads to the formation of occlusion bodies that contain a mixture of viruses from the parental strains.

The interaction of mixed baculoviruses could be additive, synergistic or antagonistic. The presence of one baculovirus may or may not enhance the replication or virulence of another. In the case of XcenGV isolated from *Xestia c-Nigrum*, a low percentage of GV correlates with a synergistic effect, while an increased presence of GV tends to promote a neutral or antagonistic effect (Goto, 1990). In other case such as *Agrotis segetum* nucleopolyhedrovirus B (AgseNPV-B) and *Agrotis segetum* granulovirus (AgseGV), *A. segetum* larvae exhibited granuloses at lower concentrations of AgseNPV-B, while higher concentrations of AgseNPV-B led to polyhdroses. Both viruses acted independently within the larval host, resulting in an additive interation (Wennmann et al., 2015). The larval dosage, larval developmental stage, or the presence of synergistic factors in baculovirus occlusion bodies play crucial roles in determining the nature of virus interactions (Ferrelli & Salvador, 2023).

7.2 Coinfection with Other Biological Control Agents

Coinfection of baculoviruses with other entomopathogens can have diverse implications in the context of pest management and ecological balance. Magholli et al. (2013) conducted an interaction study on the *Helicoverpa armigera* single nucleopolyhedrovirus (HaSNPV) and *B. thuringiensis* subsp. *kurstaki,* and a synergism effect was observed on the second instar larvae of *Plutella xylostella* at the lowest *B. thuringiensis* and HaSNPV concentrations. Feeding larvae on contaminated leaf disks decreased growth, developmental rates, pupation, pupal weight, and adult emergence rates. However, coinfection of a baculovirus with an entomopathogenic fungi may lead to antagonistic or additive effect. For instance, combination of *Metarhizium anisopliae* and *Diatraea saccharalis* granulovirus (DisaGV) has demonstrated additive effects in combating the target insect (Pauli et al., 2018). However, antagonistic effect was observed in the combination of more than one entomopathogenic fungus (*M. anisopliae* and *B. bassiana*) with DisaGV. The occurrence of antagonistic or additive interactions between the pathogens could be influenced by the mode of infection and the initial inoculum density employed (Wraight & Ramos, 2005).

8 Application

8.1 Commercialization

Table 3.2 provides a summary of the commercialization of viral bioinsecticides currently available in the market. Baculovirus bioinsecticides are commonly marketed as suspension concentrates (Grzywacz & Moore, 2017). The active viral component is formulated into a concentrate, which can be mixed with water for application via spraying on pest-infested crops. Although suspension concentrates offer a straightforward and effective method for delivering baculoviruses to target insects, ensuring their shelf life matches or surpasses that of dry formulations requires pH adjustment to the optimal level and the addition of a bacterial suppressant (Hunter-Fujita et al., 1998). Some commercial baculovirus bioinsecticides are also offered as wettable powders, which are generally easy to handle, store, and transport, boasting longer shelf life and greater stability in storage compared to other formulations (Grzywacz & Moore, 2017).

8.2 The Weaknesses of Baculovirus as Bioinsecticides

Despite its numerous advantages, baculovirus presents several drawbacks. It has a slow release and action period, making its spread less rapid compared to chemical pesticides (Sherwani et al., 2015). Baculoviruses are highly susceptible to solar radiation and environmental conditions, potentially diminishing their efficacy, especially in open fields during sunny weather (Anggraini et al., 2022; Moscardi et al., 2011). Additionally, their high host specificity, while beneficial for protecting non-target species, may require additional pesticides to control other pest types unaffected by baculoviruses (Song et al., 2016). Furthermore, utilizing baculoviruses in pest control demands more technical knowledge than chemical pesticides, and their production as bioinsecticides can be expensive, particularly on a large scale, given the complexity of in vitro culturing (Rai et al., 2013).

8.3 Strategy to Improve the Efficacy of Baculovirus

To address the potential weaknesses of baculovirus, various strategies can enhance its efficacy. Employing different strains of baculovirus helps overcome inherent limitations. Utilizing mixed strains increases the effectiveness of baculovirus in controlling insect pests. Optimization of application methods, like spray formulations, enhances the persistence and distribution of baculoviruses on treated surfaces. Additionally, incorporating adjuvants like UV protectants shield baculoviruses from sunlight degradation, while feeding stimulants encourage target insects to ingest the virus, ultimately enhancing its overall impact on pest control (Cefali et al., 2016; Wu et al., 2015). Improved formulations by encapsulating baculoviruses in microcapsules can increase their environmental persistence, protect them from degradation, and enhance their overall performance (Luo et al., 2021).

9 Conclusions and Future Perspectives

Baculoviruses exhibit significant potential as bioinsecticides in the future, owing to their distinctive traits such as specificity, safety, and natural occurrence. These viruses exclusively target certain insect species, predominantly those in the orders Lepidoptera, Hymenoptera, and Diptera, ensuring minimal harm to non-target organisms like beneficial insects, humans, and wildlife. Baculoviruses are generally safe for humans, animals, and plants, posing no zoonotic or plant infection risks. This

Table 3.2 Commercial baculovirus bioinsecticides

Product name (formulation)	A.I	Target insect	Company
Virosoft CP4™ (Suspension concentrate)	*Cydia pomonella* granulovirus	Codling moth larvae	Certis Biologicals
Gemstar® (Suspension concentrate)	*Helicoverpa zea* NPV	Corn earworm, cotton bollworm, and tobacco budworm	Certis Biologicals
Madex® (Suspension concentrate)	*Cydia pomonella* granulovirus	Codling moth	Certis Biologicals
Spod-X® (Suspension concentrate)	*Spodoptera exigua* NPV	Beet armyworms	Certis Biologicals
CYD-X® (Suspension concentrate)	*Cydia pomonella* granulovirus	Codling moth larvae	Certis Biologicals
Gypchek (Wettable Powder)	*Lymantria dispar* MNPV	Gypsy moth	USDA Forest Service
Lepigen® (Suspension concentrate)	*Autographa californica* MNPV	Diamondback moth	AgBiTech
VPN-ULTRA® (Wettable Powder)	*Autographa californica* MNPV + *Spodoptera albula* NPV	Multiple insect pest	Agricola El Sol
GRAP Baculovirus (Wettable Powder)	*Anticarsia gemmatalis* MNPV	Velvet bean caterpillar	Agrocete
Carpovirusine® (Suspension concentrate)	*Cydia pomonella* GV	*Cydia pomonella,Cydia molesta*	Natural Plant Protection (NPP)
H-NPV (Suspension concentrate)	*Helicoverpa armigera* NPV	*Helicoverpa armigera*	Agrilabh
SUN BIO HANPV (Suspension concentrate)	*Helicoverpa armigera* SNPV	*Helicoperva* complex	Sonkul Agroindustries Private LTD
SUN BIO SlNPV (Suspension concentrate)	*Spodoptera litura* SNPV	*Spodoptera litura*	Sonkul Agroindustries Private LTD
HzNPV-CCAB®	*Helicoperva zea* SNPV	*Helicoverpa* complex	CCAB Agro
Mamestrin®	*Mamestra brassicae* MPNPV	Multiple insect pest	NPP
CAPEX® (Suspension concentrate)	*Adoxophyes orana* GV	*Adoxophyes orana*	Andermatt

(continued)

Table 3.2 (continued)

Product name (formulation)	A.I	Target insect	Company
HELICOVEX® (Suspension concentrate)	*Helicoverpa armigera* NPV	*H. armigera*	Andermatt
CRYPTEX® (Suspension concentrate)	*Crypthophlebia leucotreta* GV	*Crypthophlebia leucotreta*	Andermatt
SPEXIT® (Suspension concentrate)	*Spodoptera exigua* NPV	*Spodoptera exigua*	Andermatt
LITTOVIR® (Suspension concentrate)	*Spodoptera littoralis* NPV	*Spodoptera littolaris*	Andermatt
ABIETIV® (Suspension concentrate)	*Neodiprion abietis* NPV	Sawfly	Andermatt
Loopex (Suspension concentrate)	*Autographa californica* MNPV	*Trichoplusia ni, A. californica*	Andermatt
BoVir (Suspension concentrate)	*Lymantria dispar* NPV	*Lymantria dispar*	Andermatt
TM Biocontrol® (Suspension concentrate)	*Orgyia pseudotsugata* NPV	*Orgyia leucostigma, O. pseudotsugata*	Andermatt
Leconte Virus® (Suspension concentrate)	*Neodiprion lecontei* NPV	*Neodiprion lecontei*	Andermatt
Tutavir (Suspension concentrate)	*Phtorimaea operculella* GV	*Phtorimaea operculella*	Andermatt
Helicop™ (Suspension concentrate)	*Helicoverpa armigera* NPV	*H. armigera*	Indore Biotech
HaNPV (Suspension concentrate)	*Helicoverpa armigera* NPV	*H. armigera*	Om Agro Organics
HaNPV (Wettable Powder)	*Helicoverpa armigera* NPV	*H. armigera*	UNIOASIS

safety profile allows their use without concerns about cross-contamination, adverse health effects, or disruptions to ecosystems.

As naturally occurring pathogens, baculoviruses align with the principles of integrated pest management and sustainable agriculture when utilized as bioinsecticides. Their distinct mode of action offers a chance to address insecticide resistance in pests. Baculoviruses naturally degrade over time, considering environmental factors like sunlight and temperature fluctuations. This characteristic prevents environmental residue accumulation, supporting the sustainable, long-term use of these products. Consequently, baculovirus bioinsecticides present promising advancements in pest control.

References

Anggraini, E., Lau, W. H., Vadamalai, G., Kong, L. L., & Mat, M. (2022). The effect and possible mitigation of UV radiation on baculoviruses as bioinsecticides. *Biodiversitas Journal of Biological Diversity, 23*(7), 3721–3735.

Babu, K. R., Ramakrishna, S., Reddy, Y. H. K., Lakshmi, G., Naidu, N. V., Basha, S. S., & Bhaskar, M. (2009). Metabolic alterations and molecular mechanism in silkworm larvae during viral infection: A review. *African Journal of Biotechnology, 8*(6), 899907.

Becnel, J. J., & White, S. E. (2007). Mosquito pathogenic viruses—The last 20 years. *Journal of the American Mosquito Control Association, 23*(2), 36–49.

Benz, G. A. (1986). Introduction: Historial perspectives. In R. R. Granados & B. A. Frederici (Eds.), *The biology of baculoviruses* (Vol. 1, pp. 1–31). CRC.

Beperet, I., Simón, O., López-Ferber, M., van Lent, J., Williams, T., & Caballero, P. (2021). Mixtures of insect-pathogenic viruses in a single virion: Towards the development of custom-designed insecticides. *Applied and Environmental Microbiology, 87*(3), e02180-e2220. https://doi.org/10.1128/AEM.02180-20

Bergold, G. (1947). Die isolierung des polyeder-virus und die natur der polyeder. *Zeitschrift Für Naturforschung B, 2*(3–4), 122–143.

Blissard, G. W., & Rohrmann, G. F. (1990). Baculovirus diversity and molecular biology. *Annual Review of Entomology, 35*(1), 127–155.

Bonning, B. C., & Hammock, B. D. (1996). Development of recombinant baculoviruses for insect control. *Annual Review of Entomology, 41*(1), 191–210.

Boucias, D. G., & Pendland, J. C. (1998). Baculoviruses. In D. G. Boucias & J. C. Pendland (Eds.), *Principles of insect pathology* (pp. 111–146). Springer.

Capinera, J. L. (2008). *Encyclopedia of entomology*. Springer.

Cefali, L. C., Ataide, J. A., Moriel, P., Foglio, M. A., & Mazzola, P. G. (2016). Plant-based active photoprotectants for sunscreens. *International Journal of Cosmetic Science, 38*(4), 346–353.

Chen, E., Kolosov, D., O'Donnell, M. J., Erlandson, M. A., McNeil, J. N., & Donly, C. (2018). The effect of diet on midgut and resulting changes in infectiousness of AcMNPV baculovirus in the cabbage loope, *Trichoplusia Ni. Frontiers in Physiology, 9*, 1348.

Chikhalya, A., Stephens, K. D., Archie, J. W., & Haas-Stapleton, E. J. (2013). Virulence and pathogenesis of the baculovirus *Autographa californica* M nucleopolyhedrovirus (AcMNPV) in *Pseudoplusia includens* larvae. *Biological Control, 65*(1), 101–108.

Choi, J. B., Lee, J. K., Bae, S. M., Shin, T. Y., Koo, H. N., Kim, J. I., Kwon, M., & Woo, S. D. (2009). Pathogenicity and production of *Mamestra brassicae* nucleopolyhedrovirus (MabrNPV)-K1. *International Journal of Industrial Entomology, 19*(2), 237–241.

Chopade, P., Raghavendra, C. G., & Bhaskar, R. N. (2021). Assessment of diseases in *Bombyx mori* silkworm—A survey. *Global Transitions Proceedings, 2*(1), 133–136.

Clem, R. J., & Passarelli, A. L. (2013). Baculoviruses: Sophisticated pathogens of insects. *PLoS Pathogens, 9*(11), 11–15.

Consigli, R. A., Tweeten, K. A., Anderson, D. K., & Bulla, L. A. J. (1983). Granulosis viruses, with emphasis on the GV of the Indian meal moth, *Plodia interpunctella. Advances in Virus Research, 28*, 141–173. https://doi.org/10.1016/s0065-3527(08)60723-x

Cory, J. S., & Myers, J. H. (2003). The ecology and evolution of insect baculoviruses. *Annual Review of Ecology, Evolution, and Systematics, 34*, 239–272. https://doi.org/10.1146/annurev.ecolsys.34.011802.132402

Cotmore, S. F., Agbandje-McKenna, M., Chiorini, J. A., Mukha, D. V., Pintel, D. J., Qiu, J., Soderlund-Venermo, M., Tattersall, P., Tijssen, P., & Gatherer, D. (2014). The family Parvoviridae. *Archives of Virology, 159*, 1239–1247.

Cuartas-Otálora, P. E., Gómez-Valderrama, J. A., Ramos, A. E., Barrera-Cubillos, G. P. & Villamizar-Rivero, L. F. (2019). Bio-insecticidal potential of nucleopolyhedrovirus and granulovirus mixtures to control the fall armyworm *Spodoptera frugiperda* (J.E. Smith, 1797) (Lepidoptera: Noctuidae). *Viruses, 11*, 684.

D'Amico, V., & Slavicek, J. (2012). Interactions between nucleopolyhedroviruses and polydnaviruses in larval lepidoptera. In M. P. Adoga (Eds.), *Molecular virology*. IntechOpen. https://doi.org/10.5772/35180

da Costa, V. H. D., Soares, M. A., Dimate, F. A. R., de Sá, V. G. M., Zanuncio, J. C., & Valicente, F. H. (2019). Genetic identification and biological characterization of baculovirus isolated from *Helicoverpa armigera* (Lepidoptera: Noctuidae) in Brazil. *Florida Entomologist, 102*(1), 59–64.

Elderd, B. D. (2013). Developing models of disease transmission: Insights from ecological studies of insects and their baculoviruses. *PLoS Pathogens, 9*(6), e1003372. https://doi.org/10.1371/journal.ppat.1003372

Erlandson, M., Newhouse, S., Moore, K., Janmaat, A., Myers, J., & Theilmann, D. (2007). Characterization of baculovirus isolates from *Trichoplusia ni* populations from vegetable greenhouses. *Biological Control, 41*(2), 256–263.

Federici, B. A. (1997). Baculovirus pathogenesis. In L. K. Miller (Ed.), *The baculoviruses* (pp. 33–56). Plenum Press.

Federici, B. A. (1999). Naturally occurring baculoviruses for insect pest control. In F. R. Hall & J. J. Menn (Eds.), *Biopesticides: Use and delivery* (pp. 301–320). Humana Press.

Federici, B. A., & Stern, V. M. (1990). Replication and occlusion of a granulosis virus in larval and adult midgut epithelium of the western grapeleaf skeletonizer, *Harrisina brillians. Journal of Invertebrate Pathology, 56*(3), 401–414.

Ferrelli, M. L., & Salvador, R. (2023). Effects of mixed baculovirus infections in biological control: A comprehensive historical and technical analysis. *Viruses, 15*, 1838. https://doi.org/10.3390/v15091838

Glaser, R. W., & Chapman, J. W. (1913). The wilt disease of gipsy moth caterpillars. *Journal of Economic Entomology, 6*(6), 479–488.

Glaser, R. W., & Chapman, J. W. (1916). The nature of the polyhedral bodies found in insects. *The Biological Bulletin, 30*(5), 367–390.

Gençer, D. (2023). Isolation characterization and pathogenicity of *Helicoverpa armigera* single nucleopolyhedrovirus isolate from Türkiye. *Artvin Çoruh Üniversitesi Orman Fakültesi Dergisi, 24*(1), 187–195.

Goto, C. (1990). Enhancement of a nuclear polyhedrosis virus (NPV) infection by a granulosis virus (GV) isolated from the spotted cutworm, *Xestia c-Nigrum* L.: Lepidoptera: Noctuidae. *Applied Entomology and Zoology, 25*(1), 135–137.

Goulson, D. (1997). Wipfelkrankheit: Modification of host behaviour during baculoviral infection. *Oecologia, 109*(2), 219–228.

Granoff, A., & Webster, R. G. (1999). *Encyclopedia of virology* (2nd ed.). Elsevier Science.

Grzywacz, D., & Moore, S. (2017). Production, formulation, and bioassay of baculoviruses for pest control. In L. A. Lacey (Ed.), *Microbial control of insect and mite pests: From theory to practice*. Academic Press. https://doi.org/10.1016/B978-0-12-803527-6.00007-X

Haase, S., Sciocco-Cap, A., & Romanowski, V. (2015). Baculovirus insecticides in Latin America: Historical overview, current status and future perspectives. *Viruses, 7*(5), 2230–2267.

Harrison, R., & Hoover, K. (2012). Baculoviruses and other occluded insect viruses. In F. E. Vega, & H. K. Kaya (Eds.), *Insect pathology* (2nd ed.). Elsevier. https://doi.org/10.1016/B978-0-12-384984-7.00004-X

Harrison, R. L., Herniou, E. A., Jehle, J. A., Theilmann, D. A., Burand, J. P., Becnel, J. J., Krell, P. J., van Oers, M. M., Mowery, J. D., & Bauchan, G. R. (2018). ICTV virus taxonomy profile: Baculoviridae. *Journal of General Virology, 99*(9), 1185–1186. https://doi.org/10.1099/jgv.0.001107

Hatem, A. E.-S., Aldebis, H. K., & Vargas Osuna, E. (2011). Effects of the *Spodoptera littoralis* granulovirus on the development and reproduction of cotton leafworm *S. littoralis*. *Biological Control, 59*(2), 192–199. https://doi.org/10.1016/j.biocontrol.2011.07.004

Hefferon, K. L., Oomens, A. G. P., Monsma, S. A., Finnerty, C. M., & Blissard, G. W. (1999). Host cell receptor binding by baculovirus GP64 and kinetics of virion entry. *Virology, 258*(2), 455–468.

Herniou, E. A., Arif, B. M., Becnel, J. J., Blissard, G. W., Bonning, B., Harrison, R., Jehle, J. A., Theilmann, D. A., & Vlak, J. M. (2011). Baculoviridae. In M. J. Adams, A. M. Q. King, E. B. Cartsens, & E. J. Lefkowitz (Eds.), *Virus taxonomy: Ninth report of the International Committee on Taxonomy of Viruses* (pp. 163–173). Elsevier.

Hill, C. L., Booth, T. F., Prasad, B. V., Grimes, J. M., Mertens, P. P., Sutton, G. C., & Stuart, D. I. (1999). The structure of a cypovirus and the functional organization of dsRNA viruses. *Nature Structural Biology, 6*(6), 565–568.

Hughes, K. (1953). The development of an insect virus within cells of its host. *Hilgardia, 22*(12), 391–406.

Hunter-Fujita, F. R., Entwistle, P. F., Evans, H. F., & Crook, N. E. (1998). *Insect viruses and pest management.* John Wiley & Sons Ltd.

Javier-Hila, A. M. V., & Caoili, B. L. (2020). Cross-infectivity of a putative *Spodoptera picta* nucleopolyhedrovirus to *Spodoptera litura* fabricius (Lepidoptera: Noctuidae). *Philippine Journal of Science, 149*(3), 887–896.

Kalagi, A., Raghavan, D., Raghavendra, C. G., Bajannavar, S., & Bhavani, V. S. (2022). Sericulture technology towards sustainable management. *IEEE International Conference on Distributed Computing and Electrical Circuits and Electronics (ICDCECE), 2022*, 1–5.

Khorramnejad, A., Perdomo, H. D., Palatini, U., Bonizzoni, M., & Gasmi, L. (2021). Cross talk between viruses and insect cells cytoskeleton. *Viruses, 13*(8). https://doi.org/10.3390/v13081658

Krell, P. J. (2008). Baculoviruses: General features. In B. W. J. Mahy, & M. H. V. van Regenmortel (Eds.), Encyclopedia of virology (Third Edition) (pp. 247–254). Academic Press.

Lauzon, H. A. M., Lucarotti, C. J., Krell, P. J., Feng, Q., Retnakaran, A., & Arif, B. M. (2004). Sequence and organization of the *Neodiprion lecontei* nucleopolyhedrovirus genome. *Journal of Virology, 78*(13), 7023–7035.

Li, L. M. (1981). *China's silk trade: Traditional industry in the modern world 1842–1937 (Issue 97).* Harvard University Press.

Luo, M., Zhu, D., Lin, J., Zhou, X., Zheng, C., & Pu, X. (2021). Preparation and performance of insect virus microcapsules. *Egyptian Journal of Biological Pest Control, 31*, 104.

Maciel-Vergara, G., & Ros, V. I. D. (2017). Viruses of insects reared for food and feed. *Journal of Invertebrate Pathology, 147*, 60–75.

Miller, L. K. (1997). *The baculoviruses.* Springer.

Millward, J. A. (2013). *The silk road: A very short introduction.* Oxford University Press.

Magholli, Z., Marzban, R., Abbasipour, H., Shikhi, A., & Karimi, J. (2013). Interaction effects of *Bacillus thuringiensis* subsp. *kurstaki* and single nuclear polyhedrosis virus on *Plutella xylostella. Journal of Plant Diseases and Protection, 120*, 173–178.

Monteiro, F., Carinhas, N., Carrondo, M. J. T., Bernal, V., & Alves, P. M. (2012, November). Toward system-level understanding of baculovirus-host cell interactions: From molecular fundamental studies to large-scale proteomics approaches. *Frontiers in Microbiology, 3*, 1–16. https://doi.org/10.3389/fmicb.2012.00391

Moscardi, F., De Souza, M. L., De Castro, M. E. B., Lara Moscardi, M., & Szewczyk, B. (2011). Baculovirus pesticides: Present state and future perspectives. In I. Ahmad, F. Ahmad, & J. Pichtel (Eds.), *Microbes and microbial technology: Agricultural and environmental applications* (pp. 415–445). Springer.

Mukhopadhyay, A., & De, D. (2009). Pathogenecity of a baculovirus isolated from *Arctornis submarginata* (Walker) (Lepidoptera:Lymantriidae), a potential pest of tea growing in the Darjeeling foothills of India. *Journal of Invertebrate Pathology, 100*(1), 57–60. https://doi.org/10.1016/j.jip.2008.08.006

Murphy, F. A., Fauquet, C. M., Bishop, D. H. L., Ghabrial, S. A., Jarvis, A. W., Martelli, G. P., Mayo, M. A., & Summers, M. D. (1995). *Virus taxonomy—The classification and nomenclature of viruses: Sixth report of the International Committee on Taxonomy of Viruses*. Springer-Verlag.

Nazli-Huda, Sajap, A. S., & Lau, W. H. (2012). Stability of *Spodoptera litura* nucleopolyhedrovirus in sodium dodecyl sulphate. *African Journal of Biotechnology, 11*(16), 3877–3881.

Paillot, A. (1926). Sur une nouvelle maladie du noyau ou grasserie des chenilles de *P. brassicae* et un nouveau groupe de microorganismes parasites. *Comptes Rendus De L'académie Des Sciences, 182*, 180–182.

Pauli, G., Mascarin, G. M., Eilenberg, J., & Delalibera, J. I. (2018). Within-host competition between two entomopathogenic fungi and a granulovirus in *Diatraea saccharalis* (Lepidoptera: Crambidae). *Insects, 9*(2), 64.

Rahul, K., Moamongba, K. S., Rabha, M., & Sivaprasad, V. (2019). Identification and characterization of bacteria causing flacherie in mulberry silkworm, *Bombyx mori* L. *Journal of Crop and Weed, 15*(3), 178–181.

Rai, D., Sharma, G., & Pandey, A. K. (2013). Baculovirus biopesticides—An ecofriendly approach for insect-pest management. *Journal of Hill Agriculture, 4*(1), 1–7.

Ravishankar, B. S., & Venkatesha, M. G. (2010). Effectiveness of SlNPV of *Spodoptera litura* (Fab.) (Lepidoptera: Noctuidae) on different host plants. *Journal of Biopesticides, 3*(1), 168–171.

Reid, S. (1994). *Inventions and trade: The silk and spice routes*. James Lorimer & Company.

Rohrmann, G. (2019). *Baculovirus molecular biology* (4th ed.). Bethesda (MD). http://www.ncbi.nlm.nih.gov/books/NBK138304/

Rohrmann, G. F. (2014). Baculovirus nucleocapsid aggregation (MNPV vs SNPV): An evolutionary strategy, or a product of replication conditions? *Virus Genes, 49*(3), 351–357.

Sajap, A., Kotulai, J. R., Bakir, M. A., Lau, W. H., Samad, N. A., & Kadir, H. A. (2000). Pathogenicity and characteristics of *Spodoptera litura* nucleopolyhedrovirus from Peninsular Malaysia. *Pertanika Journal of Tropical Agricultural Science, 23*(1), 23–28.

Sanjaya, Y., Machmudin, D., & Kurniawati, N. (2010). Histological study of SlNPV infection on body weight and peritrophic membrane damage of *Spodoptera litura* larvae. *Nusantara Bioscience, 2*(3), 135–140.

Saxena, A., Byram, P. K., Singh, S. K., Chakraborty, J., Murhammer, D., & Giri, L. (2018). A structured review of baculovirus infection process: Integration of mathematical models and biomolecular information on cell–virus interaction. *Journal of General Virology, 99*(9), 1151–1171.

Schoeser, M. (2007). *Silk*. Yale University Press.

Senthil-Nathan, S. (2015). A Review of biopesticides and their mode of action against insect pests. In P. Thangavel, & G. Sridevi (Eds.), *Environmental sustainability: Role of green technologies* (pp. 49–62). Springer; New Delhi, India.

Sherwani, S. I., Khan, H. A., Nollet, L., & Rathore, H. (2015). Modes of action of biopesticides. In L. M. L. Nollet, & H. S. Rathore (Eds.), *Biopesticides handbook* (pp. 51–68). Taylor & Francis Group.

Shu, R., Meng, Q., Miao, L., Liang, H., Chen, J., Xu, Y., Cheng, L., Jin, W., Qin, Q., & Zhang, H. (2020). Genome analysis of a novel clade b betabaculovirus isolated from the legume pest *Matsumuraeses phaseoli* (Lepidoptera: Tortricidae). *Viruses, 12*(10), 1068.

Sivakumar, G., Kannan, M., Babu, V. R., Mohan, M., Kumari, S., Rangeshwaran, R., Venkatesan, T., & Ballal, C. R. (2020). Characterization and field evaluation of tetrahedral and triangular nucleopolyhedrovirus of *Spilosoma obliqua* (SpobNPV) strain NBAIR1 against jute hairy caterpillar. *Egyptian Journal of Biological Pest Control, 30*, 82.

Slack, J., & Arif, B. M. (2006). The baculoviruses occlusion-derived virus: Virion structure and function. *Advances in Virus Research, 69*(6), 99–165. https://doi.org/10.1016/S0065-3527(06)69003-9

Slavicek, J. M., & Popham, H. J. R. (2005). The *Lymantria dispar* nucleopolyhedrovirus enhancins are components of occlusion-derived virus. *Journal of Virology, 79*(16), 10578–10588.

Song, J., Wang, X., Hou, D., Huang, H., Liu, X., Deng, F., Wang, H., Arif, B. M., Hu, Z., & Wang, M. (2016). The host specificities of baculovirus *per os* infectivity factors. *PLoS ONE, 11*(7), e0159862. https://doi.org/10.1371/journal.pone.0159862

Steinhaus, E. A. (1947). A new disease of the variegated cutworm, *Peridroma margaritosa* (Haw.). *Science, 106*(2753), 323. https://doi.org/10.1126/science.106.2753.323

Steinhaus, E. A. (1975). *Disease in a minor chord: Being a semihistorical and semibiographical account of a period in science when one could be happily yet seriously concerned with the diseases of lowly animals without backbones, especially the insects*. The Ohio State University Press.

Tanada, Y., & Hess, R.T. (1984). The cytopathology of baculovirus infections in insects. In R. C. King & H. Akai (Eds.), *Insect ultrastructure*. Springer. https://doi.org/10.1007/978-1-4613-2715-8_14

Terra, W. R. (2001). The origin and functions of the insect peritrophic membrane and peritrophic gel. *Archives of Insect Biochemistry and Physiology, 47*(2), 47–61. https://doi.org/10.1002/arch.1036

Trager, W. (1935). The relation of cell size to growth in insect larvae. *Journal of Experimental Zoology, 71*(3), 489–508.

Wahl, U. (1909). Uber die Polyederkrankheit der Nonne (*Limantria monacha* L.). *Zentralbl. ges Forstwirtsch, 35*, 164.

Wang, M., & Hu, Z. (2019). Cross-talking between baculoviruses and host insects towards a successful infection. *Philosophical Transactions of the Royal Society b: Biological Sciences, 374*(1767), 20180324.

Wennmann, J. T., Köhler, T., Gueli Alletti, G., & Jehle, J. A. (2015). Mortality of cutworm larvae is not enhanced by *Agrotis segetum* granulovirus and *Agrotis segetum* nucleopolyhedrovirus B coinfection relative to single infection by either virus. *Applied and Environmental Microbiology, 81*(8), 2893–2899.

Williams, T., Virto, C., Murillo, R., & Caballero, P. (2017). Covert infection of insects by baculoviruses. *Frontiers in Microbiology, 8*, 1337.

Wraight, S. P., & Ramos, M. E. (2005). Synergistic interaction between *Beauveria bassiana* and *Bacillus thuringiensis tenebrionis* based biopesticides applied against field populations of Colorado potato beetle larvae. *Journal of Invertebrate Pathology, 90*(3), 139–150.

Wu, Z. W., Fan, J. B., Yu, H., Wang, D., & Zhang, Y. L. (2015). Ultraviolet protection of the *Cydia pomonella* granulovirus using zinc oxide and titanium dioxide. *Biocontrol Science and Technology, 25*(1), 97–107.

Yasser, N., Sayed, S., Ghanem, N., Thabit, A., Abdelal, A., Elmenofy, W., & Osman, E. (2022). Determination of virulence and genetic variability of three *Spodoptera littoralis* nucleopolyhedrovirus isolates from Egypt. *Egyptian Journal of Biological Pest Control, 32*(1), 15.

Zhang, Y., Enden, G., Wei, W., Zhou, F., Chen, J., & Merchuk, J. C. (2020). Baculovirus transit through insect cell membranes: A mechanistic approach. *Chemical Engineering Science, 223*, 115727.

Zhou, Y., Qin, T., Xiao, Y., Qin, F., Lei, C., & Sun, X. (2014). Genomic and biological characterization of a new cypovirus isolated from *Dendrolimus punctatus*. *PLoS ONE, 9*(11), e113201.

Nature-Based Active Ingredients: Harnessing Plants and Microbes for Enhanced Biopesticide Efficacy Against Insect Pests

Norhayu Asib, Mohamad Firdaus Ahmad, Masdah Mawi, and Dzolkhifli Omar

Abstract The aggressive attack of insect pests has led farmers to use of synthetic chemical pesticide as the main option to manage the infestation in most countries for decades mostly because its rapid effect. However, despite its effectiveness, there has been an increasing awareness on the harm caused by extensive and prolong use of chemical pesticide to non-target organisms and the environment as well as their highly cost price for the past few years. Recently, the use of natural substances as an alternative is gaining attention of researchers because the effect produced by this nature-based substance has been proven affective in laboratory and field trials. Thus, this review paper will contribute the knowledge and understanding of the potential of nature-based active ingredients for enhanced biopesticide efficacy against insect pests.

Keywords Plant-based biopesticides · Microbial-based biopesticides · Mechanism of action · Biopesticide formulation · Active ingredient

1 Introduction

Insect pests pose significant challenges to agricultural productivity and food security worldwide. The use of synthetic pesticides has long been the dominant approach for pest control, but concerns over their adverse effects on the environment, human health, and non-target organisms have led to the search for safer and more sustainable alternatives. Nature-based active ingredients derived from plants and microbes have emerged as promising candidates for biopesticide development. These natural sources offer a diverse array of compounds that can target insect pests effectively while minimizing ecological impacts. Exploring the potential of nature-based active

N. Asib (✉) · M. F. Ahmad · M. Mawi · D. Omar
Department of Plant Protection, Faculty of Agriculture, Universiti Putra Malaysia, Serdang, Malaysia
e-mail: norhayuasib@upm.edu.my

M. Wong (ed.), *Advances in Tropical Crop Protection*,
https://doi.org/10.1007/978-3-031-59268-3_4

ingredients and understanding their mechanisms of action can contribute to the development of improved biopesticides and sustainable pest management strategies.

The objectives of this review paper are twofold. Firstly, we aim to comprehensively examine the potential of plants and microbes as active ingredients for formulating biopesticides. Specifically, we will explore the effectiveness of botanical extracts, essential oils, allelochemicals, secondary metabolites, and plant-based bioinsecticides derived from plants. Additionally, we will delve into the utilization of Bacillus thuringiensis and its toxins, entomopathogenic fungi, and microbial secondary metabolites as active ingredients derived from microbes. Secondly, we will investigate the mechanisms of action underlying the efficacy of these nature-based active ingredients against insect pests. By understanding these mechanisms, we can gain insights into how to optimize formulation techniques for enhanced efficacy. Moreover, we will examine the field applications and efficacy of plant-based and microbial-based biopesticides through case studies. Lastly, we will discuss the challenges and prospects associated with the utilization of nature-based active ingredients for improved biopesticide efficacy, including regulatory considerations, formulation optimization, ecological impacts, and the integration of biopesticides into integrated pest management (IPM) strategies. By addressing these objectives, this review paper aims to contribute to the knowledge and understanding of the potential of nature-based active ingredients for enhanced biopesticide efficacy against insect pests.

2 Mechanisms of Action

2.1 Plants

2.1.1 Antifeedant and Deterrent Effects

Plants have natural compounds that repel insects and discourage feeding. Bitter compounds like tannins and alkaloids create an unpleasant taste, while volatile compounds like terpenoids and essential oils disrupt feeding behavior and protect crops (Lucía et al., 2018; Nwokafor et al., 2020; Tewari & Tiwari, 2018). Azadirachtin, found in *Azadirachta indica*, is the main insecticidal ingredient with antifeedant and growth-disrupting effects (Tewari & Tiwari, 2018). *Azadirachta indica* is better at lysing bacteria than *Moringa oleifera* at lower concentrations (Lucía et al., 2018). The aqueous extract of *A. indica* is particularly effective against *Escherichia coli* compared to *M. oleifera* and ginger extracts (Nwokafor et al., 2020). However, *M. oleifera* (1%) had faster mortality than *A. indica* (1%), suggesting *A. indica* inhibits *N. lugens* at a slower rate (Fariq et al., 2021). Additionally, the methanolic extract of *M. oleifera* had stronger antibacterial effects than *A. indica* (Lucía et al., 2018). These results show the antifeedant and deterrent effects of plant compounds, including Azadirachtin, *A. indica*, and *M. oleifera*, against insects and bacteria.

2.1.2 Disruption of Insect Growth and Development

Certain compounds derived from plants can interfere with the growth and development of insect pests. For example, some plant allelochemicals or secondary metabolites can mimic insect molting hormones or interfere with hormonal regulation, leading to developmental abnormalities or disruption of the insect's life cycle (Gaur & Kumar, 2020). These compounds can affect processes such as egg hatching, larval development, pupation, and metamorphosis, ultimately reducing the population growth and survival of insect pests (Gad et al., 2019).

2.1.3 Toxic Effects on Insect Physiology

Plants produce various toxic compounds that can directly target and disrupt the physiological processes of insects. These compounds may interfere with the functioning of the insect's nervous system, digestive system, or other vital physiological processes (Britto et al., 2021). For instance, certain alkaloids, such as nicotine or pyrethroids, can disrupt the nervous system of insects, leading to paralysis or death. Other toxic compounds may disrupt the insect's cellular membranes, interfere with enzyme activity, or induce oxidative stress, resulting in physiological dysfunction and mortality (Jaffar et al., 2022).

2.2 Microbes

2.2.1 Insecticidal Proteins and Toxins

Microbes, such as *Bacillus thuringiensis* (Bt) and *Cordyceps fumosorosea*, produce insecticidal proteins and toxins that selectively target specific insect pests (Wang et al., 2022). Microbe toxins, commonly used in biopesticides, are ingested by the insect and bind to receptors in their gut, initiating a sequence of events that leads to the formation of pores in the insect's gut membrane. This process causes cell lysis, gut paralysis, and ultimately, the death of the insect (Bowling et al., 2017). The specificity of these toxins toward certain insect orders or families ensures minimal harm to non-target organisms. In addition, the development of mycoinsecticides utilizing *C. fumosorosea* as an active ingredient offers an alternative approach to control the population of *Metisa plana*, reducing reliance on chemical insecticides (Nasir et al., 2023). In a trial, mycoinsecticide formulations developed with *C. fumosorosea* effectively reduced the bagworm population by over 95% (Nasir et al., 2023). Application of these mycoinsecticides in infested oil palm areas resulted in a significant reduction of the M. plana population by more than 95% within 30 days after treatment. Furthermore, the formulations did not cause a significant increase in mortality among the oil palm pollinator, *Elaeidobius kamerunicus*, indicating the potential of *C. fumosorosea* for bagworm management without harming pollinators on oil palm plantations (Nasir et al., 2023).

2.2.2 Enzymatic Degradation of Insect Cuticle

Some microbes, particularly entomopathogenic fungi, possess enzymatic capabilities to degrade the insect's cuticle, the protective outer layer of the exoskeleton. Fungal enzymes, such as chitinases and proteases, break down the structural components of the cuticle, allowing the fungi to penetrate and invade the insect's body (Huang et al., 2020). Once inside, the fungi proliferate and release toxins or cause physical damage, leading to the insect's demise.

2.2.3 Pathogenicity and Host Invasion

Microbes can exhibit pathogenicity toward insects, causing diseases that result in host mortality. Entomopathogenic bacteria, viruses, and other microorganisms have evolved mechanisms to invade and infect the insect host (Bondarchuk et al., 2021). These mechanisms may include the production of toxins, manipulation of the insect's immune response, or colonization of host tissues. The successful invasion and proliferation of the microbes within the insect host ultimately lead to the demise of the pest population.

Understanding the diverse mechanisms of action employed by nature-based active ingredients, whether derived from plants or microbes, is crucial for the development of effective biopesticides. By targeting specific aspects of insect biology and physiology, these active ingredients provide sustainable and eco-friendly solutions for insect pest management (Deka et al., 2021).

3 Formulation Techniques

3.1 Plant-Based Biopesticide Formulations

3.1.1 Extracts and Emulsions

Plant-based biopesticides can be formulated as liquid or semi-solid formulations using plant extracts. Extracts are obtained by extracting active compounds from plant materials using suitable solvents. These extracts can be formulated as emulsions, where the active ingredients are dispersed in water or oil-based emulsions for better stability and ease of application (Giunti et al., 2022). Emulsions provide uniform distribution of the active compounds and facilitate their adhesion to plant surfaces or penetration into insect pests (Fariq et al., 2021).

3.1.2 Encapsulation and Microencapsulation

Encapsulation and microencapsulation techniques involve coating plant-based active ingredients with protective materials to improve stability, controlled release, and efficacy (Arzani & Santo, 2022). Encapsulation can be achieved using natural polymers, such as starch or alginate, or synthetic polymers. The encapsulation process provides physical protection to the active compounds, preventing degradation and enhancing their longevity. Microencapsulation further reduces the risk of phytotoxicity and enhances the adhesion and absorption of the biopesticide.

3.1.3 Nanoparticles and Nanotechnology

Nanotechnology has opened new avenues for the formulation of plant-based biopesticides. Nanoparticles, such as liposomes, solid lipid nanoparticles, or polymeric nanoparticles, can be used to encapsulate plant-derived active compounds. Nanoparticles provide controlled release properties, protect the active ingredients from environmental degradation, and enhance their stability and bioavailability (Kumar et al., 2019). Additionally, nanotechnology enables targeted delivery and improved penetration into insect pests.

3.2 *Microbial-Based Biopesticide Formulations*

3.2.1 Formulations Incorporating Spores and Conidia

Microbial-based biopesticides, particularly those derived from entomopathogenic fungi and bacteria, can be formulated using spores or conidia. These formulations involve optimizing the concentration of spores or conidia, adding suitable additives for stability, and developing carrier materials for ease of application (Nasir et al., 2023). The spores or conidia, when applied to target pests, germinate and infect the insects, leading to pest control.

3.2.2 Biofilms and Gels

Biofilms and gels are alternative formulation approaches for microbial-based biopesticides. Biofilms consist of microbial cells embedded in a matrix, which provides protection, adhesion, and prolonged activity. Gels, on the other hand, are semi-solid formulations that contain microbial cells dispersed in a gel matrix. Both biofilms and gels enhance the stability and efficacy of the microbial-based biopesticides by prolonging the activity of the microorganisms and improving their adherence to plant surfaces (Flemming et al., 2016).

3.2.3 Fermentation and Bioencapsulation

Microbial-based biopesticides can also be formulated through fermentation processes. Fermentation involves the growth of microbial cultures in suitable nutrient media, allowing them to produce and accumulate bioactive metabolites. The resulting fermented broth can then be used as a formulation or further processed into solid or liquid formulations. Bioencapsulation techniques, like those used for plant-based biopesticides, can also be employed to protect and deliver the active microbial cells or metabolites.

Formulation techniques play a critical role in optimizing the stability, efficacy, and delivery of nature-based active ingredients in biopesticides. By selecting the appropriate formulation method, researchers can enhance the bioavailability, shelf life, and target specificity of the biopesticides, facilitating their integration into pest management programs. Continued advancements in formulation technologies will further contribute to the development of efficient and sustainable nature-based biopesticides for insect pest control.

4 Field Applications and Efficacy

4.1 Plant-Based Biopesticides

Several case studies have demonstrated the efficacy of plant-based biopesticides in controlling insect pests. For example, neem-based biopesticides derived from the neem tree (*Azadirachta indica*) have shown effectiveness against a wide range of pests, including aphids, caterpillars, and beetles (Fariq et al., 2021). Neem extracts containing azadirachtin, a potent bioactive compound, have exhibited antifeedant, growth-regulating, and repellent effects on insect pests (Fariq et al., 2021). Similarly, pyrethrum-based biopesticides, derived from the flowers of *Chrysanthemum cinerariaefolium*, have been successful in managing pests such as mosquitoes, flies, and moths. The pyrethrins present in pyrethrum extracts act as neurotoxins, targeting the nervous systems of insects.

4.2 Microbial-Based Biopesticides

Microbial-based biopesticides have also demonstrated efficacy in pest control. One prominent example is the use of *Bacillus thuringiensis* (Bt) and *Cordyceps fumosorosea*-based biopesticides. The formulations have been employed to manage pets, such as the bagworm in oil palm (Nasir et al., 2023). The specificity of toxins ensures that they primarily affect target pests while minimizing harm to beneficial organisms. Another case study involves the use of entomopathogenic fungi, such as *Beauveria bassiana*, for controlling pests like whiteflies, aphids, and thrips (Sayed et al., 2021). These fungi infect insects and cause disease, leading to reduced pest populations in field conditions.

4.3 Synergistic Approaches: Plant–Microbe Biopesticides

Synergistic approaches that combine plant-based and microbial-based biopesticides have shown great potential in pest management (Suriani et al., 2021). For instance, the use of plant extracts or essential oils in combination with entomopathogenic microbes has been explored to enhance efficacy. The antimicrobial and insecticidal properties of plant compounds can augment the pathogenic effects of the microbes, resulting in improved pest control outcomes. Such synergistic approaches have been reported in studies targeting pests like the diamondback moth and agricultural mites.

These case studies highlight the practical applications and effectiveness of nature-based biopesticides in pest management. Field trials and experiments have demonstrated their ability to control a wide range of insect pests while minimizing environmental impact and non-target effects (Rahman et al., 2019). The integration of these biopesticides into integrated pest management (IPM) programs offers sustainable and eco-friendly alternatives to synthetic pesticides, reducing reliance on chemical inputs and promoting ecological balance in agricultural systems (Rahman et al., 2019). Further research and field studies will continue to refine and optimize the efficacy of nature-based biopesticides, leading to their wider adoption and implementation.

5 Challenges and Future Prospects

5.1 Regulatory Hurdles and Commercialization

One of the significant challenges for nature-based biopesticides is navigating the regulatory requirements for their registration and commercialization. The process often involves extensive testing and documentation to ensure safety, efficacy, and environmental compatibility. Regulatory agencies have specific guidelines and standards that need to be met, which can be time-consuming and costly. Streamlining the regulatory processes and providing support for small-scale producers and researchers can facilitate the development and market access of nature-based biopesticides.

5.2 Ecological Considerations and Non-target Effects

While nature-based biopesticides are generally considered safer and more environmentally friendly than synthetic pesticides, it is important to consider their potential ecological impacts. Careful evaluation of the non-target effects on beneficial organisms, such as pollinators and natural enemies, is necessary. Additionally, understanding the potential for resistance development in target pests and implementing strategies to mitigate resistance is crucial. Research focusing on the ecological risk assessment and mitigation of nature-based biopesticides will contribute to their sustainable use and integration into pest management strategies.

5.3 Integrated Pest Management (IPM) Strategies

The prospects of nature-based biopesticides lie in their integration into holistic pest management approaches, such as integrated pest management (IPM). IPM combines various pest control methods, including cultural practices, biological control, and judicious pesticide use. Nature-based biopesticides can play a vital role as a component of IPM strategies, offering effective pest control while minimizing the reliance on synthetic pesticides. Integrating biopesticides with other pest management tactics, such as crop rotation, habitat manipulation, and monitoring techniques, can lead to sustainable and resilient pest management systems.

6 Conclusions

Nature-based active ingredients derived from plants and microbes hold significant promise for enhancing biopesticide efficacy against insect pests. The diverse mechanisms of action, formulation techniques, and field applications demonstrate their potential as sustainable alternatives to synthetic pesticides. However, several challenges need to be addressed to fully harness the benefits of these biopesticides. Overcoming regulatory hurdles and facilitating commercialization processes will support their wider adoption. Optimization of formulations and delivery systems will enhance stability and efficacy. Attention to ecological considerations and non-target effects will ensure their environmental compatibility. Finally, integrating biopesticides into IPM strategies will promote sustainable pest management practices. Continued research, collaboration, and support are essential for realizing the full potential of nature-based biopesticides and achieving a more sustainable and environmentally friendly approach to insect pest control.

References

Arzani, F. A., & Dos Santos, J. H. Z. (2022). Biocides and techniques for their encapsulation: A review. *Soft Matter, 18*(29), 5340–5358. https://doi.org/10.1039/d1sm01114f

Bondarchuk, E. Y., Tsygichko, A. A., Tomashevich, N. S., & Asaturova, A. (2021). Entomopathogenic activity of bacterial and viral strains from the bioresource collection "State Collection of Entomoacariphages and Microorganisms". *E3S Web of Conferences, 285*, 03004. https://doi.org/10.1051/e3sconf/202128503004

Bowling, A. J., Pence, H. E., Li, H., Tan, S. Y., Evans, S. L., & Narva, K. E. (2017). Histopathological effects of Bt and TcdA Insecticidal proteins on the midgut epithelium of western corn rootworm larvae (*Diabrotica virgifera*). *Toxins, 9*(5), 156. https://doi.org/10.3390/toxins9050156

Britto, I. O., Araújo, S. H. C., Toledo, P. F. S., Lima, G. D. A., Salustiano, I. V., Alves, J. R., Mantilla-Afanador, J. G., Kohlhoff, M., Oliveira, E. E., & Leite, J. P. V. (2021). Potential of *Ficus carica* extracts against *Euschistus heros*: Toxicity of major active compounds and selectivity against beneficial insects. *Pest Management Science, 77*(10), 4638–4647. https://doi.org/10.1002/ps.6504

Deka, B., Baruah, C., & Babu, A. (2021). Entomopathogenic microorganisms: Their role in insect pest management. *Egyptian Journal of Biological Pest Control, 31*, 1–8.

Fariq, M. N., Mohkheri, M., Mawi, M., Ishak, I., & Asib, N. (2021). Synergism activity between *Moringa oleifera* and *Azadirachta indica* crude extract against *Nilaparvata lugens* (Hemiptera: Delphacidae). *Serangga, 26*(4), 72–83.

Flemming, H. C., Wingender, J., Szewzyk, U., Steinberg, P., Rice, S. A., & Kjelleberg, S. (2016). Biofilms: An emergent form of bacterial life. *Nature Reviews Microbiology, 14*(9), 563–575.

Gad, M. A., Elwassimy, M. M., Aref, S. A., & Abdelhamid, A. A. (2019). Chemical design and effects of new insect growth regulators as potential insecticidal agents on *Spodoptera littoralis* (Boisd.). *Novel Research in Science, 2*(3).

Gaur, S. K., & Kumar, K. (2020). Toxicity and insect growth regulatory activities of medicinal plant, Withania somnifera, in flesh fly, Sarcophaga ruficornis (Diptera: Sarcophagidae). *The Journal of Basic and Applied Zoology, 81*.

Giunti, G., Laudani, F., Lo, P. E., Bacchi, M. R., Palmeri, V., & Campolo, O. (2022). Contact toxicity and ovideterrent activity of three essential oil-based nano-emulsions against the olive fruit fly *Bactrocera oleae*. *Horticulturae, 8*(3), 240. https://doi.org/10.3390/horticulturae8030240

Huang, W., Huang, Y., Hao, Y., Huang, S., Gao, T., Keyhani, N. O., & Huang, Z. (2020). Host-dependent contributions of the Cfcdp1 protease gene to virulence in the entomopathogenic fungus *Cordyceps fumosorosea*. *Pest Management Science, 76*(2), 575–588. https://doi.org/10.1002/ps.5549

Jaffar, S., Ahmad, S., & Lu, Y. (2022). Contribution of insect gut microbiota and their associated enzymes in insect physiology and biodegradation of pesticides. *Frontiers in Microbiology, 13*, 979383. https://doi.org/10.3389/fmicb.2022.979383

Kumar, A., Singh, P., & Yadav, R. N. (2019). Nanotechnology: A promising tool for plant-based insect pest management. In *Nanotechnology in agriculture* (pp. 235–252). Springer.

Lucía, A., Miguel, A. A., Stefany, C., Néstor, G., & Juana, V. (2018). Antibacterial and cytotoxic effects of *M. oleifera* (Moringa) and *Azadirachta indica* (Neem) methanolic extracts against strains of *Enterococcus faecalis*. *International Journal of Dentistry, 2018*, 1–5.

Nasir, A. A., Syarif, N. Y., Omar, D., & Asib, N. (2023). Effectiveness of *Cordyceps fumosorosea* wettable powder formulation against *Metisa plana* (Walker) and its side effects on *Elaeidobius kamerunicus* in oil palm plantation. *Journal of Pesticide Science, 48*(2), 54–60. https://doi.org/10.1584/jpestics.D22-072

Nwokafor, C., Udensi, C., Ogbonna, H. N., Udekwu, C., Nwankpa, U., Amanze, E., Chibuzor, W., & Okeke, K. (2020). Antimicrobial activities of M.oleifera, A.indica and ginger plant extracts against bacteria associated with the spoilage of fruit juice. *South Asian Journal of Research in Microbiology, 7*, 21–30.

Rahman, M., Howlader, M. T., Islam, K., & Morshed, N. (2019). Efficacy of three biopesticides against cucurbit fruit fly, Bactrocera cucurbitae Coquillett (Diptera: Tephritidae) and yield of bitter gourd. *Journal of the Bangladesh Agricultural University, 17*, 483–489.

Sayed, S., Al-Otaibi, S., El-Shehawi, A., Elarnaouty, S. A., El-Shazly, S., Gaber, A., & Ibrahim, R. (2021). Field Evaluation of native fungus, *Beauveria bassiana* (Bals.) vuillemin against some piercing-sucking insects on the grapevine. *Pakistan Journal of Biological Sciences, 24*(1), 158–164. https://doi.org/10.3923/pjbs.2021.158.164

Suriani, N. L., Suprapta, D. N., Indrayani, A. W., Herlambang, S., Resiani, N. M., Al-Shwaiman, H. A., Al Khulaifi, M. M., Elgorban, A. M., Datta, R., Gunawan, S., & Zuan, A. T. (2021). The synergistic action of three piper plant extracts and biofertilizer for growth promotion and biocontrol of blast disease in red rice. *Sustainability, 13*(18), 10412. https://doi.org/10.3390/su131810412

Tewari, A., & Tiwari, S. (2018). *Synthesis of Medicinal Agents from Plants*. Elsevier.

Wang, W., Zhang, D., Zhao, S., & Wu, K. (2022). Susceptibilities of the invasive fall armyworm (*Spodopterafrugiperda*) to the insecticidal proteins of Bt maize in China. *Toxins, 14*(8), 507. https://doi.org/10.3390/toxins14080507

Pink Pineapple Mealybug, *Dysmicoccus Brevipes* (Hemiptera: Pseudococcidae): An Overview

Siti Mahsuri Talib and Anis Syahirah Mokhtar

Abstract Small yet disastrous is the best word to describe *Dysmicoccus brevipes* (Cockerell). Known as pink pineapple mealybug (PPM), this polyphagous pest has infested more than 159 plant genera and spread across the globe. This chapter reviews the life cycle and infestation of *D. brevipes* and its management.

Keywords *Dysmicoccus brevipes* · Pineapple · Mealybug · PPM · Pest

1 Introduction

Dysmicoccus brevipes (Cockerell, 1893), also known as pink pineapple mealybug (PPM), is a major global insect pest. It primarily affects pineapple and other potential hosts like citrus, nuts, cocoa, sugarcane, and other crops (Bertin et al., 2013). These mealybugs were named when first discovered in Jamaica on pineapples (*Ananas comosus*) in 1893. They are believed to be native to tropical America (Carter, 1932; Waterhouse, 1998) and widely distributed in South America (Granara de Willink, 2009). Even though it has a common name, this troublesome insect is polyphagous, meaning it can feed on a variety of plants. The small sap-sucking insect belongs to the Pseudococcidae family under the suborder of Sternorrhyncha of Hemiptera and is known to cause damage directly and indirectly to the host plants. *Dysmicoccus brevipes* undergoes asexual reproduction through obligate parthenogenesis. Nevertheless, in some instances, *D. brevipes* reproduces sexually with biparental parents. However, there is no detailed report on the condition and environment that might affect the parthenogenesis of PPM.

S. M. Talib · A. S. Mokhtar (✉)
Department of Plant Protection, Faculty of Agriculture, Universiti Putra Malaysia, 43400 UPM Serdang, Selangor, Malaysia
e-mail: anissyahirah@upm.edu.my

M. Wong (ed.), *Advances in Tropical Crop Protection*,
https://doi.org/10.1007/978-3-031-59268-3_5

1.1 Life Cycle

Known for having incomplete metamorphosis, *D. brevipes* has three stages throughout its life cycle: eggs, nymph and adult. They are also a parthenogenetic insect species, where they undergo asexual reproduction without egg fertilization by males.

Adult: PPM adults reproduce through ovoviviparity, a method in which the adult retains the eggs within its body. This means the eggs develop and hatch internally before being born (Pandey & Johnson, 2006). The newborn crawlers, also known as first instar nymphs, come out from the female's body, as illustrated in Fig. 1. Ito (1938) has recorded that female adults produce about 234 crawlers per individual or 19–137 crawlers, according to Waterhouse (1998). On average, an adult female can produce about 250 crawlers, but this number can be up to 1000, as described by Egelie and Gillett-Kaufman (2015), during certain conditions.

In the condition where parthenogenesis reproduction occurs, males are absent. Adult females of *D. brevipes* can live for up to 56 days according to Mamahit et al. (2008). The distinct life cycles of female and male mealybugs are evident in biparental PPM. The female goes through three nymphal instars before transitioning into the adult phase. Conversely, the male follows a sequence of two nymphal instars, a prepupa stage, pupa, and eventually attains adulthood (Colen et al., 2000; Lim, 1973; Santa-Cecília et al., 2004). The male has a shorter life cycle than the female, around 32 days (Santa-Cecília et al., 2004), while the female is around 56 days, approximately (Petty et al., 2002). According to Waterhouse (1998), the lifespan of adult females ranges from 7 to 49 days. Before giving birth to crawlers, a female adult has a preoviposition period lasting around 27 days, followed by an oviposition period of about 25 days. The adult females live approximately five more days after parturitions of the progeny before dying (Mau & Kessing, 2007). The PPM has an average

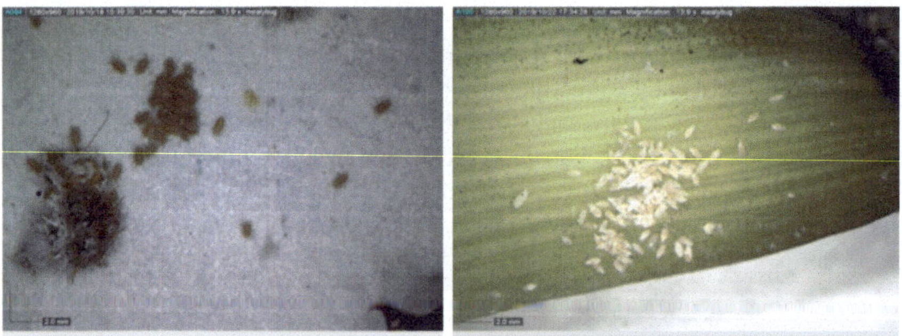

(a) (b)

Fig. 1 **a** First instar nymphs (crawlers) of *D. brevipes* emerged from an adult female's body. **b** The crawlers on pineapple leaf. Both **a** and **b** were observed using a Dino-Lite digital microscope (Dino-Lite Premier Digital Microscope AM-7013M2T with 13.9× magnification). *Source* Mahsuri, UPM

lifespan of 95 days, ranging from 78 to 111 days (Mau & Kessing, 2007; Petty et al., 2002). In comparison, Manjushree et al. (2018) reported that the average duration required to complete a life cycle of PPM in Kerala ranged from 60 to 70 days. The life cycle, survival, reproduction and development of PPM vary depending on the host plant (Santa-Cecília et al., 2004; Mamahit et al., 2008; Bua et al., 2013).

Dysmicoccus brevipes exhibits a notably small adult female size appearance, approximately three millimeters long and one millimeter wide (Mamahit et al., 2008). Oval-shaped and convex, they have a plum-like appearance with a pink-orange hue. In addition to being covered in a white waxy and powdery secretion, these structures are encircled by 17 pairs of waxy filaments. With eight antenna segments and discoidal pores around the eyes, the hind femur and tibia of the metacoxae display clear pores that allow light to pass through. Additionally, they have a circulus, a well-developed ostiole, and a lack of oral rim tubular ducts. The dorsal setae located on the eighth abdominal segment, positioned before the anal ring, exhibit a greater length compared to other dorsal setae. Moreover, these setae take on a quadrate shape within the sclerotization of the ventral anal lobe (Williams & Watson, 1988; Beardsley, 1992; Williams & Granara de Willink, 1992; Petty et al., 2002; Williams, 2004; Mau & Kessing, 2007; Santa-Cecília et al., 2007; Suh & Hyun, 2015). The adult females, as well as the immature nymphs, lack wings. In contrast, if males are present, they have wings and possess antennas consisting of eight segments, along with a pair of caudal filaments resembling those found in females (Beardsley, 1965).

Nymph: The nymphal stage of PPM females includes three instars, with the first instar nymph (crawler) being the most active stage, particularly in movement and feeding. Mau and Kessing (2007) reported crawlers as the primary dispersal stage of PPM infestation. Crawlers can move far from their original location and can be spread by wind and animals. The nymphs look like scaled-down adults, featuring compact, white bodies coated in wax and adorned with lengthy wax filaments. They are very similar except for a slight variation in size. As Mamahit et al. (2008) revealed, the nymph's development period spans from 32 to 36 days. The duration of the first instar spans from 10 to 26 days, followed by the second instar lasting between 6 and 22 days, and finally, the third instar takes place for a period ranging from 7 to 24 days. Besides that, research conducted by Mau and Kessing (2007) revealed that the overall duration required for the transition from the nymph to the adult stage (see Fig. 2) falls within the range of approximately 26–55 days, with an average of 34 days.

1.2 Host Preference

PPM live in colonies and can be found in all parts of the plant, especially in protected places (Luz et al., 2005), such as hidden underneath the plant part or base area of the plant host, as shown in Fig. 3. They are often found feeding on sap from roots and leaf joints but can also be spotted on fruits, upper leaf surfaces, crowns, and seedlings (Sanches & Matos, 1999). González-Hernández (1995) reported that on pineapples,

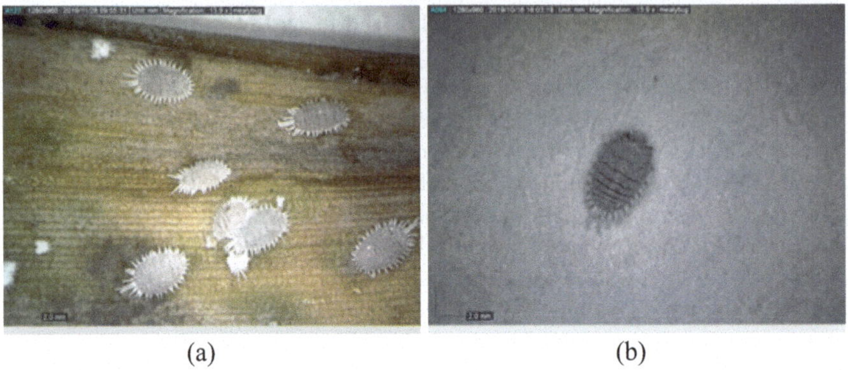

Fig. 2 **a** Nymphs and female adults of *D. brevipes* on pineapple leaf. **b** Female adult of *D. brevipes*. Both **a** and **b** were observed using a Dino-Lite digital microscope (Dino-Lite Premier Digital Microscope AM-7013M2T with 13.9× magnification). *Source* Mahsuri, UPM

Fig. 3 **a** *D. brevipes* colony on pineapple's crown. **b** *D. brevipes* also infest other parts of the pineapple, including the stem. **c** Colony of *D. brevipes* infesting pumpkin fruit. *Source* Mahsuri, UPM

the PPM is commonly found both below ground on the roots and above ground at various levels, including the aerial parts of the plant. Besides that, they were also found attached to the leaves, fruits, blossom cups and crowns (González-Hernández et al., 1999; Mamahit et al., 2010).

1.3 Association of Ants

Dysmicoccus brevipes are often found in association with ants in the field. Among the ants that show a close association with PPM are the red fire ant, *Solenopsis* sp. (Formicidae), *Camponotus mitis* (Formicinae) and the white-footed ant, *Technomyrmex albipes* (Dolichoderinae) (Hasibuan, 2005; Manjushree et al., 2018). The ants harvest honeydew as a source of carbohydrates from PPM, offer protection against natural enemies, and transport PPM to the nearest location or new host plants (González-Hernández et al., 1999; Jahn & Beardsley, 2000). Ants feeding on honeydew could also help mealybugs by preventing the buildup of honeydew on their bodies.

2 Infestation and Damage

The infestation and damages done by *D. brevipes* are discussed below:

2.1 Infestation

Primarily documented in tropical and subtropical regions, PPM can be found in various areas, such as South Pacific islands, Central and South America, Micronesia and Southern Asia, as indicated by research from Williams and Watson (1988), Williams and Granara de Willink (1992), Huang et al. (2002), Sether (2002), Granara de Willink (2009) and Dey et al. (2018). As reported by Scalenet (2022), its presence has now extended to over 126 countries. PPM has spread worldwide as a cosmopolitan pest associated with plant transportation and cultivation where later, PPM has been reported in Hawai (Beardsley et al., 1982), Colombia (Kondo et al., 2008), Cuba (Blanco et al., 2003), Uganda (Kabi et al., 2016; Kubiriba et al., 2001), Brazil (da Silva et al., 2016; Daane et al., 2012) and Hungary (Kozár et al., 2013).

Meanwhile, in Asia, PPM was recorded in many countries, such as India (Rishi et al., 2019), Sri Lanka (Sirisena et al., 2013), Taiwan (Huang et al., 2002; Mau & Kessing, 2007), China (He et al., 2012), South Korea (Suh & Hyun, 2015) and Iran (Moghaddam, 2004). Williams (2004) reported PPM occurrence in almost all Southeast Asia countries: Philippines, Thailand, Indonesia, Brunei, Singapore, Laos and Malaysia. In 2015, PPM was intercepted in Korea on a Ficus plant imported from Malaysia (Suh & Hyun, 2015). To date, PPM is recognized as a significant pest of pineapple and other potential host plants in Malaysia, but information about PPM remains limited.

2.2 Damages

Pineapple plants affected by infestation exhibit signs such as yellowing leaves, stunted growth, wilting, reduced leaf count, diminished root length and weight, and root decay. These symptoms ultimately lead to lower yields and a diminished plant population (Bua et al., 2013; Colen et al., 2000; Lim, 1973; Santa-Cecília et al., 2004). Infestations of PPM also lead to a decline in sugar content, hindered development, and weakened plant vigor (Godfrey et al., 2002; Bertin et al., 2013). On mangroves, PPM sucks the sap on the soft tissues of the leaves and injects saliva that causes curling and contortion of leaves (Rishi et al., 2019). The whole plant displayed stunted growth, and the shoot tips seemed dried out. Moreover, mealybugs secrete honeydew, which fosters the growth of sooty mold. This mold adversely affects the photosynthesis rate, quality and commercial value of the fruits. The existence of mealybugs and the resulting contamination from their feeding can cause aesthetic harm to the fruits, rendering them unsuitable for commercial purposes and consequently diminishing their market appeal (Daane et al., 2008; González & Volosky, 2004; Kishino et al., 2007).

PPM directly weakens plants and causes indirect damage as major vectors by transmitting the virus. PPM serves as a carrier for various plant viruses, including the pineapple mealybug wilt-associated virus (PMWaV) responsible for mealybug wilt disease in pineapples (Hughes & Samita, 1998; Sether & Hu, 2002; Sether et al., 2005). Additionally, PPM transmits the grapevine plant virus (Daane et al., 2012; Walton & Pringle, 2004) and Banana Streak Virus (BSV) (Kubiriba et al., 2001). Furthermore, PPM functions as a vector for the transmission of the Cacao Mottle Leaf virus (CMLV) and Cacao Trinidad Virus (CTV) (Harris, 1981). The virus transmitted will cause leaf symptoms, including dieback and discoloration due to root damage, making the plant lose rigidity (Rohrbach & Johnson, 2003; Sether & Hu, 2002). The infected plant will show a reduction in root length, weight and leaf size (Egelie & Gillett-Kaufman, 2015; Rohrbach & Schmitt, 2003). In Cuba, Hernández et al. (2010) documented a 40% decrease in yield caused by the transmission of PMWaV-3 (Pineapple Mealybug Wilt-associated Virus-3) disease. Meanwhile, in Central Uganda in 2013, the prevalence of pineapple mealybug wilt disease in the production fields varied between 15 and 100% (Bua et al., 2013).

3 Management

Managing *D. brevipes* infestation is quite challenging due to their secretive hiding habit and the wax coating covering their body. Preplanting treatment should be practiced as a prevention method. The selection of healthy sucker for planting is crucial (Hepton, 2003; Mamahit et al., 2008). The sucker can be dipped in insecticidal solutions like malathion and diazinon (Dey et al., 2018; Rohrbach & Johnson, 2003; Sipes, 2000) or botanical extracts (Joy et al., 2012) right before planting. The insecticides, such as imidacloprid, chlorpyriphos and quinalphos, can be applied to kill PPM (Geiger & Daane, 2001; Joy et al., 2012). As an alternative to chemical control, studies were conducted to understand and find a better management plan to control PPM using botanical-based insecticides. Azadirachtin and neem oil are potentially effective in causing PPM mortality (Memon et al., 2017; Rishi et al., 2019). When sprayed once on pineapple plants, a 10% concentration of *Azadirachta indica* extract and 5–10% *Cyperus iria* extracts may negatively impact the PPM population. This suggests its potential use in Integrated Pest Management (IPM), either alone or in conjunction with other control methods (Lira et al., 2020). The use of natural and predatory ladybird beetles can also contribute to suppressing the population of sap-sucking insects (Rishi et al., 2019). Saranya et al. (2010) and Manjushree et al. (2018) showed the employment of microbial entomopathogens, including *Metarhizium anisopliae*, *Beauveria bassiana* and *Lecanicillium lecanii*. So far, *L. lecanii* has been reported as the best entomopathogen to control the PPM at a certain concentration, with 90.04% mortality (Saranya et al., 2010). Zart et al. (2021) reported recent findings on the potential of entomopathogenic nematodes as a biological control of PPM. In addition, *Anagyrus ananatis* (González-Hernández, 1995) and *Spalgis epeus* (Manjushree et al., 2018) can be potential natural enemies for biocontrol of PPM. Besides that, an excellent cultural system, sanitation and management of crops can be good measures to control and prevent the infestation of PPM, as suggested by Sulaiman (2000). For example, fumigation treatment with ethyl formate is considered an effective and potential phytosanitary treatment to manage PPM (Abdullah et al., 2020; Petty et al., 2002). It is also believed that PPM can be minimized by suppressing the ant population and eradicating the weeds and alternate hosts nearby in the field (Egelie & Gillett-Kaufman, 2015; Joy et al., 2012). In brief, IPM should be implemented to control the infestation of *D. brevipes* in the field to get maximum control without causing any negative impact on humans and the environment.

4 Perspectives

While there is a considerable amount of literature on the insect diversity of *D. brevipes*, there is limited information and current updates on this pest in Malaysia. PPM has the potential to cause significant damage to both pineapple and other host plants. PPM not only inflicts harm on crops directly and indirectly but also poses a quarantine threat. This situation leads to elevated production costs in preventing or eradicating PPM from plants and agricultural products. Thus, further studies on these species and their management are crucial to prevent the spread and infestation of PPM.

Acknowledgements The authors would like to express gratitude to Universiti Putra Malaysia for the financial support provided under Geran Putra—Inisiatif Putra Muda (GP-IPM) Grant, VOT number: 9677200.

References

Abdullah, S. A., Yusof, S., Osman, M. S., Tajuddin, M. K., Ariffin, S. I., & Rosali, N. (2020). Simulation study of vapor mate fumigation treatment against pineapple mealy bug (*D. brevipes*) and Pineapple Var. MD2 quality evaluation meant for China export market. *International Journal of Agriculture, Forestry and Plantation, 10*, 183–186.

Beardsley, J. W. (1965). Notes on the pineapple mealybug complex, with descriptions of two new species (Homoptera: Pseudocoecidae). *Proceedings of the Hawaiian Entomological Society, 19*(1), 55–68.

Beardsley, J. W. (1992). The pineapple mealybug complex; taxonomy, distribution and host relationships. *International Pineapple Symposium, 334*, 383–386.

Beardsley, J. W., Su, T. H., McEwen, F. L., & Gerling, D. (1982). Field investigations on the interrelationships of the big- headed ant, the gray pineapple mealybug, and pineapple mealybug wilt disease in Hawaii. *Proceedings of the Hawaiian Entomological Society, 24*, 51–67.

Bertin, A., Bortoli, L. C., Botton, M., & Parra, J. R. P. (2013). Host plant effects on the development, survival, and reproduction of *D. brevipes* (Hemiptera: Pseudococcidae) on grapevines. *Annals of the Entomological Society of America, 106*, 604–609.

Blanco, R. E., Pérez, V. I., & Rodríguez, C. A. M. (2003). Encuesta de los pseudocóccidos de Cuba. Resultados del período 2001–2002. *Fitosanidad, 7*(2), 31–36.

Bua, B., Karungi, J., & Kawube, G. (2013). Occurrence and effects of pineapple mealybug wilt disease in central Uganda. *Journal of Agricultural Science and Technology A, 3*(5), 410–416.

Carter, W. (1932). The pineapple mealybug (*Pseudococcus brevipes*) and wilt of pineapples (abstract). *Phytopathology, 20*(12), 996–997.

Cockerell, T. D. A. (1893). The west Indian species of Dactylopius. *The Entomologist., 26*, 177–179.

Colen, K. G. F., Santa-Cecília, L. V. C., Moraes, J. C., & Reis, P. B. (2000). Efeitos de diferentes temperaturas sobre a biologia da cochonilha pulverulenta *D. brevipes* (d) (Hemiptera: Pseudococcidae). *Revista Brasileira De Fruticultura, 22*, 248–252.

da Silva, P. V. C., Kaydan, M. B., Germain, J., Malausa, T., & Botton, M. (2016). Three new species of mealybug (Hemiptera, Coccomorpha, Pseudococcidae) on persimmon fruit trees (*Diospyros kaki*) in southern Brazil. *ZooKeys, 584*, 61–82.

Daane, K. M., Almeida, R. P. P., Bell, V. A., Walker, J. T. S., Botton, M., Fallahzadeh, M., Mani, M., Miano, J. L., Sforza, R., Walton, V. M., & Zaviezo, T. (2012). Biology and management of mealybugs in vineyards. In N. J. Bostanian, C. Vincent, & R. Isaacs (Eds.), *Arthropod management in vineyards: Pests, approaches, and future directions* (pp. 271–307). Springer.

Daane, K. M., Cooper, M. L., Triapitsyn, S. V., Walton, V. M., Yokota, G. Y., Haviland, D. R., Bentley, W. J., Godfrey, K. E., & Wunderlich, L. R. (2008). Vineyard managers and researchers seek sustainable solutions for mealybugs, a changing pest complex. *California Agriculture, 62*, 167–176.

Dey, K., Green, J., Melzer, M., Borth, W., & Hu, J. (2018). Mealybug wilt of pineapple and associated viruses. *Horticulturae, 4*(4), 52.

Egelie, A. A., & Gillett-Kaufman, J. L. (2015). Pineapple mealybug, *D. brevipes* (Cockerell) (Insecta: Hemiptera: Pseudococcidae). *Entomology* and Nematology Department, University of Florida. UF/IFASExtension Publication Number EENY-635. https://edis.ifas.ufl.edu/pdffiles/IN/IN110600.pdf. Accessed 25 May 2021.

Geiger, C. H., & Daane, K. M. (2001). Seasonal movement and distribution of grape mealybug (Homoptera: Pseudococcidae): Developing sampling program for San Joaquin Valley vineyards. *Journal of Economic Entomology, 94*(1), 291–301.

Godfrey, K. E., Daane, K. M., Bentley, W. J., Gill, R. J., & Malakar-Kuenen, R. (2002). *Mealybugs in California vineyards*. University of California Agriculture and Natural Resources (UC ANR) Publication.

González, R. H., & Volosky, C. (2004). Chanchitos blancos y polillas de la fruta: Problemas cuarentenarios de la fruticultura de exportación. *Revista Fruticola, 25*, 41–62.

González-Hernández, H. (1995). *The status of biological control of pineapple mealybugs in Hawaii.* Ph.D. Dissertation, Department of Entomology, College of Tropical Agriculture and Human Resources, University of Hawaii.

González-Hernández, H., Reimer, N. J., & Johnson, M. W. (1999). Survey of the natural enemies of Dysmicoccus mealybugs on pineapple in Hawaii. *BioControl, 44*, 47–58.

Granara de Willink, M. C. (2009). Dysmicoccus de la Región Neotropical (Hemiptera: Pseudococcidae). *Revista De La Sociedad Entomológica Argentina, 68*, 11–95.

Harris, K. F. (1981). Arthropod and nematode vectors of plant viruses. *Annual Review of Phytopathology, 19*(1), 391–426.

Hasibuan, R. (2005). Hubungan kutu dompolan *D. brevipes* (Ckll.) (Homoptera: Pseudococcidae) dan semut api *Solenopsis* sp. (Hymenoptera: Formicidae) pada dua cara bertanam nenas. *Jurnal Hama dan Penyakit Tumbuhan Tropika, 5*(1), 17–23.

He, Y. B., Wan, X. W., Liu, Y. H., Sun, G. M., & Zhan, R. L. (2012). Mitochondrial COI from *D. brevipes* (Hemiptera: Pseudococcidae) suggests cryptic lineage and pinpoints the source of the introduction to China. *Florida Entomologist, 95*(1), 183–191.

Hepton, A. (2003). *Cultural system, in the pineapple botany, production and uses* (pp. 109–142). CAB International Publication.

Hernández, L., Ramos, P. L., Rodríguez, M., Peña, I., & Pérez, J. M. (2010). First report of Pineapple mealybug wilt associated virus-3 infecting pineapple in Cuba. *New Disease Reports, 22*, 18.

Huang, S. H., Wong, C. Y., & Cheng, C. H. (2002). A newly recorded insect pest, pink pineapple mealybug (*D. brevipes* (Cockerell)) (Hemiptera: Pseudococcidae), infecting on the roots of peanut in Taiwan. *Plant Protect Bulletin, 44*(2), 141–146.

Hughes, G., & Samita, S. (1998). Analysis of patterns of pineapple mealybug wilt disease in Sri Lanka. *Plant Disease, 82*, 885–890.

Ito, K. (1938). Studies on the life history of the pineapple mealybug, *D. brevipes* (Ckll.). *Journal of Economic Entomology, 31*, 291–298.

Jahn, G. C., & Beardsley, J. W. (2000). Interactions of ants (Hymenoptera: Formicidae) and mealybugs (Homoptera: Pseudococcidae) on pineapple. *Proceedings of the Hawaiian Entomological Society, 34*, 181–185.

Joy, P. P., Anjana, R., & Sournya, K. K. (2012). *Pest of pineapple and their management.* Pineapple Research Station (Kerala Agricultural University).

Kabi, S., Karungi, J., Sigsgaard, L., & Ssebuliba, J. M. (2016). *Dysmicoccus brevipes* (Cockerell) occurrence and infestation behaviour as influenced by farm type, cropping systems and soil management practices. *Agriculture, Ecosystems and Environment, 222*, 23–29.

Kishino, A. Y., Carvalho, S. L. C., & Roberto, S. R. (2007). *Viticultura tropical: O sistema de produção do Paraná.* Instituto Agronômico do Paraná.

Kondo, T., Ramos Portilla, A. A., & Vergara Navarro, E. V. (2008). Updated list of mealybugs and putoids from Colombia (Hemiptera: Pseudococcidae and Putoidae). *Boletín Del Museo De Entomología De La Universidad Del Valle, 9*(1), 29–53.

Kozár, F., Benedicty, Z. K., Fetykó, K., Kiss, B., & Szita, E. (2013). An annotated update of the scale insect checklist of Hungary (Hemiptera, Coccoidea). *ZooKeys, 309*, 49–66.

Kubiriba, J., Legg, J. P., Tushemereirwe, W., & Adipala, E. (2001). Vector transmission of Banana streak virus in the screenhouse in Uganda. *Annals of Applied Biology, 139*, 37–43.

Lim, W. H. (1973). Studies on the bisexual race of *D. brevipes* Ckll.: Its bionomics and economic importance. *Malaysian Agricultural Journal, 49*, 254–267.

Lira, J. A., Pereira, R. R. D. C., Costa, L. D. O., Nogueira, R. C. C., & Pereira, C. E. (2020). Insecticidal activity of plant extracts on *D. brevipes* in pineapple. *Revista De Ciências Agrárias, 43*(1), 82–90.

Luz, P. B., Bonani, J. P., & Santa-Cecília, L. V. C. (2005). Primeira ocorrência de *D. brevipes* (Cockerell, 1893) (Hemiptera: Pseudococcidae) na palmeira *Rhapis excelsa* (Thunberg) Henry ex. Rehder no Brasil. *Arquivos Do Instituto Biológico, 72*, 391–393.

Mamahit, J. M., Manengkey, G. S. J., & Makal, H. V. G. (2010). Kelimpahan populasi kutu putih *Dysmicoccus* sp. (Hemiptera: Pseudococcidae) pada tanaman nenas di Kabupaten Bolaang, Mongondow. *Eugenia, 16*(3), 11500.

Mamahit, J. M., Manuwoto, S., Hidayat, P., & Sobir, S. (2008). Biologi kutu putih *D. brevipes* Cockerell (Hemiptera: Pseudococcidae) pada tanaman nenas dan kencur. *Buletin Penelitian Rempah Dan Obat XIX, 2*, 164–166.

Manjushree, G., Mani, C., & Madhu, S. (2018). Studies on natural enemies of pink pineapple mealybug, *D. brevipes* (Cockerell) (Hemiptera: Pseudococciade) in Kerala. *Journal of Biological Control, 33*, 53–59.

Mau, R. F. L., & Kessing, J. L. M. (2007). *Dysmicoccus brevipes* (Cockerell). In J. M. Diez (Ed.), *The crop master program, extension entomology and UH-CTAHR integrated pest management program.* http://www.extento.hawaii.edu/kbase/crop/Type/d_brevip.htm. Accessed 19 May 2021.

Memon, S. A., Rajput, A. A., Rajput, S., Narejo, M., Rashid, M., Mastoi, A. H., & Azizullah, S. (2017). Efficacy of different botanical pesticide concentrations against mealybug, *D. brevipes* (Cockerell) of banana. *Lasbela University Journal of Science and Technology, 6*, 130–135.

Moghaddam, M. (2004). New record on three mealybugs (three species and two genera) (Hem. Coccoidea: Pseudococcidae) in Iran. *Applied Entomology and Phytopathology, 72*(1), 23.

Pandey, R. R., & Johnson, M. W. (2006). Weeds adjacent to Hawaiian pineapple plantings harboring pink pineapple mealybugs. *Environmental Entomology, 35*(1), 68–74.

Petty, G. J., Stirling, G. R., & Barholomew, D. P. (2002). *Pest of pineapple, in tropical fruit pests and pollinators* (p. 430). CAB International Publication.

Rishi, R. R., Sundararaj, R., Shwetha, V., Sunayana, T. P., Rao, S. P., & Karnat, N. M. (2019). Record of *D. brevipes* Cockerell (Hemiptera: Pseudococcidae) as pest in mangroves of Maharashtra and its management. *Journal of Entomology and Zoology Studies, 7*(2), 176–179.

Rohrbach, K. G., & Johnson, M. W. (2003). *Pests, disease and weed, in the pineapple: Botany, production and uses* (pp. 203–251). CAB International.

Rohrbach, K. G., & Schmitt, D. (2003). Disease of pineapple. In R. C. Ploetz (Ed.), *Disease of tropical fruit crops* (pp. 443–464). CAB International.

Sanches, N. F., & Matos, A. P. (1999). Murcha associada à cochonilha *D. brevipes* (Cockerell 1893). In: *O abacaxizeiro: Cultivo, agroindústria e economia* (pp. 343–366). EMBRAPA-CNPMF.

Santa-Cecília, L. V. C., Souza, B., Souza, J. C., Prado, E., Moino Junior, A., Fornazier, M. J., & Carvalho, G. A. (2007). *Cochonilhas-farinhentas em cafeeiros: bioecologia, danos e métodos de controle* (p. 48). EPAMIG.

Santa-Cecília, L. V. C., Bueno, V. H. P., & Prado, E. (2004). Desenvolvimento de *D. brevipes* (Cockerell) (Hemiptera:Pseudococcidae) em duas cultivars de abaxi. *Ciencia e Agrotecnologia, 28*(5), 1015–1020.

Saranya, S., Ushakumari, R., Jacob, S., & Philip, B. M. (2010). Efficacy of different entomopathogenic fungi against cowpea aphid, *Aphis craccivora* (Koch). *Journal of Biopesticides, 31*(1), 138–142.

Scalenet. (2022). *Dysmicoccus brevipes* (Cockerell). http://scalenet.info/catalogue/Dysmicoccus%20brevipes/. Accessed 23 April 2022.

Sether, D. M. (2002). *Pineapple mealybug wilt associated viruses: Vectors, impacts, and dynamics [doctoral dissertation].* University of Hawaii.

Sether, D. M., & Hu, J. S. (2002). Closterovirus infection and mealybug exposure are necessary for the development of mealybug wilt of pineapple disease. *Phytopathology, 92*(9), 928–936.

Sether, D. M., Melzer, M. J., Busto, J., Zee, F., & Hu, J. S. (2005). Diversity and mealybug transmissibility of ampeloviruses in pineapple. *Plant Disease, 89*(5), 450–457.

Sipes, S. B. (2000). *Crop profile for pineapple in Hawaii, NSF centre for integrated pest management.* North Carolina State University.

Sirisena, U. G. A. I., Watson, G. W., Hemachandra, K. S., & Wijayagunasekara, H. N. P. (2013). Mealybugs (Hemiptera: Pseudococcidae) species on economically important fruit crops in Sri Lanka. *Tropical Agricultural Research, 25*(1), 69–82.

Suh, S. J., & Hyun, I. H. (2015). A preliminary checklist of scale insects (Hemiptera: Coccoidea) intercepted in Korea on dracaena and ficus plants (Asparagales: Asparagaceae, Rosales: Moraceae) imported from southern Asia. *Insecta Mundi, 14*, 937.

Sulaiman, S. F. M. (2000). Effect of pesticidal pre-treatments of pineapple plants on the incidence of mealybug wilt disease. *Acta Horticulturae, 529*, 273–288.

Walton, V., & Pringle, K. (2004). A survey of mealybugs and associated natural enemies in vineyards in the Western Cape Province, South Africa. *South African Journal for Enology and Viticulture, 25*(1), 35–96.

Waterhouse, D. F. (1998). *Biological control of insect pests: Southeast Asian prospects* (p. 548). Australian Center for International Agricultural Research.

Williams, D. J., & Watson, G. W. (1988). *Scale insects of the tropical South Pacific region. Part 2. Mealybugs (Pseudococcidae)* (p. 260). CAB International.

Williams, D. J., & Granara de Willink, M. C. (1992). *Mealybugs of Central and South America* (p. 635). CAB International.

Williams, D. J. (2004). *Mealybugs of Southern Asia* (p. 896). The Natural History Museum.

Zart, M., de Macedo, M. F., Simões Santos Rando, J., Doneze, G. S., Brito, C. P., de Souza Poletto, R., & Alves, V. S. (2021). Performance of entomopathogenic nematodes on the mealybug, *D. brevipes* (Hemiptera: Pseudococcidae) and the compatibility of control agents with nematodes. *Journal of Nematology, 53*(1), 1–10.

Advances in Tropical Plant Disease Management

Emerging Rubber Disease and Potential Factors Contributing

Syd Ali Nusaibah, Syed Sagaff Sharifah Aliya, and Zaiton Sapak

Abstract *Hevea brasiliensis* or rubber tree has been one of the most indispensable commodity crops since World War I. The necessity for latex in the production of various human essentials has continued till to date. The recent pandemic has again boosted the prerequisite of latex for the production of gloves and other medical items. Therefore, a sustainable yield from rubber trees to cater the expanding demand is crucial. However, newly emerging plant diseases have been hampering our plantation crops that results in decreased yield. Discovery of new plant diseases may require disease symptom characterization, identification of the causal pathogen, disease mechanism and factors that have contributed to the disease incidences to determine the curative and preventive measures. Herein, we present the most recent findings on a newly emerging rubber disease designated as new leaf spot disease caused by fungal pathogen/s. In addition, a summary on the potential factors that may have been contributing to emerging plant diseases as whole is included.

Keywords Rubber tree · Emerging plant disease · Fungal pathogens

1 Introduction

An indigenous tropical tree of South America's Amazon Forest is the *H. brasiliensis* or rubber (family *Euphorbiaceae*). The tree was cultivated actively for its natural latex since World War I. At present, the top five natural rubber-producing nations are Thailand, Indonesia, Malaysia, India and Vietnam based on ranking in terms of production.

S. A. Nusaibah (✉) · S. S. S. Aliya
Department of Plant Protection, Faculty of Agriculture, Universiti Putra Malaysia, 43400 Serdang, Selangor, Malaysia
e-mail: nusaibah@upm.edu.my

Z. Sapak
Faculty of Plantation and Agrotechnology, Universiti Teknologi MARA (UiTM), Jasin Campus, 77300 Merlimau, Melaka, Malaysia
e-mail: zaiton3338@uitm.edu.my

M. Wong (ed.), *Advances in Tropical Crop Protection*,
https://doi.org/10.1007/978-3-031-59268-3_6

To date, different goods, including medical gloves, safety boots, aeroplanes and vehicle tires, are made using latex as a base material. However, the production of natural rubber is currently beset by a number of difficulties, including price swings, a shortage of skilled tappers, and the threat of damaging epidemics (Heng & Joo, 2017). Rubber-producing nations have recently been reported to be experiencing a significant epidemic of new leaf disease. On mature rubber leaves, the disease manifests as circular spots with a brown necrotic halo. Approximately 90% of the leaves are lost due to this disease. A complex disease caused by several fungal pathogens, according to recent study, was reported to be the cause of this disease. *Colletotrichum siamense*, a main fungal pathogen, and *P. jesteri*, a secondary fungal pathogen, were determined to be the causal pathogens (Aliya et al., 2022). This disease was named as complex circular leaf spot (CCLS) disease. The aforementioned disease has significance because they have an impact on physiological functions and latex production, which in turn has an impact on the economy of the nation that produces them. Thus, the purpose of this chapter is to introduce the newly emergent rubber tree epidemic as well as any potential factors that may have triggered it.

2 Complex Circular Leaf Spot (CCLS) Disease of Rubber

In 2016, North Sumatera, Indonesia, had the first reports of the illness. In 2017, South Sumatera saw the disease's expansion. Four other nations that produce rubber have now been infected. In Kottayam, India, in 2017, an outbreak was reported. This disease was discovered in Johor, Malaysia, that same year, according to the Malaysian Rubber Board (MRB). According to Pornsuriya et al. (2020), Narathiwat Province in Thailand had the first case of this disease in 2019. Sri Lanka experienced CCLS disease in 2019 as well, according to the International Rubber Research and Development Board. Nearly all rubber clones that were planted were susceptible to the new leaf disease, according to reports from the afflicted rubber-producing nations.

2.1 Symptoms of Complex Circular Leaf Spot Disease (CCLS) of Rubber

Irregular circular spots with brown necrotic lesion were seen on mature rubber leaves as CCLS disease signs (see Fig. 1). Mature rubber trees between 10 and 15 years old are affected by the causing fungal infection. Figure 2 shows that the disease can defoliate up to 90% of the canopy.

Fig. 1 Symptoms of CCLS disease **a** identified as circular spots with brown necrotic lesion on matured rubber leaf and **b** colour changes observed on the leaves and spot (darker lesion with yellow halo) prior to defoliation. Photos were taken at Sembawa, Palembang, Indonesia in 2019

Fig. 2 90% of defoliation observed in severely affected plantations with CCLS disease. Photos were taken at Sembawa, Palembang, Indonesia in 2019

2.2 Disease Causal Pathogen

At present many hypotheses have been hypothesized on the culprit causing this devastating disease of rubber trees. However, the first published report was from Malaysia in 2022. The findings concluded that, when *C. siamense* injures rubber leaves, *P. jesteri* may be the opportunistic pathogen that penetrates to the injured leaves. Opportunistic pathogens will result in subsequent infections and worsen the condition. It is possible to draw the conclusion that *P. jesteri* is the secondary pathogen and *C. siamense* is the major pathogen responsible for CCLS disease of rubber.

2.2.1 *Colletotrichum siamense* as Plant Fungal Pathogen

Colletotrichum siamense is from the genus *Colletotrichum*, which is traditionally recognized as an asexual genus of fungi that comes from the family *Glomerellaceae* (Sharma and Shenoy, 2016). Based on observations reported by Prasannath et al. (2020) and Aliya et al. (2022), *C. siamense* isolate produces greyish and cottony aerial mycelium with visible orange conidial masses when they sporulate (see Fig. 3). Generally, it is a widespread plant pathogenic fungus that damages a variety of economically valuable plant hosts because it adapts effectively to different environmental conditions (Yan et al., 2015). It can also persist in a latent state in the tissues of the host as a hemibiotrophic pathogen. This privilege is due to its conidium that develops into an appressorium on the surface of the plant, followed by penetration and invasion of the epidermal cells and the production of primary hyphae (Chung et al., 2022).

However, *C. siamense* shows a hemibiotrophic lifestyle that includes intracellular hemibiotrophy and a predominance of necrotrophic life cycles (De Silva et al., 2017). As a result, primary infection vesicles are created without the cells being killed during the early infection of the epidermal cells. Recently, *C. siamense* was identified as the cause of leaf spots on redbud (Ji et al., 2019) and areca palms (Chou et al., 2019) in China. It also affects high-value fruit crops such as strawberry, mango, citrus, avocado, and banana (Cannon et al., 2007). Consequently, *Colletotrichum* spp. were ranked as the world's eighth most important phytopathogenic fungal genus (Dean et al., 2012).

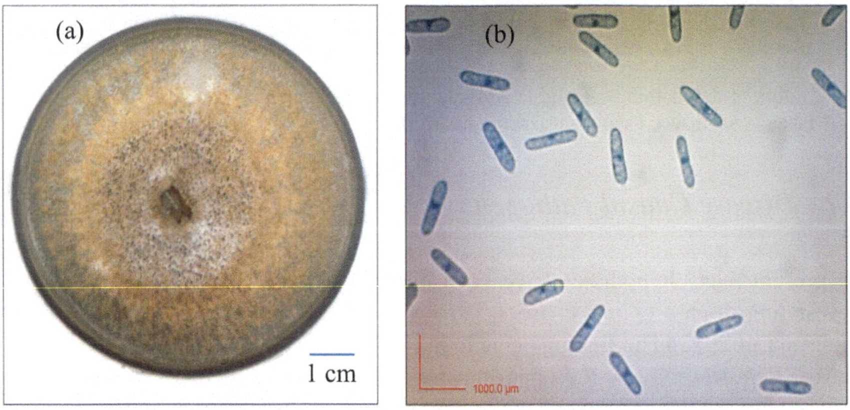

Fig. 3 Cultural view of *C. siamense* fungus on potatodextrose agar (PDA) plate (**a**) and spores of *C. siamense* viewed under a light microscope at 40× magnification (**b**) (Aliya et al., 2022)

2.2.2 *Pestalotiopsis jesteri* as Plant Fungal Pathogen

Pestalotiopsis jesteri is an anamorphic fungus with appendage bearing conidia that belongs to *Pestalotiopsis* genus and Amphisphaeriaceae family (Lee et al., 2011; Maharachchikumbura et al., 2014). Generally, *Pestalotiopsis* spp. synthesize yellowish-white mycelia with an aerial texture, producing black conidiomata (see Fig. 4). Based on microscopic examination, it produces septate hyphae and generate conidia with light concolorous median cells, bearing two to three apical appendages which are long and unbranched (Aliya et al., 2022). Commonly, *Pestalotiopsis* spp. are frequently isolated as endophytes. However, they also play the role as plant pathogens that causes numerous infections such as diseases of foliage, stems and roots which significantly reduces commercial production (Debbab et al., 2013).

In certain, *Pestalotiopsis* species could lead to post-harvest damage in addition to diseases including leaf spots, grey blights, leaf blights and fruit rots (Crous et al., 2011; Zhang et al., 2013). Chelong et al. (2020) also described that the fungal pathogen, *Pestalotiopsis* sp. affects rubber trees by causing their leaves to fall easily from their branches since the leaf-falling disease is easily propagated by the wind. However, Agrios (2005) stated that *Pestalotiopsis* are mild pathogens that could penetrate the host through naturally occurring openings such as stomas, lenticels and hydathodes. Apart from that, pruning wounds or other physical techniques also play major roles in the dissemination of the disease caused by *Pestalotiopsis* spp. (Keith et al., 2006) as *Pestalotiopsis* genus only infect stressed or injured plants (Wright et al., 1998). In addition, *Pestalotiopsis* spp. favors wet environments with high humidity since the pathogen spreads or transmits mainly through rain splash and air. Thus, the disease transmits more easily in favorable temperatures and on vulnerable rubber clones (Aliya et al., 2022).

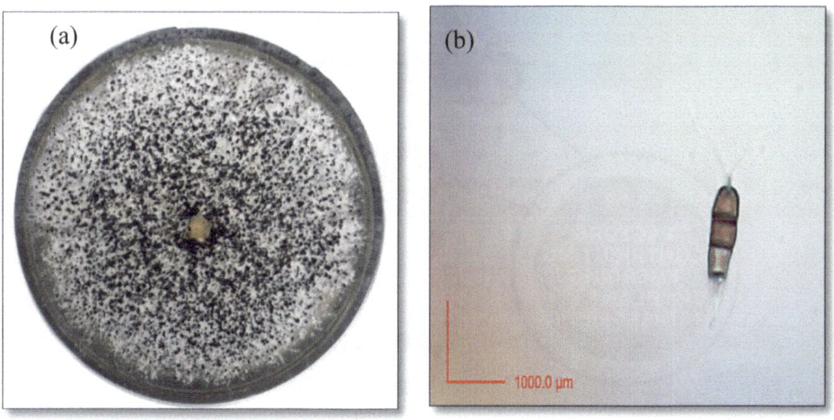

Fig. 4 Cultural view of *P. jesteri* fungus on potato dextrose agar (PDA) plate (**a**) and spore of *P. jesteri* viewed under a light microscope at 40× magnification (**b**) (Aliya et al., 2022)

3 Factors Contributing to Endemic and Emerging Plant Diseases

Emerging plant diseases are those that have altered their pathogenesis, increased the frequency or range of hosts for a pathogen, or have been discovered and recognised for the first time in a particular region (Ristaino et al., 2021). The following are some hypotheses regarding how new harmful microorganisms emerge: (1) The organism may already be endemic in crop areas, but a new host has just been identified; (2) The organism developed pathogenicity after becoming endemic, either as a result of an increase in virulence or a decrease in the host's defenses; (3) The organism may have recently been introduced into a new area and previously unexposed hosts, and the organism is pathogenic to new plants; or (4) Insect vectors feed on new plants that are retaining the pathogenic microorganism and disseminate it to other plants (Avila-Quezada et al., 2018). Figure 5 lists the established factors that contribute to the emergence of new plant diseases.

3.1 *Climate Change*

Climate change may affect crop growth and quality as well as cause plant mortality, as shown by changes in average temperatures, annual rainfall reductions, irregular rainfall distributions, and extended dry periods (Avila-Quezada et al., 2018). In previously non-infested areas, increased climate adaptability may expand the range of plant pathogens, their vectors, and hosts, potentially promoting the spread of new emerging diseases (Bebber, 2015; Ristaino et al., 2021). In general, when plants are

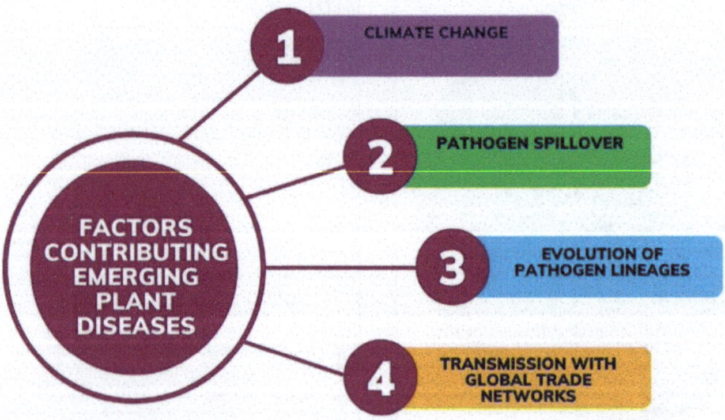

Fig. 5 There are evidence that endemic and newly emerging plant diseases develop due to factors like pathogen spillover, the formation of novel pathogen lineages, and transmission through international trade networks

weak or stressed due to environmental causes, microbes could effortlessly colonize them, establishing infection and subsequently causing plant mortality.

3.2 *Pathogen Spillover*

Pathogen spillover is the process by which a pathogen is transferred from a reservoir host species to a recipient host species (Becker et al., 2019). According to the "spillover effect" mechanism, the pathogen typically multiplies rapidly in the reservoir before spreading to the other host (Power & Mitchell, 2004).

3.3 *Evolution of New Pathogen Lineages*

Gene functionality can vary as a result of mutations, translocations and gene creep, which can happen when several genes are replicated (Avila-Quezada et al., 2018). Numerous fungi, including many crop diseases, are known to reproduce clonally or to perform sexual recombination only in very rare instances (McDonald & Stukenbrock, 2016). The majority of plant diseases have the capacity to engage in genetic recombination, which can lead to the emergence of pathogenic lineages and the establishment of new genetic diversity, which may lead to the origin of novel infections with increased pathogenicity (De Silva et al., 2017). In addition, to swiftly overcome new host resistance or develop resistance to new fungicides, fungal crop diseases could demonstrate enough genetic diversity even in the absence of sexual recombination (Taylor et al., 2015).

3.4 *Transmission with Global Trade Networks*

International trade is essential in supplying agricultural raw materials in the form of substances or materials that are utilized to make commodities. As a result, the risk of disease transmission is concentrated in places with a high level of trade (Ristaino et al., 2021). Plant pathogens are dispersed by both long-distance and short-distance dispersion mechanisms. Weather (rain and wind) and inter-regional, domestic, and continental trade are two major factors that cause long-distance pathogen transmission. Additionally, human practices like pruning and grafting, as well as vectors carried by insects or people, are major drivers of short-distance dispersion (Ashraf et al., 2014). In fact, spores could move independently over land, air, or water from a few meters to 1000 km. Nonetheless, storms have been shown to rapidly disseminate spores over long distances, such as from one continent to another or the southern hemisphere to the northern (Isard et al., 2005). Consequently, new plant diseases have emerged as a result of these circumstances.

4 Conclusions

This chapter presents on the emergence of a new devastating disease of rubber trees encountered by almost all the rubber producing countries in Asia. A recent study in Malaysia has identified the disease as a complex infection caused by more than one fungal pathogen. The study reported *C. siamense* and *P. jesteri* as the primary and secondary fungal pathogens respectively. In addition, this chapter also reviews several factors that could be responsible for the emergence of new plant diseases reported in less than a decade. In order to tackle on disease prevention, humans need to prevent or minimize the associated factors to mitigate the occurrence of these newly emerging plant diseases.

Acknowledgements The authors would like to acknowledge the assistance and cooperations from the Rubber Research Institute Malaysia (RRIM) and International Rubber Research Development Board (IRRDB).

References

Agrios, G. N. (2005) *Plant pathology*, 5th Edn (p. 922). Elsevier Academic Press.

Aliya, S. S. S., Nusaibah, S. A., Mahyudin, M. M., Yun, W. M., & Yusop, M. R. (2022). *Colletotrichum siamense* and *P. jesteri* as potential pathogens of new rubber leaf spot disease via detached leaf assay. *Journal of Rubber Research, 25*(3), 195–212. https://doi.org/10.1007/s42464-022-00157-4

Ashraf, S., Khan, G. A., Ali, S., Iftikhar, M., & Mehmood, N. (2014). Managing insect pests and diseases of citrus: On farm analysis from Pakistan. *Pakistan Journal of Phytopathology, 26*(2), 301–307.

Avila-Quezada, G. D., Esquivel, J. F., Silva-Rojas, H. V., Leyva-Mir, S. G., Garcia-Avila, C. D., Noriega-Orozco, L., Rivas-Valencia, P., Ojeda-Barrios, D., & Melgoza-Castillo, A. (2018). Emerging plant diseases under a changing climate scenario: Threats to our global food supply. *Emirates Journal of Food and Agriculture, 30*(6), 443–450. https://doi.org/10.9755/ejfa.2018.v30.i6.1715

Bebber, D. P. (2015). Range-expanding pests and pathogens in a warming world. *Annual Review of Phytopathology, 53*, 335–356. https://doi.org/10.1146/annurev-phyto-080614-120207

Becker, D. J., Washburne, A. D., Faust, C. L., Pulliam, J. R. C., Mordecai, E. A., Lloyd-Smith, J. O., & Plowright, R. K. (2019). Dynamic and integrative approaches to understanding pathogen spillover. *Philosophical Transactions of the Royal Society B, 374*, 1–7. https://doi.org/10.1098/rstb.2019.0014

Cannon, P. F., Damm, U., Johnston, P. R., & Weir, B. S. (2007). *Colletotrichum*–current status and future directions. *Studies in Mycology, 59*(1), 129–145. https://doi.org/10.3114/sim0014

Chelong, I. A., Moye, J. J., Adair, A., & Bonwanno, S. (2020). First report of efficacy study of bioextract to control *Pestalotiopsis* sp. affecting para rubber leaf disease (*H. brasiliensis* Muell. Arg) under climate variability. *International Journal of Advanced Research in Engineering and Technology (IJARET), 11*(10), 209–217.

Chung, P. C., Wu, H. Y., Chen, Y. C., Hung, T. H., & Chung, C. L. (2022). Development of a nested PCR assay for detecting *C. siamense* and *Colletotrichum fructicola* on symptomless strawberry plants. *PLoS ONE, 17*(6), 687. https://doi.org/10.1371/journal.pone.0270687

Crous, P. W., Summerell, B. A., Swart, L., Denman, S., Taylor, J. E., Bezuidenhout, C. M., & Groenewald, J. Z. (2011). Fungal pathogens of Proteaceae. *Persoonia-Molecular Phylogeny and Evolution of Fungi, 27*(1), 20–45. https://doi.org/10.3767/003158511X606239

De Silva, D. D., Crous, P. W., Ades, P. K., Hyde, K. D., & Taylor, P. W. (2017). Life styles of *Colletotrichum* species and implications for plant biosecurity. *Fungal Biology Reviews, 31*(3), 155–168. https://doi.org/10.1111/PPA.13351

Dean, R., Van Kan, J. A., Pretorius, Z. A., Hammond-Kosack, K. E., Di Pietro, A., Spanu, P. D., & Foster, G. D. (2012). The Top 10 fungal pathogens in molecular plant pathology. *Molecular Plant Pathology, 13*(4), 414–430. https://doi.org/10.1111/j.1364-3703.2011.00783.x

Debbab, A., Aly, A. H., & Proksch, P. (2013). Mangrove derived fungal endophytes: A chemical and biological perception. *Fungal Diversity, 61*(1), 1–27. https://doi.org/10.1007/s13225-013-0243-8

Heng, T. S., & Joo, G. K. (2017). Rubber. *Encyclopedia of Applied Plant Sciences, 3*(2), 402–409. https://doi.org/10.1016/B978-0-12-394807-6.00175-1

Isard, S. A., Gage, S. H., Comtois, P., & Russo, J. M. (2005). Principles of the atmospheric pathway for invasive species applied to soybean rust. *BioScience, 55*(10), 851–861. https://doi.org/10.1641/0006-3568(2005)055[0851:POTAPF]2.0.CO;2

Ji, J., Wang, T., Xu, X., Wang, X. Y., Wu, Q. Q., Li, W. Y., & Yao, L. G. (2019). First report of *C. siamense* causing leaf spot on redbud in China. *Plant Disease, 103*(3), 585–585. https://doi.org/10.1094/PDIS-08-18-1331-PDN

Keith, L. M., Velasquez, M. E., & Zee, F. T. (2006). Identification and characterization of *Pestalotiopsis* spp. causing scab disease of guava, *Psidium guajava*, in Hawaii. *Plant Disease, 90*(1), 16–23. https://doi.org/10.1094/PD-90-0016

Lee, S., Crous, P. W., & Wingfield, M. J. (2011). Pestalotioid fungi from *Restionaceae* in the Cape floral kingdom. *Studies in Mycology, 68*(1), 211–235. https://doi.org/10.3114/sim.55.1.175

Maharachchikumbura, S. S., Hyde, K. D., Groenewald, J. Z., Xu, J., & Crous, P. W. (2014). *Pestalotiopsis* revisited. *Studies in Mycology, 79*(1), 121–186. https://doi.org/10.1016/j.simyco.2014.09.005

McDonald, B. A., & Stukenbrock, E. H. (2016). Rapid emergence of pathogens in agro-ecosystems: Global threats to agricultural sustainability and food security. *Philosophical Transactions B, 371*, 1–9. https://doi.org/10.1098/rstb.2016.0026

Pornsuriya, C., Chairin, T., Thaochan, N., & Sunpapao, A. (2020). Identification and characterization of *Neopestalotiopsis* fungi associated with a novel leaf fall disease of rubber trees (*H. brasiliensis*) in Thailand. *Journal of Phytopatholology, 121*, 1–12. https://doi.org/10.1111/jph.12906

Power, A. G., & Mitchell, C. E. (2004). Pathogen spillover in disease epidemics. *The American Naturalist, 164*(S5), S79–S89. https://doi.org/10.1086/424610

Prasannath, K., Galea, V. J., & Akinsanmi, O. A. (2020). Characterisation of leaf spots caused by *Neopestalotiopsis clavispora* and *C. siamense* in macadamia in Australia. *European Journal of Plant Pathology, 156*(4), 1219–1225. https://doi.org/10.1007/s10658-020-01962-6

Ristaino, J. B., Anderson, P. K., Bebber, D. P., Brauman, K. A., Cunniffe, N. J., Fedoroff, N. V., Finegold, C., Garret, K. A., Gilligan, C. A., Jones, C. M., Martin, M. D., MacDonald, G. K., Neenan, P., Records, A., Schmale, D. G., Tateosian, L., & Wei, Q. (2021). The persistent threat of emerging plant disease pandemics to global food security. *PNAS, 118*(23), 1–9. https://doi.org/10.1073/pnas.2022239118

Sharma, G., & Shenoy, B. D. (2016). *Colletotrichum* systematics: Past, present and prospects. *Mycosphere, 7*(8), 1093–1102. https://doi.org/10.5943/mycosphere/si/2c/2

Taylor, J. W., Hann-Soden, C., Branco, S., Sylvain, I., & Ellison, C. E. (2015). Clonal reproduction in fungi. *PNAS, 112*(29), 8901–8908. https://doi.org/10.1073/pnas.1503159112

Wright, E. R., Rivera, M. C., & Flynn, M. J. (1998). First report of *Pestalotiopsis guepini* and *Glomerella cingulata* on blueberry in Buenos Aires (Argentina). *EPPO Bulletin, 28*(1–2), 219–220. https://doi.org/10.1111/j.1365-2338.1998.tb00725.x

Yan, J. Y., Jayawardena, M. M. R. S., Goonasekara, I. D., Wang, Y., Zhang, W., Liu, M., & Li, X. H. (2015). Diverse species of *Colletotrichum* associated with grapevine anthracnose in China. *Fungal Diversity, 71*(1), 233–246. https://doi.org/10.1007/s13225-014-0310-9

Zhang, Y., Maharachchikumbura, S. S., Tian, Q., & Hyde, K. D. (2013). *Pestalotiopsis* species on ornamental plants in Yunnan Province, China. *Sydowia, 65*(1), 113–128. https://doi.org/10.1016/j.simyco.2014.09.005

Common Diseases in Pineapple and Their Management

Zaiton Sapak and Syd Ali Nusaibah

Abstract Pineapple (*Ananas comosus* L. Merr) is a tropical fruit with sweet taste, vibrant tropical flavour, and contains enormous health benefits. Currently, pineapple is ranked as the second in the list of commercial tropical fruits for worldwide production after mangoes. Like other crops, pineapple is prone to various diseases caused by different groups of plant pathogens. Several common pineapple diseases have been reported particularly in Malaysia and other countries including bacterial and fungal heart rots, fruit collapse, butt and black rots, fruitlet core rot, internal black spot, fusariosis, mealybug wilt-associated virus, yeasty, stem end rot, leaf spot disease, and parasitic nematodes–associated disease. These diseases can cause huge losses to pineapple production by reducing quantity and quality of fruit. Knowledge on the causal pathogen, disease symptoms and dissemination of pathogen in pineapple farms are important to prevent and control these diseases. Therefore, this chapter presents more than 10 common diseases in pineapple with the descriptions of causal pathogen and symptoms, as well as the available control methods to manage the diseases.

Keywords Pineapple · Plant diseases · Bacteria · Fungi · Virus · Nematode · Control methods

Z. Sapak (✉)
Faculty of Plantation and Agrotechnology, Universiti Teknologi MARA (UiTM), Jasin Campus, 77300 Merlimau, Melaka, Malaysia
e-mail: zaiton3338@uitm.edu.my

S. A. Nusaibah
Department of Plant Protection, Faculty of Agriculture, Universiti Putra Malaysia, 43400 Serdang, Selangor, Malaysia
e-mail: nusaibah@upm.edu.my

M. Wong (ed.), *Advances in Tropical Crop Protection*,
https://doi.org/10.1007/978-3-031-59268-3_7

1 Introduction

Pineapple (*Ananas comosus*) has distinctive qualities that makes it highly valuable to both small-scale farmers and large-scale commercial growers worldwide. A growing trend of pineapple farming in various countries can be seen as a result of the crop's potential for great profitability. Along with avocado and mango, pineapple is set to continue to be the most significantly traded tropical fruit in terms of export quantity in 2022 (FAO, 2023). With an estimation of 3.2 million tonnes of global exports in 2023, pineapples will continue to dominate the market. The majority of the world's pineapple plantations is located in tropical areas. More than 82 different countries with more than 2.1 million acres are devoted for pineapple production (Hossain, 2016). The Philippines is the world's biggest pineapple producer with 2.7 million metric tonnes of pineapples per year. Pineapple plantation areas in the country are estimated to be over 40,000 ha, with most production is controlled by big companies such as Dole Philippines and Del Monte. Meanwhile, Costa Rica is the second largest global pineapple producers with 2.6 metric tonnes of fruits per year, followed by Brazil with 2.4 million metric tonnes of fruits per year. Additionally, other countries including Indonesia, China, India, Thailand, Nigeria, Mexico, and Colombia are listed as the top 10 pineapple producing countries. Malaysia is on small scale as compared to other top producing countries and was ranked 23rd in 2020 with global export share of 0.22%. Yet, the pineapple industry in Malaysia has huge potential to be escalated.

Pineapple has various cultivars that differ in terms of their morpho agronomic characteristics, phytochemical profiles, and antioxidant properties. Through research and analysis, scientists have identified and classified different pineapple varieties based on several key attributes. Generally, commercial cultivars of pineapple are placed in five groups which are Cayenne, Queen, Spanish, Pernambuco, and Mordilona. As for the international trade, the pineapple cultivars are grouped into four main classes including Smooth Cayenne, Red Spanish, Queen, and Abacaxi, despite many variations in the types within each class. In Malaysia, 16 varieties have been registered and promoted by the Malaysian Pineapple Industrial Board (MPIB) under several incentive programs for growers to adopt and to expend their farms and boost their income. Among these varieties, MD2 has been acknowledged as the key crop and becomes economically important for the country. This MD2 variety is considered as the most preferable variety due to its outstanding qualities including bright gold colour, sweeter taste, thinner skin, higher vitamin C, lower acidity, less fibrous, higher total soluble solids, and longer shelf life (Thalip et al., 2015). Both pulp and juice of pineapple are commonly consumed in various forms of fresh and processed products such as beverages, jam, confectionery jelly, canned in syrup, fruit roll/leather, breaded, dehydrated candied, pickle, sauce, fruit fillings and powder (Zainun, 2011).

Despite the growing trend of pineapple farming, the growers face various challenges including marketing problems, lack of capital, pests, and diseases affecting pineapple yields (Barrett et al., 2012; Sapak et al., 2021). Pests and diseases can cause significant yield losses if necessary control methods are not properly conducted.

Several insect pests such as mealybugs, mites, symphylids, bud moth, scales, beetles, termites, white grubs, and weevils are associated with pineapple losses in farms (Petty et al., 2002). Some insects such as mealybug have been recognized as vectors that transmit viruses and cause diseases such as wilt and mosaic. The importance of these insect pests in pineapple cannot be ignored; however, this chapter focuses on common diseases in pineapple and available control measures for these diseases only. Types of diseases, causal pathogens, signs and symptoms of infected pineapple and available control methods for the diseases are also discussed in this chapter.

2 Pineapple Diseases

Pineapple diseases are economically important due to their impact on the production of pineapple fruits and planting materials. The diseases can be caused by various groups of pathogenic microorganisms including fungi, bacteria, viruses, and nematodes. All parts of pineapple plants could be attacked by these pathogenic microorganisms. Some of the most economically significant diseases include bacterial and fungal heart rots, fruit collapse, butt rot, black rot, fruitlet core rot, internal black rot, stem end rot, fusariosis, yeasty, mealybug wilt of pineapple associated with virus (PMWaV) and leaf spot. These diseases can cause huge losses to pineapple productivity, for instance, bacterial heart rot and fruit collapse that can reduce the yield productivity up to 40% (Young et al., 2022). Meanwhile, Sether and Hu (2002) reported that PWMaV can cause reduction in pineapple fruit yield by 30–55%, depending on the age of the plant at the onset of disease. Other diseases also can cause huge losses if the diseases are not properly managed.

2.1 *Bacterial Heart Rot*

Bacterial heart rot (BHR) is a significant disease affecting pineapple production worldwide. This disease is primarily caused by pathogenic bacterium of *Dickeya zeae* syn. *Erwinia chrysanthemi*. The outbreak of BHR has been reported in various pineapple-growing regions, including Malaysia (Lim & Lowings, 1979; Sapak et al., 2021), Hawaii (Marrero and Alvarez, 2011), Indonesia (Cano Reinoso et al., 2021), Puerto Rico (Veléz-Negrón et al., 2023), and Australia (Young et al., 2022). *Dickeya zeae* has a broad range of host plants and able to attack different plant parts. It can cause significant losses of various economically important crops including banana, potato, rice, taro, and maize (Aeny et al., 2020; Boluk et al., 2021; Martinez-Cisneros et al., 2014). This pathogenic bacterium is responsible for most of soft rot disease that is characterized by the decay of plant tissues, resulting in visual symptoms such as water-soaked lesions, tissue maceration, and foul odour (Tan et al., 2022). In pineapple, the disease can be characterized by water-soaked lesions on the white basal part of leaves in the centre whorl (Rohrbach & Johnson, 2003; Sipes & de Matos,

Fig. 1 Bacterial heart rot in MD2 pineapple plant caused by *D. zeae*. **a** The disease shows symptoms of water-soaked lesion and rotten tissues of basal part of leaves in central whorl and **b** spread into the green part of the leaves. **c** The infected leaves bloat and **d** can easily be pulled out from the plant. Photos were taken in Kelantan, Malaysia in 2017

2018). These water-soaked lesions spread into the green part of the leaf then change the leaf colour to olive-green to brown (see Fig. 1a–c). As infection progresses, the infected leaves are easily to be pulled out from the plant. Normally, the infected leaves bloat with bad odour gas formed as the leaves are fermented by the pathogen (see Fig. 1c–d).

Successful infection of pineapple plants by BHR pathogen involves its movement towards the plant surface, following penetration and proliferation inside the host tissues immediately after the entry (Campilho et al., 2020). As mentioned by several authors, the penetration by *D. zeae* into the host plant is commonly through plant natural openings, injuries, and wounds (Cano Reinoso et al., 2021; Huang, 1986). Some insects such as ants could be responsible to transport the disease inoculum into the healthy plants (Lim & Lowings, 1979). Besides BHR caused by *D. zeae*, heart rot disease of pineapple can also be attacked by several species of *Phytophthora*. *Phytophthora nicotianae* has been reported as a main causal pathogen that attacks pineapple and causes severe losses in various countries such as China (Shen et al., 2013), Ecuador (Ratti et al., 2018), Cuba (Rodríguez et al., 2002), Uganda (Oculi et al., 2020), and Mexico (Espinosa-Rodríguez et al., 2015). Meanwhile, in Australia, *P. cinnamomi* has been reported as a main causal pathogen of heart rot disease particularly in MD2 variety (Anderson et al., 2011). Disease symptoms of heart rot disease caused by pathogenic *Phytophthora* sp. are similar to BHR including soft rotting of the basal white tissues of the youngest leaves at the heart of the apical meristem. Infected leaves are easily to be pulled from the plant, and as disease progresses sufficiently, plants die. On fruit-bearing plants of susceptible varieties, the infection can move up through the fruit peduncle and rot the fruit (Green & Nelson, 2015).

2.2 *Fruit Collapse*

Fruit collapse is also known as Ghost disease caused by the same bacterial pathogen of BHR, *D. zeae*. Disease symptoms of fruit collapse commonly visible on mature fruits 2–3 weeks before ripening and can be characterized by an exudation of juice and release of rotten-smelling gas bubbles (Kaneshiro et al., 2008; Rohrbach & Johnson, 2003). As the gas escapes, it releases audible hissing sound, particularly in the quiet night. This leads to ghost disease name, as it frightens many field workers as they return to their homes at night (Young et al., 2022). As the disease progresses, the pineapple fruit shell changes colour to olive green (see Fig. 2a). Dissection of the infected fruit shows rotten cavities within the skeletal fibres within the fruit (see Fig. 2b). Insects such as ants, beetles, mites, and flies are known to be transporters of the bacterial inoculum from collapsed fruit or infected plants with BHR into healthy fruits. The pathogen enters fruits via open flowers or injuries (Rohrbach & Johnson, 2003). Besides that, windblown and rain splash also help in the transmission of the pathogen. In Malaysia, this disease is considered as a common and serious disease in pineapple farms and often caused a significant reduction of fruit quantity and quality (Lim & Lowings, 1979).

Fig. 2 Fruit collapse in pineapple plant variety MD2 caused by *D. zeae*. **a** Fruit shell changes colour to olive green, **b** rotten cavities within the skeletal fibres within the fruit. Photos were taken in Johor, Malaysia, in Jun 2017

Fig. 3 Butt rot disease in MD2 pineapple plant caused by *T. paradoxa*. **a** Symptom of rotten sucker base tissues and **b** the infected tissues turn grey-black and decay. These photos were taken in Johor, Malaysia, in June 2017

2.3 Butt Rot

Butt rot disease is caused by *Thielaviopsis paradoxa*, a fungus that penetrates the plant roots and spreads to other parts of the plant. This fungal pathogen is an Ascomycetes fungus and possess both asexual and sexual reproductive stages. The sexual stage of this pathogen is known as *Ceratocystis paradoxa*. Normally, *T. paradoxa* (asexual state) produces infective spores that responsible for the blackening of the infected plant tissues. This fungal pathogen is well known as a wound parasite that seizes opportunity through the wound into hosts. This pathogen spreads the disease through wind, water, soil, insects, and rodents (Elliott, 2019). Meanwhile, the sexual stage (*C. paradoxa*) is seldom found during growing season but only attack the host plant under stress conditions such as overwintering (Álvarez et al., 2012). Disease symptoms of butt rot can be detected as rotten tissues on sucker base (see Fig. 3a and b), crown, and slip stem (Joy & Sindhu, 2012). As the infection progresses, the infected tissues turn grey-black and then decay (Rohrbach & Johnson, 2003). Consequently, the plant fails to grow, stunted, hence limiting the production of pineapples in the field.

2.4 Black Rot

Black rot in pineapple fruit is a common postharvest disease and the disease is caused by similar pathogen of butt rot, *T. paradoxa*. Normally, the disease infection starts from a broken peduncle during or after harvesting then the fungal pathogen gets into fruit (Joy & Sindhu, 2012; Swaroop Kumar, 2007). However, the fungal pathogen may also get into pineapple fruit through wound and natural opening. Reyes et al. (2004) reported pineapple fruit that suffer from translucency will be more susceptible to the disease infection as translucent fruit is susceptible to injury.

Furthermore, leaking juice from broken peduncle or damaged fruit will encourage the fungal pathogen to attack and grow on the broken or damaged area.

Initially, disease symptoms of circular and water-soaked spots can be seen on the infected peduncle and fruit. As the disease progresses, the spots enlarge extending over the flesh region, making fruit watery, soft, and disintegrated with black patches. Normally, the disease takes 3–4 days to develop after harvest (Joy & Sindhu, 2012). In Malaysia, the varieties that are susceptible to this disease include Morris, Sarawak, and MD2 (Singh et al., 2018).

2.5 *Fruitlet Core Rot*

Fruitlet core rot (FCR) is known to be associated with various fungal pathogen species such as *Fusarium ananatum*, *F. guttiforme*, *F. subglutinans* and *Penicillium funiculosum* (Barral et al., 2020; Gu et al., 2015; Lim, 1980; Vignassa et al., 2021). In Malaysia, the causal pathogen of FCR has been identified as *P. funiculosum* by Lim (1985) but recently, several species of *Fusarium* including *F. proliferatum*, *F. sacchari*, and *F. verticillioides* have been reported to be the causal pathogens of FCR in pineapple (Ibrahim et al., 2017).

Disease symptoms of FCR can be described as affected fruitlets become brown to black and sunken (see Fig. 4). Normally, the browning flesh remains quite firm, dry rot type, and limited to a single infected fruitlet (Gu et al., 2015). This browning is the result of the phenolic compounds oxidation into quinones by the enzymes polyphenol oxidase and laccase (Avallone et al., 2003). Several studies revealed that soluble and cell wall-bound phenolic acids in the pineapple fruit parenchyma increase significantly after the onset of the first symptoms of FCR (Barral et al., 2019, 2020; Steingass et al., 2015).

2.6 *Fusariosis*

Fusariosis of pineapple is one of the well-known diseases in pineapple. This disease is associated with several fungal species of *Fusarium*. *Fusarium guttiforme* has been reported to cause the disease with an estimated loss of about 30–40% in Brazil (Carnielli-Queiroz et al., 2019). Meanwhile, Jacobs et al. (2010) reported that *F. ananatum* as a causal pathogen of fusariosis in South Africa. Stępień et al. (2013) discovered that more species of *Fusarium* fungi could be associated with fusariosis including *F. concentricum*, *F. fujikuroi*, *F. incarnatum*, *F. oxysporum*, *F. polyphialidicum*, *F. proliferatum*, *F. temperatum* and *F. verticillioides* in Costa Rica, Ecuador, Honduras, Hawaii, Indonesia, and Vietnam. In Malaysia, Ibrahim et al. (2016) reported that *Fusarium fujikuroi*, *F. semitectum* and *F. chlamydosporum* are responsible for the disease. Disease symptoms of fusariosis can be recognized by development of lesion and brown discolouration of the affected fruitlets, rotten or

Fig. 4 Fruitlet core rot disease in MD2 pineapple plant caused by *P. funiculosum*. **a** Symptom of rotten sucker base tissues and **b** the infected tissues turn grey-black and decay. These photos were taken in Johor, Malaysia, in June 2017

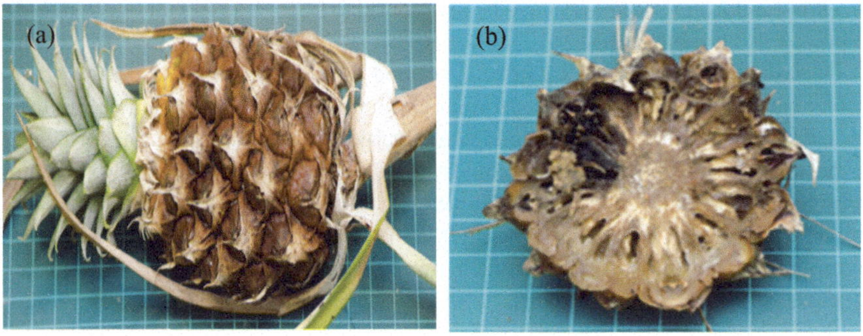

Fig. 5 Fusariosis disease in MD2 pineapple plant caused by *F. fujikuroi*. **a** Symptom of brown and rotten fruit skin and **b** affected flesh displays dry rot and gum exudation. These photos were taken in Johor, Malaysia, in June 2017

sunken fruit skin (see Fig. 5a) and stem, gum exudation on some fruits (see Fig. 5b), dry rot on the leaf, stem bending, chlorosis, increased number of leaves per spiral and natural cracks on the fruits (Ibrahim et al., 2016; Sapak et al., 2021).

2.7 *Wilt Virus Associated with Mealybug*

Incidences of pineapple mealybug wilt virus associated with mealybug disease have been reported in various cultivated areas of pineapple (Hutahayan, 2017; Massé et al., 2021; Nyarko & Asare-Bediako, 2019). The disease has an intriguing aetiology

Fig. 6 Pineapple wilt virus associated with mealybug disease. **a** Symptoms of colour changes in leaves of MD2 pineapple from green to yellow, loss of turgidity, dieback, curling, and drying up, and **b** malformed shape of fruit produced in Sarawak pineapple. Photos were taken in Johor and Kelantan, Malaysia in 2017

and involves a unique relationship between virus, mealybug, and ant (Dey et al., 2018; Jahn et al., 2003). Mealybug is a vector of the virus and ant plays as an agent to protect and transport the vector in pineapple farms. Several distinct virus species have been identified to be associated with this wilt disease in pineapple. These viruses include pineapple mealybug wilt-associated virus subgroup I (PMWaV-1), subgroup II (PMWaV-2), subgroup III (PMWaV-3) and two putative of PMWaV-4 and PMWaV-5 (Dey et al., 2018). Recently, PMWaV-3 is identified as the most reported causal virus for this disease in many countries (Massé et al., 2021; Nyarko & Asare-Bediako, 2019). Two species of mealybug have been identified as the vectors which are *Dysmicoccus brevipes* and *D. neobrevipes* (Dey et al., 2018; Jahn et al., 2003). Wilt disease caused by PMWaV can be grouped into two types which are slow wilt and quick wilt (Dey et al., 2018). The slow wilt normally occurs after mealybugs feed on the pineapple plants for many months and causes destruction of leaf tissues. Meanwhile, quick wilt is observed approximately two months after a sudden period of feeding by high population of mealybugs. The symptoms of quick wilt in plants up to six months old include the colour change in the inner leaves either too yellow or pink (varies by pineapple varieties), loss of turgidity, browning of the leaf tips (dieback), curling, and drying up (see Fig. 6a). In both wilt types, the plant may die, fail to produce fruits, or produce malformed shape (see Fig. 6b) or smaller fruits than the normal fruits if the plant does not recover. In a severe infection, the disease gradually spreads from the leaves to the roots that consequently leads to root collapse.

2.8 Stem End Rot and Internal Black Spot

Fungal pathogen, *Neoscytalidium dimidiatum* has been reported to cause various diseases in various important crops. In 2020, this fungal pathogen was found to cause

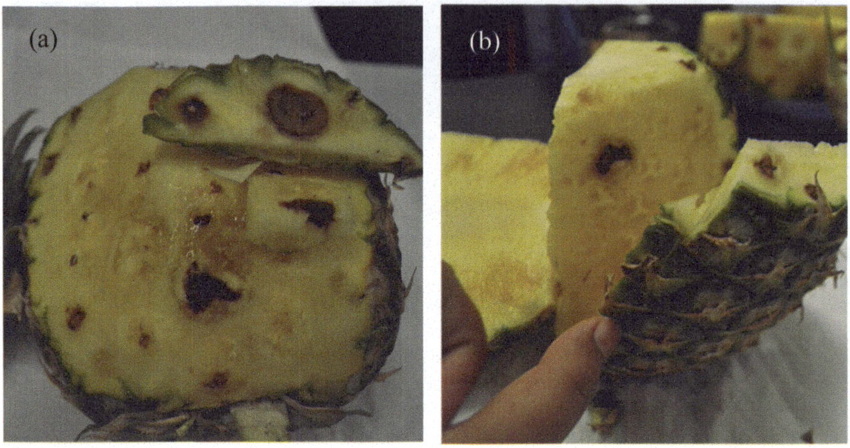

Fig. 7 Internal black spot in MD2 pineapple variety. **a** Symptom of brown to black lesion on fruit flesh and **b** no visible disease symptom on fruit skin. These samples of diseased fruits were recorded from Johor, Malaysia in Jun 2018

a disease of stem-end rot in MD2 pineapple in Malaysia (Kuruppu et al., 2021). Sapak et al. (2021) also found that *N. dimidiatum* is able to cause internal black spot disease in pineapple fruit without damaging fruit skin. Most likely, the fungal pathogen enters the fruit at the early stage of fruit development through nectary ducts. The pathogen then induces browning of an individual fruitlet with various degrees of severity during its growth period. As described by Kuruppu et al. (2021), pineapple stem end rot disease displays symptoms of brown and rotten fruit skin near the stem end region or darker skin with black discolouration. Furthermore, internal black spot can be recognized as brown to black lesion on flesh tissues, occurring immediately beneath the floral cavity, and generally limited to a single fruitlet. This disease is not easily recognized as there are no external symptoms on the fruit skin (see Fig. 7).

2.9 Leaf Spot

Generally, leaf spot disease of pineapple is considered less important unless the disease severely attacks the whole plant. Various fungal pathogens have been associated to the leaf spot including *Exserohilum rostratum* (Luo et al., 2012), *Curvularia clavata* (Zhong et al., 2016) and *Chalara paradoxa* (Joy & Sindhu, 2012). In Malaysia, leaf spot disease has been associated with several *Fusarium* species as reported by Ibrahim et al. (2017). Recently, Sapak et al. (2021) reported that leaf spot disease caused by *N. dimidiatum* is found on pineapple leaves in Perak. Leaf spot disease caused *E. rostratum* can be recognized as greyish white to yellowish white spots emerged on the leaf surfaces that ranged from 1.0 to 2.4 cm × 0.3 to 0.7 cm in the early stage of infection and later, these spots would merge to form

Fig. 8 Brown leaf in Josapine pineapple. **a** Colour of the infected area turns cream or almost white, but the margins of the spot often remain brown and **b** evidence of fungal pathogen can be seen on the lesion. The photos were taken in Perak, Malaysia, in September 2017

larger lesions, covering more than 67% of the leaf surface, which can lead to plant death (Luo et al., 2012). Additionally, leaf spot symptoms caused by *C. clavata* can be described as water-soaked, yellow, chlorotic, or necrotic spots and greyish-brown irregular lesions (0.8–4.2 cm), predominantly on the leaf margin and main veins (Zhong et al., 2016). Symptoms of leaf spot caused by *Fusarium* spp. have been described as small necrotic spots that change to brown and dark brown lesions as the disease developed (Ibrahim et al., 2017). Furthermore, leaf spot disease caused by *N. dimidiatum* starts with small brown spots then eventually these spots lengthen rapidly to more than 20 cm. Normally, the colour of the infected area turns cream or almost white, but the margins of the spot often remain brown (see Fig. 8a) and sometimes, sign of fungal pathogen dark mycelium and spores can be seen on the lesion (see Fig. 8b).

2.10 *Yeasty*

Yeasty disease of pineapple is caused by several species of yeast such as *Hanseniaspora valbyensis*, *Saccharomyces* spp. and *Candida* spp. (Korres et al., 2011; Rohrbach & Johnson, 2003). In Malaysia, this disease is caused *Candida* sp. (Sapak et al., 2021). Disease symptoms of yeasty can be described as bubbling exudation of gas and juice through the crack or injury on the infected area (see Fig. 9a). The skin then turns olive green, the pulp becomes spongy, and finally, only the shell and spongy tissue remain (see Fig. 9b).

Fig. 9 Yeast of MD2 pineapple caused by *Candida* spp. **a** Symptoms of bubbling gas and juice produced by affected fruit and **b** spongy flesh. The photos were taken in Selangor, Malaysia in Jun 2019

2.11 Diseases Associated with Parasitic Nematodes

Several species of parasitic nematodes that are responsible for the losses in pineapple farms have been recorded in Costa Rica (Araya, 2019), Peru (Vera et al., 2017), France (Soler et al., 2021), Africa (Coyne et al., 2018), Nigeria (Daramola et al., 2013) and Malaysia (Masdek et al., 2007). These nematodes species include *Meloidogyne javanica*, *Meloidogyne incognita*, *Pratylenchus brachyurus* and *Rotylenchulus reniformis*. Normally, pineapple plants infected by *Meloidogyne* (root-knot nematodes) display the symptoms of galls that develop on the roots because of feeding by active second-stage juveniles (Rabie, 2017). A high number of galls and branching are indications of severe root-knot nematode infection. This condition results in a stunted root system which makes the plants more susceptible to nutrient and moisture stress and diseases (Back et al., 2002). Meanwhile, symptoms of root damaged by *Pratylenchus* (lesion nematodes) can be seen as black and necrotic lesions where the nematodes penetrate. However, the lesion can extend across the entire root surface (Rabie, 2017). Above-ground symptoms of nematode infected pineapple plants resemble the symptoms of nutrient deficiency. Severe root injury will disturb the nutrient uptake by the plant, especially potassium uptake, which will subsequently affect the crop growth and finally result in low pineapple farm production and low fruit quality (Masdek et al., 2007).

3 Disease Controls

3.1 Cultural Practices

Maintaining pineapple farms by practicing good cultural practices is highly recommended and the advantages of these practices towards pineapple farms have been discussed by many authors (Aristoteles et al., 2011; Baiswar et al., 2021; Rohrbach & Johnson, 2003). Good practices on soil preparation, planting system and spacing, production and management of planting material, fertilization, artificial floral induction, pre- and post-harvest managements of fruits, as well as weed, pest, and disease controls, could enhance high productivity of pineapple and improve fruit quality. In managing pests and diseases in farms, practises such as destroying crop residues in pineapple farms to reduce disease inoculums and avoiding the emergence of resistant structures of fungal pathogens including chlamydospore and sclerotia that could remain a long time in the soil, are highly encouraged (Ventura et al., 2009). For instance, destroying affected pineapple plant and fruit with *D. zeae* could prevent the disease spread. Use of sanitized farm tools during fruit harvesting and minimize incidence that cause fruit injury can reduce some post-harvest diseases such as black rot, stem end rot, yeasty and fruitlet core rot. The implementation of proper irrigation system could improve the plant health and avoid the disease attack because pineapple plants require adequate water for growth, and water stress can predispose them to pest and disease attacks (Soler et al., 2018). Furthermore, an appropriate nutrient management is crucial to be practiced in pineapple farms to improve plant health and to avoid plant stress due to nutrient deficiency. Plants under stress are more susceptible to disease infestations. Crop rotation with non-related crops between pineapple cycles is a crucial practice to prevent diseases in pineapple as it can help to disrupt the disease cycle and reduces the accumulation of soil borne pathogen. Effective weed control in pineapple farms is also essential to avoid competition for water, nutrient and space between weeds and pineapple plants, and also to reduce disease spread because weeds could become alternative hosts for pathogens. Besides that, artificial floral induction in pineapple farm plantation is also suggested as it helps to synchronize the flowering period and facilitates harvesting, promoting the marketability of the fruits (Andrade et al., 2021).

3.2 Biological Control

Biological control agents can be utilized as an alternative to chemical pesticides. It is an effective method in managing diseases such as bacterial heart rot, black rot, and fusariosis (Sapak et al., 2021). Several microorganisms have been recognized as potential biological control agents for pineapple diseases, including *Bacillus cereus*, *Bacillus subtilis*, *Trichoderma harzianum*, and *Pseudomonas fluorescens* (Husin & Sapak, 2022; Wijesinghe et al., 2010; Yao Fulgence et al., 2021). For instance, the use of *Trichoderma* to control some pineapple diseases such as fruit collapse, butt and black rots has been reported by multiple researchers (Cano-Reinoso et al., 2022; Ishak et al., 2021; Korres et al., 2011). *Trichoderma* has capability to suppress the growth of plant pathogens by engaging in antibiosis and mycoparasitism, and synthesizing specific metabolites such growth regulators, enzymes, antibiotics, and siderophores. *Trichoderma* can produce iron chelating compounds which absorbs soil iron compounds and makes the compounds unavailable for pathogens. However, the success of biological control agents depends on the timing and the dosage of application, as well as the host–pathogen–microorganism interaction.

3.3 Chemical Control

The use of chemical pesticides for managing pests and diseases in pineapple is still the main preference among growers despite increasing concerned among consumers about the chemical residues in their food. As stated by Darnaudery et al. (2016), pineapple is a conventional monocropping systems with high levels of agrochemical inputs for fertilization, weed management, pest and disease control and flowering induction. In managing diseases, several chemical pesticides such as triadimenol, propiconazole, 1% Bordeaux mixture, and 0.25% copper oxychloride are used prior to the planting by dipping, spraying, or drenching to control butt and black rots (Joy & Sindhu, 2012). In some places, insecticides such as chlorpyrifos, imidacloprid, methyl parathion, malathion and diazinon are used in controlling mealybug, a vector of wilt virus disease in pineapple (Jahn et al., 2003; Joy & Sindhu, 2012). Meanwhile, although benomyl is regularly used to control BHR in pineapple, a recent study by Sidek and Sapak (2021) showed that benomyl is no longer effective. Heavy use of these chemical pesticides will encourage the development of pesticide resistance. Therefore, their application should be well-regulated to maintain ecological sustainability. Proper timing, dosage, and selecting a safer chemical are also crucial for effective pest and disease management.

3.4 Integrated Disease Management

Integrated disease management (IDM) is an approach that combines various control methods to manage diseases effectively while minimizing environmental and health hazards. Generally, IDM strategies for pineapple cultivation include the use of disease-resistant cultivars, crop rotation, habitat modification, biological control, cultural practices, and chemical control (Sipes & Matos, 2018). Some pineapple-producing countries such as Brazil have developed an integrated disease management in managing fusariosis disease by implementing a disease strategy of using disease-free planting material and resistant cultivars to the pathogen, inducing the flowering of plants in months with low rainfall, protecting the inflorescences by covering them with paper bags and applying chemical pesticides when necessary (Ventura et al., 2009). Some countries promote organic production of pineapple by emphasizing the use of natural control measures such as applying biological control products, disease-resistant cultivar, crop rotation, and habitat modification such as the use of mulch to reduce the occurrence of soil-borne diseases (Sipes & Matos, 2018; Soler et al., 2018). In Malaysia, the implementation of integrated pineapple disease management (IPDM) in managing pineapple diseases is encouraged by the government to be considered by the growers to reduce the application of chemical pesticides and also to promote the sustainable pineapple production with reduced disease incidence.

4 Conclusions

Generally, this chapter concludes that pineapple is susceptible to various diseases including heart rot caused by bacterial and fungal pathogen, fruit collapse, butt and black rots, fruitlet core rot, fusariosis, mealybug wilt-associated virus, yeasty, stem end rot and leaf spot. These diseases can lead to devastating production losses and economic impacts in pineapple cultivation. Some diseases are able to kill the whole plants while some are able to reduce the marketability of fruits. Early detection and proper diagnosis of these diseases based on visible symptoms and pathogen signs that can be observed in pineapple farms could help growers to prevent the disease infection and spread. The disease outbreak can be prevented by implementing good agriculture practices including sanitation, applying disease-free planting material and resistant cultivar, implementing balanced nutrient inputs, suitable irrigation, and IPDM strategies. Several strategies of IPDM such as the use of organic fertilizers, green manure and biological products could help to reduce the reliance on chemical pesticides and promotes sustainable pineapple cultivation. Furthermore, continued research and monitoring of disease outbreaks are essential to develop effective management strategies and to ensure the long-term sustainability of pineapple production.

References

Aeny, T. N., Suharjo, R., Ginting, C., Hapsoro, D., & Niswati, A. (2020). Characterization and host range assessment of *D. zeae* associated with pineapple soft rot disease in East Lampung, Indonesia. *Biodiversitas Journal of Biological Diversity, 21*(2), 587–595.

Alvarez, E., Llano, G. A., John, B. L., & Maria, A. C. (2012). Characterization of *T. paradoxa* isolates from oil palms in Colombia, Ecuador and Brazil. *Journal of Phytopathology, 160*(11–12), 690–700. https://doi.org/10.1111/jph.12012

Anderson, J., Pegg, K. G., Ronald, S. C., & Drenth, A. (2011). Phosphonate applied as a pre-plant dip controls *Phytophthora cinnamomi* root and heart rot in susceptible pineapple hybrids. *Australasian Plant Pathology, 41*(1), 59–68. https://doi.org/10.1007/s13313-011-0090-6

Andrade, R. A., Brito, R. S., Mendes, R. F., & Andrade Neto, R. (2021). Cultural treatments in pineapple crop management for high yield: Review. *Scientific Electronic Archives, 14*(12), 1487. https://doi.org/10.36560/14122021l487

Araya, M. (2019). Frequencies and population densities of parasitic nematodes in Costa Rican pineapple plantations. *Acta Horticulturae, 1239*, 153–166.

Aristoteles, P. M., Sanches, N. F., Teixeira, F. A., & Simão, A. H. (2011). Pineapple integrated pest management: An overview. *Acta Horticulturae, 902*, 339–347.

Avallone, S., Guiraud, J. P., Brillouet, J. M., & Teisson, C. (2003). Enzymatic browning and biochemical alterations in black spots of pineapple [*A. comosus* (L.) Merr.]. *Current Microbiology, 47*(2), 113–118. https://doi.org/10.1007/s00284-002-3958-4

Back, M. A., Haydock, P. P. J., & Jenkinson, P. (2002). Disease complexes involving plant parasitic nematodes and soilborne pathogens. *Plant Pathology, 51*(6), 683–697. https://doi.org/10.1046/j.1365-3059.2002.00785.x

Baiswar, P., Borah, T. R., & Singh, A. R. (2021). Postharvest diseases of pineapple and their management. In D. Singh, R. R., Sharma, V. Devappa, & D. Kamil (Eds.), *Postharvest handling and diseases of horticultural produce* (pp. 287–294). CRC Press. https://doi.org/10.1201/9781003045502

Barral, B., Chillet, M., Doizy, A., Grassi, M., Ragot, L., Léchaudel, M., Durand, N., Rose, L. J., Viljoen, A., & Schorr-Galindo, S. (2020). Diversity and toxigenicity of fungi that cause pineapple fruitlet core rot. *Toxins, 12*(5), 339. https://doi.org/10.3390/toxins12050339

Barral, B., Chillet, M., Léchaudel, M., Lugan, R., & Schorr-Galindo, S. (2019). Coumaroyl-isocitric and caffeoyl-isocitric acids as markers of pineapple fruitlet core rot disease. *Fruits, 74*(1), 11–17.

Barrett, C. B., Bachke, M. E., Bellemare, M. F., Michelson, H. C., Narayanan, S., & Walker, T. F. (2012). Smallholder participation in contract farming: Comparative evidence from five countries. *World Development, 40*(4), 715–730. https://doi.org/10.1016/j.worlddev.2011.09.006

Boluk, G., Arizala, D., Dobhal, S., Zhang, J., Hu, J., Alvarez, A. M., & Arif, M. (2021). Genomic and phenotypic biology of novel strains of *D. zeae* isolated from pineapple and taro in Hawaii: Insights into genome plasticity, pathogenicity, and virulence determinants. *Frontiers in Plant Science, 12*, 851. https://doi.org/10.3389/fpls.2021.663851

Campilho, A., Nieminen, K., & Ragni, L. (2020). The development of the periderm: The final frontier between a plant and its environment. *Current Opinion in Plant Biology, 53*, 10–14. https://doi.org/10.1016/j.pbi.2019.08.008

Cano Reinoso, D. M., Soesanto, L., Kharisun, K., & Wibowo, C. (2021). Review: Fruit collapse and heart rot disease in pineapple: Pathogen characterization, ultrastructure infections of plant and cell mechanism resistance. *Biodiversitas Journal of Biological Diversity, 22*(5), 504. https://doi.org/10.13057/biodiv/d220504

Cano-Reinoso, D. M., Soesanto, L., Kharisun, K., & Wibowo, C. (2022). Fruit collapse incidence and quality of pineapple as affected by biopesticides based on *P. fluorescens* and *T. harzianum*. *Acta Agriculturae Slovenica, 118*(3), 2485. https://doi.org/10.14720/aas.2022.118.3.2485

Carnielli-Queiroz, L., Fernandes, P. M. B., Fernandes, A. A. R., & Ventura, J. A. (2019). A rapid and reliable method for molecular detection of *F. guttiforme*, the etiological agent of pineapple Fusariosis. *Brazilian Archives of Biology and Technology, 62*, 591. https://doi.org/10.1590/1678-4324-2019180591

Coyne, D. L., Cortada, L., Dalzell, J. J., Claudius-Cole, A. O., Haukeland, S., Luambano, N., & Talwana, H. (2018). Plant-parasitic nematodes and food security in Sub-Saharan Africa. *Annual Review of Phytopathology, 56*(1), 381–403. https://doi.org/10.1146/annurev-phyto-080417-045833

Daramola, F., Afolami, S., Idowu, A., & Nwanguma, E. (2013). Studies on the occurrence and distribution of plant-parasitic nematodes in some pineapple-producing states in Nigeria. *Asian Journal of Crop Science, 5*(2), 190–199. https://doi.org/10.3923/ajcs.2013.190.199

Darnaudery, M., Fournier, P., & Léchaudel, M. (2016). Low-input pineapple crops with high quality fruit: Promising impacts of locally integrated and organic fertilisation compared to chemical fertilisers. *Experimental Agriculture, 54*(2), 286–302. https://doi.org/10.1017/s0014479716000284

Dey, K., Green, J., Melzer, M., Borth, W., & Hu, J. (2018). Mealybug wilt of pineapple and associated viruses. *Horticulturae, 4*(4), 52. https://doi.org/10.3390/horticulturae4040052

Elliott, M. L. (2019). Thielaviopsis trunk rot of palm. *EDIS, 2006*(1), 143.

Espinosa-Rodríguez, C. J., Nieto-Angel, D., León-García de Alba, C. D., Villegas-Monter, Á., Aguilar-Pérez, L. A., & Ayala-Escobar, V. (2015). Etiology of the heart rot of pineapple (*A. comosus* L. Merril) MD2 cultivar in Isla, Veracruz, México. *Revista Mexicana de Fitopatología, 33*(1), 104–115.

FAO. (2023). *Major tropical fruits market review: Preliminary results 2022.* Rome. https://www.fao.org/3/cc3939en/cc3939en.pdf

Green, J., & Nelson, S. (2015). Heart and root rots of pineapple. *Plant Disease, 106*, 1–7.

Gu, H., Zhan, R. L., Zhang, L. B., Gong, D. Q., & Jia, Z. W. (2015). First report of *F. ananatum* causing pineapple fruitlet core rot in China. *Plant Disease, 99*(11), 1653–1653. https://doi.org/10.1094/pdis-03-15-0279-pdn

Hossain, F. (2016). World pineapple production: An overview. *African Journal of Food, Agriculture, Nutrition and Development, 16*(4), 11443–11456. https://doi.org/10.18697/ajfand.76.15620

Huang, J. (1986). Ultrastructure of bacterial penetration in plants. *Annual Review of Phytopathology, 24*(1), 141–157. https://doi.org/10.1146/annurev.py.24.090186.001041

Husin, N., & Sapak, Z. (2022). *Bacillus cereus* for controlling bacterial heart rot in pineapple var. MD2. *Tropical Life Sciences Research, 33*(1), 77–89. https://doi.org/10.21315/tlsr2022.33.1.5

Hutahayan, A. J. (2017). Identification of distribution the pineapple mealybug wilt disease in the pineapple plant in North Tapanuli. *International Journal of Environment, Agriculture and Biotechnology, 2*(5), 2473–2480. https://doi.org/10.22161/ijeab/2.5.26

Ibrahim, N. F., Mohd, M. H., Mohamed Nor, N. M. I., & Zakaria, L. (2017). Characterization of *Fusarium* spp. associated with pineapple fruit rot and leaf spot in Peninsular Malaysia. *Journal of Phytopathology, 165*(11–12), 718–726. https://doi.org/10.1111/jph.12611

Ibrahim, N. F., Mohd, M. H., Mohamed Nor, N. M. I., & Zakaria, L. (2016). *Fusarium fujikuroi* causing fusariosis of pineapple in peninsular Malaysia. *Australasian Plant Disease Notes, 11*(1), 206. https://doi.org/10.1007/s13314-016-0206-5

Ishak, A. A., Zulkepli, F. R. A., Hayin, N. F. M., Zain, N. H. M., & Sapak, Z. (2021). Effect of high inlet temperature of spray dryer on viability of microencapsulated *Trichoderma asperellum* conidia. *IOP Conference Series: Earth and Environmental Science, 757*(1), 012023. https://doi.org/10.1088/1755-1315/757/1/012023

Jacobs, A., Van Wyk, P. S., Marasas, W. F., Wingfield, B. D., Wingfield, M. J., & Coutinho, T. A. (2010). *Fusarium ananatum* sp. nov. in the *Gibberella fujikuroi* species complex from pineapples with fruit rot in South Africa. *Fungal Biology, 114*(7), 515–527. https://doi.org/10.1016/j.funbio.2010.03.013

Jahn, G. C., Beardsley, J. W., & González-Hernández, H. (2003). A review of the association of ants with mealybug wilt disease of pineapple. *Proceedings of the Hawaiian Entomology Society, 36*, 9–28.

Joy, P. P., & Sindhu, G. (2012). Disease of pineapple (*A. comosus*): Pathogen, symptoms, infection, spread and management (pp. 23–25). *Pineapple Research Station (Kerala Agricultural University).*

Kaneshiro, W. S., Burger, M., Vine, B. G., de Silva, A. S., & Alvarez, A. M. (2008). Characterization of *E. chrysanthemi* from a bacterial heart rot of pineapple outbreak in Hawaii. *Plant Disease, 92*(10), 1444–1450. https://doi.org/10.1094/pdis-92-10-1444

Korres, A. M., Buss, D. S., Ventura, J. A., & Fernandes, P. M. (2011). *Candida krusei* and *Kloeckera apis* inhibit the causal agent of pineapple fusariosis *F. guttiforme. Fungal Biology, 115*(12), 1251–1258. https://doi.org/10.1016/j.funbio.2011.09.001

Kuruppu, M., Siddiqui, Y., Kong, L. L., Ahmed, K., & Ali, A. (2021). First report of postharvest stem end rot disease on md2 pineapple fruits caused by *N. dimidiatum* in Malaysia. *Plant Disease, 105*(5), 1564. https://doi.org/10.1094/pdis-10-20-2318-pdn

Lim, T. K. (1980). Role of *P. funiculosum* strains in the development of pineapple fruit diseases. *Phytopathology, 70*(7), 663. https://doi.org/10.1094/phyto-70-663

Lim, W. (1985). Pathogenicity of *P. funiculosum* strains from pineapple fruit blemishes. *MARDI Research Bulletin,* 13(2), 166–171.

Lim, W. H., & Lowings, P. H. (1979). Pineapple fruit collapse in peninsular Malaysia: Symptoms and varietal susceptibility. *Plant Disease Reporter, 63,* 170–174.

Luo, Z. W., He, F., Fan, H. Y., Wang, X. H., Hua, M., Hu, F. C., Li, X. H., Liu, Z. X., & Yu, N. T. (2012). First report of leaf spot disease caused by *E. rostratum* on pineapple in Hainan Province, China. *Plant Disease, 96*(3), 458–458. https://doi.org/10.1094/pdis-11-11-0979

Marrero, G. & Alvarez, A.M. (2011). Hawaiian strains of *E. chrysanthemi* (*Dickeya* sp.) associated with pineapple heart rot disease. *Pineapple Working Group News,* 14, 57.

Martinez-Cisneros, B. A., Juarez-Lopez, G., Valencia-Torres, N., Duran-Peralta, E., & Mezzalama, M. (2014). First report of bacterial stalk rot of maize caused by *D. zeae* in Mexico. *Plant Disease, 98*(9), 1267–1267. https://doi.org/10.1094/pdis-02-14-0198-pdn

Masdek, H. N., Ismail, A. B., Zulkifli, M., & Malip, M. (2007). *Paratylenchus* sp. associated with pineapple yield decline. *Journal of Tropical Agriculture and Food Science, 35*(1), 191.

Massé, D., Cassam, N., Hostachy, B., Iskra-Caruana, M. L., Darnaudery, M., Lefeuvre, P., & Lett, J. M. (2021). First report of three pineapple mealybug wilt-associated viruses in Queen Victoria pineapples in Reunion Island. *Plant Disease, 105*(3), 715–715. https://doi.org/10.1094/pdis-05-20-1068-pdn

Nyarko, J., & Asare-Bediako, E. (2019). First report of Pineapple mealybug wilt-associated virus-1 and -3 in Ghanaian pineapple. *New Disease Reports, 40*(1), 18–18. https://doi.org/10.5197/j.2044-0588.2019.040.018

Oculi, J., Bua, B., & Ocwa, A. (2020). Reactions of pineapple cultivars to pineapple heart rot disease in central Uganda. *Crop Protection, 135,* 105213. https://doi.org/10.1016/j.cropro.2020.105213

Petty, G. J., Stirling, G. R., & Bartholomew, D. P. (2002). Pests of pineapple. In J. E. Peña, J. L. Sharp, & M. Wysoki (Eds.), *Tropical fruit pests and pollinators: Biology, economic importance, natural enemies and control* (pp. 157–195). CABI Publishing.

Rabie, E. C. (2017). Nematode pests of pineapple. In H. Fourie, D. D. Waele, V. W. Spaull, R. K. Jones, & M. S. Daneel (Eds.), *Nematology in South Africa: A view from the 21st century.* Springer International Publishing. https://doi.org/10.1007/978-3-319-44210-5_18

Ratti, M. F., Ascunce, M. S., Landivar, J. J., & Goss, E. M. (2018). Pineapple heart rot isolates from Ecuador reveal a new genotype of *P. nicotianae. Plant Pathology, 67*(8), 1803–1813. https://doi.org/10.1111/ppa.12885

Reyes, M. E. Q., Rohrbach, K. G., & Paull, R. E. (2004). Microbial antagonists control postharvest black rot of pineapple fruit. *Postharvest Biology and Technology, 33*(2), 193–203. https://doi.org/10.1016/j.postharvbio.2004.02.003

Rodríguez, Y., Mosqueda, M., Companioni, B., Arzola, M., Borras, O., Perez, M. C., Lorenzo, J. C., & Santos, R. (2002). Bioassay for in vitro differentiation of pineapple cultivar resistance levels to heart rot disease. *In Vitro Cellular and Developmental Biology Plant, 38*(6), 613–616. https://doi.org/10.1079/ivp2002346

Rohrbach, K. G., & Johnson, M. W. (2003). Pests, diseases and weeds. In D. P. Bartholomew, R. E. Paull, & K. G. Rohrbach (Eds.), *The pineapple: Botany, production and uses* (pp. 203–251). CABI publishing. https://doi.org/10.1604/9780851995038

Sapak, Z., Mohd Faisol Mahadeven, A. N., Nurul Farhana, M. H., Norsahira, S., & Mohd Zafri, A. W. (2021). A review of common diseases of pineapple: the causal pathogens, disease symptoms, and available control measures. *Food Research, 5*(S4), 1–14. https://doi.org/10.26656/fr.2017.5(s4).004

Sether, D. M., & Hu, J. S. (2002). Yield impact and spread of pineapple mealybug wilt associated virus-2 and mealybug wilt of pineapple in Hawaii. *Plant Disease, 86*(8), 867–874. https://doi.org/10.1094/pdis.2002.86.8.867

Shen, H. F., Lin, B. R., Zhan, J. X., & Pu, X. M. (2013). First report of pineapple heart rot caused by *P. nicotianae* in Hainan Province, China. *Plant Disease*, *97*(4), 560–560. https://doi.org/10.1094/pdis-11-12-1017-pdn

Sidik, S., & Sapak, Z. (2021). Evaluation of selected chemical pesticides for controlling bacterial heart rot disease in pineapples variety MD2. In IOP Conference Series: Earth and Environmental Science (Vol. 757, No. 1, p. 012072). IOP Publishing. https://doi.org/10.1088/1755-1315/757/1/012072

Singh, H. K. B., Santhirasegaram, V., Razali, Z., & Somasundram, C. (2018). Effects of black rot on the antioxidant properties of Morris and Sarawak pineapple (*A. Comosus*). In D. Mandal, A. C. Shukla, & M. W. Siddiqui (Eds.), *Sustainable horticulture: Food, health, and nutrition* (pp. 176–184). CRC Press. https://doi.org/10.1201/b22430-14

Sipes, B., & Matos, A. D. (2018). Pests, diseases and weeds. In D. P. Bartholomew, R. E. Paull, & K. G. Rohrbach (Eds.), *The pineapple: Botany, production and uses* (pp. 269–294). CABI publishing. https://doi.org/10.1079/9781786393302.0269

Soler, A., Marie-Alphonsine, P. A., Quénéhervé, P., Prin, Y., Sanguin, H., Tisseyre, P., Daumur, R., Pochat, C., Dorey, E., Gonzalez Rodriguez, R., Portal, N., & Smith-Ravin, J. (2021). Field management of *R. reniformis* on pineapple combining crop rotation, chemical-mediated induced resistance and endophytic bacterial inoculation. *Crop Protection, 141*, 105446. https://doi.org/10.1016/j.cropro.2020.105446

Soler, A., Reinhardt, D. H., Matos, A. P. D., & Pádua, T. R. P. D. (2018). Organic production. *The Pineapple: Botany, Production and Uses, 22*, 203–221. https://doi.org/10.1079/9781786393302.0203

Steingass, C. B., Glock, M. P., Schweiggert, R. M., & Carle, R. (2015). Studies into the phenolic patterns of different tissues of pineapple (*A. comosus* [L.] Merr.) infructescence by HPLC-DAD-ESI-MS n and GC-MS analysis. *Analytical and Bioanalytical Chemistry, 407*(21), 6463–6479.https://doi.org/10.1007/s00216-015-8811-2

Stępień, U., Koczyk, G., & Waśkiewicz, A. (2013). Diversity of *Fusarium* species and mycotoxins contaminating pineapple. *Journal of Applied Genetics, 54*(3), 367–380. https://doi.org/10.1007/s13353-013-0146-0

Swaroop Kumar, H. M. (2007). *Pathological and biochemical investigations on stem end rot disease in pineapple (A. comosus (L.) MERR.)* (Doctoral dissertation, University of Mysore).

Tan, X. J., Zhang, Z. W., Xiao, J. J., Wang, W., He, F., Gao, X., Jiang, B., Shen, L., Wang, X., Sun, Y., & Zhu, G. P. (2022). Genomic and phenotypic biology of a novel *D. zeae* WH1 isolated from rice in China: Insights into pathogenicity and virulence factors. *Frontiers in Microbiology, 13*, 7486. https://doi.org/10.3389/fmicb.2022.997486

Thalip, A. A., Tong, P. S., & Ng, C. (2015). The MD2 'super sweet' pineapple (*A. comosus*): A new pineapple variety, MD2, has been selected to spearhead Malaysia's new ambitions in pineapple export. *UTAR Agriculture Science Journal (UASJ), 1*(4), 15–17.

Veléz-Negrón, Y. I., Simbaña-Carrera, L. L., Soto-Ramos, C. M., Medina, O. H., Dinkel, E., Hardy, C., Rivera-Vargas, L. I., & Ramos-Sepúlveda, L. (2023). First report of bacterial pineapple heart rot caused by *D. zeae* in Puerto Rico. *Plant Disease, 107*(1), 210. https://doi.org/10.1094/pdis-01-22-0174-pdn

Ventura, J., Costa, H., Cabral, J., & de Matos, A. (2009). "Vitória": New pineapple cultivar resistant to fusariosis. *Acta Horticulturae, 822*, 51–56. https://doi.org/10.17660/actahortic.2009.822.4

Vera, N., Maicelo, J., Guevara, E., & Oliva, S. (2017). Plant parasitic nematodes associated with pineapple cultivation (*A. comosus*) in Amazonas, Peru. *Scientia Agropecuaria, 8*, 79–84. https://doi.org/10.17268/sci.agropecu.2017.01.08

Vignassa, M., Meile, J. C., Chiroleu, F., Soria, C., Leneveu-Jenvrin, C., Schorr-Galindo, S., & Chillet, M. (2021). Pineapple mycobiome related to fruitlet core rot occurrence and the influence of fungal species dispersion patterns. *Journal of Fungi, 7*(3), 175. https://doi.org/10.3390/jof7030175

Wijesinghe, C., Wijeratnam, R. W., Samarasekara, J., & Wijesundera, R. (2010). Biological control of *T. paradoxa* on pineapple by an isolate of *T. asperellum*. *Biological Control, 53*(3), 285–290. https://doi.org/10.1016/j.biocontrol.2010.02.009

Yao Fulgence, K., Moumouny, T., Yapi Eric, Y., Ban Koffi, L., Diguta, C., Alloue-Boraud, M., & Matei, F. (2021). Biocontrol of post-harvest fungal diseases of pineapple (*A. comosus* L.) using bacterial biopesticides. *American Journal of Microbiological Research, 9*(2), 34–43. https://doi.org/10.12691/ajmr-9-2-1

Young, A. J., Pathania, N., Manners, A., & Pegg, K. G. (2022). Heart rot of Australian pineapples caused by *D. zeae*. *Australasian Plant Pathology, 51*(5), 525–533. https://doi.org/10.1007/s13313-022-00880-x

Zainun, C. (2011). Food products from *Ananas*. *Acta Horticulturae, 902*, 533–538. https://doi.org/10.17660/actahortic.2011.902.70

Zhong, L. C., Ai, Y. J., Chun, R. H., & Yi, Y. D. (2016). Identification of *C. clavata* causing leaf spot on pineapple (*A. comosus*) in China. *Canadian Journal of Plant Pathology, 38*(2), 250–253. https://doi.org/10.1080/07060661.2016.1158743

The Microbiome of Banana and Its Role in Managing Fusarium Wilt Disease

Jia Xin Ong, Nurul Shamsinah Mohd Suhaimi, and Noor Baity Saidi

Abstract Plants are associated with diverse microorganisms that collectively form a microbial community called the microbiome. Distinct microbial communities are found in different plant niches, including the endosphere, rhizosphere, and phyllosphere, and can directly or indirectly impact plant growth and health. Banana (*Musa* spp.) is an important crop in the sub-tropics and tropics and plays a crucial role in the livelihoods of millions of small-scale farmers. However, worldwide banana production is constrained by the Fusarium wilt disease, caused by the soil-borne pathogen, *Fusarium oxysporum* f. sp. *cubense* (*Foc*). Significant efforts have been made to decipher banana-associated microbiome and their correlations with the host plant and pathogens, particularly the most virulent strain of *Foc*, Tropical Race 4 (TR4). This information is critical to manipulating beneficial traits of the soil–plant microbiome to suppress the Fusarium wilt disease of bananas and ensure global food security. Herein, we summarize the current state of knowledge on the banana microbiome and detail the use of microbial consortia to improve the crop's health and overall performance. In addition, this chapter also discussed the emerging strategies toward sustainable routes to Fusarium wilt disease management using banana microbiome and the key challenges that remain.

Keywords Metagenomics · *Fusarium oxysporum* f. sp. *cubense* · *Musa* sp. · Soil-borne · Management

J. X. Ong · N. B. Saidi (✉)
Department of Cell and Molecular Biology, Faculty of Biotechnology and Biomolecular Sciences, Universiti Putra Malaysia, 43400 Serdang, Selangor, Malaysia
e-mail: norbaity@upm.edu.my

N. S. M. Suhaimi
Institute of Biological Sciences, Faculty of Science, Universiti Malaya, 50603 Kuala Lumpur, Malaysia

M. Wong (ed.), *Advances in Tropical Crop Protection*,
https://doi.org/10.1007/978-3-031-59268-3_8

1 Plant Microbiome

Diverse microorganisms that live, thrive, and interact with plant tissues are known as the plant microbiome (Berg et al., 2020; Orozco-Mosqueda et al., 2018). The microbiome can also be defined as the ecological community of microorganisms living in commensal, symbiotic, and pathogenic relationships with the host plant in a particular environment (Berg et al., 2020; Lederberg & McCray, 2001). Plant microbiomes reside in different compartments of plants, which can be mainly differentiated into the rhizosphere (root zones), phyllo-sphere (aerial surface), and endosphere (inner tissues) (Mwajita et al., 2013). Each of them hosts a specific microbial community, with bacteria and fungi being the most dominant. Among all organs, the root endosphere was recognized as the most colonized and diverse niche (Pande et al., 2021; Thomas & Sekhar, 2017). Since bacterial root endophytes hold great potential in promoting plant growth and mitigating biotic and abiotic stress (Thomas & Sekhar, 2017), the rhizosphere and endosphere were the most extensively studied microhabitat in plants.

Members of the microbiome include bacteria, archaea, fungi, algae, protists, and non-living microorganisms such as viruses and phages (Berg et al., 2020). Some of them are pathogens that can cause serious diseases, while others are decomposers that emit greenhouse gases, contributing to climate change (Jiao et al., 2018; Zolkhairi & Ismail, 2022). On the contrary, beneficial microbes can improve soil health through biogeochemical processes of preservation, carbon sequestration, and nutrient mineralization (Thakur & Geisen, 2019; Zolkhairi & Ismail, 2022). These plant-associated microbes produce phyto-hormones that stimulate the growth of the host plants and are capable of fixing nitrogen and mineralizing phosphorus, two of the most important elements for increasing soil fertility (Mwajita et al., 2013). In addition, they can enhance the resistance of plants toward biotic and abiotic stresses by inhibiting the growth of pathogens or stimulating the host's defense mechanisms (Pande et al., 2021; Turner et al., 2013). These microbial activities increase agricultural production and reduce chemical inputs, resulting in more sustainable agricultural practices.

2 Banana and Its Microbiome

2.1 Banana Plant

Banana (*Musa* spp.) is a giant monocotyledonous perennial herb found mainly in Southeast Asia and Africa (Soorianathasundaram et al., 2016). It is one of the most important cash crops around the globe that generates income and employment in banana-producing countries, mainly in the sub-tropics and tropics (Caro, 2020). Most cultivated bananas are triploid hybrids between two diploid wild types: *M. acuminata* and *M. balbisiana* (Were et Banana (*Musa* spp.) is a giant monocotyledonous perennial herb found mainly in South-East Asia and Africa (Soorianathasundaram et al., 2016). It is one of the most important cash crops around the globe that generates income and employment in banana-producing countries, mainly in the sub-tropics and tropics (Caro, 2020). Most cultivated bananas are triploid hybrids between two diploid wild types: *M. acuminata* and *M. balbisiana* (Were et al., 2022). Bananas are propagated mainly via in vitro culture techniques and planted in large-scale monoculture plantations for international trade or in small farms for local markets.

The global export of bananas reached the highest record in 2020, with more than 26 million tons of bananas being traded (FAOSTAT, 2023). However, the global production and trade of bananas were constrained by the Panama (Fusarium wilt) disease. Fusarium wilt of banana is a vascular disease caused by the soil-borne fungus *F. oxysporum* f. sp. *cubense* (*Foc*). This host-specific fungus invades and blocks the host's vascular system, causing wilting symptoms and browning of the xylem (Pérez-Vicente et al., 2014). Fusarium wilt was recognized as the most destructive disease as it nearly wiped out the banana plantations in the mid-twentieth century. Although switching susceptible Gros Michel with resistant Cavendish had solved this crisis, the emergence of the most virulent strain, *Foc* Tropical Race 4 (TR4) that infects both Cavendish and those that are susceptible to other strains, once again threatened the banana industry (Dita et al., 2018). This issue has raised global concern and calls for solution-oriented research to develop strategies for managing this disease. Apart from Fusarium wilt, several other diseases caused by different microbes were also found in banana plantations, including Black Sigatoka, Moko disease, and Xanthomonas wilt.

2.2 The Core Microbiome of Banana

Banana hosts diverse microorganism that influences their health and production. Management of these communities may provide unique solutions to overcome the various diseases affecting banana production and improve its yields. A core microbiome can be defined as a set of taxa that consistently occur within a given habitat type that are important for plant fitness (Custer et al., 2023). Common core taxa represent the relatively frequent microbiome across most individuals of a host species across diverse environmental gradients (Birt et al., 2022, 2023). These core taxa narrowed the research focus and served as potential microorganisms in disease management.

2.2.1 Bacteria

Several phyla of bacteria that are usually found associated with banana plants include Pseudomonadota (Proteobacteriota), Acidobacteriota, Actinobacteri-ota, Gemmatimonadota, Nitrospirota, Bacillota (Firmicutes), Bacteroidota and Verrucomicrobiota (Birt et al., 2022). Proteobacteria is the most abundant phylum found in banana plants. Studies have focused on members of Gammaproteobacteria under this phylum since they play a crucial role in supporting the well-being of many plant species (Köberl et al., 2015, 2017). This section will discuss the beneficial core bacteria and pathogen representatives in the banana microbiome.

Pseudomonadota

Gammaproteobacteria governs several well-known plant colonizers that provide beneficial plant-microbiome interactions. Among the most extensively studied was Pseudomonas in the Pseudomonadeles order for its phytopatho-gen suppression potential (Shen et al., 2022). Certain strains, such as *P. chlororaphis* subspecies *piscium*, *aurantiaca*, and *aureofaciens*, were shown to suppress the Fusarium wilt disease. Pseudomonas systemically induced resistance in plants (Kaushal et al., 2020). The ability to suppress disease is also achieved by the production of antimicrobial compounds, as an increase in secondary metabolite biosynthesis genes was found in disease suppressive soil of bananas (Shen et al., 2022). The species of this genus also improve plant growth via indole acetic acid (IAA) production, which stimulates plant cell elongation and proliferation, as well as efficient root system development to increase nutrient uptake (Beibei et al., 2016; Beltran-Garcia et al., 2021). Plant growth is also promoted by phosphate-solubilizing rhizobacterium *Acinetobacter rhizosphaerae* (Köberl et al., 2017). Apart from pathogen-suppressive bacteria, abiotic stress-protecting microbes are also present in the banana microbiome. For instance, Stenotrophomonas under the Xanthomonadales order comprises species that protect plants against salinity (Köberl et al., 2015). However, Erwinia was also found in relatively high abundance in the Enterobacteria class. They are mainly known as plant pathogens, such as *E. amylovora* and *E. chrysanthemi* (Köberl et al., 2015; Zolkhairi & Ismail, 2022).

Burkholderiales of phylum Betaproteobacteria are one of the core taxa in the order level found in banana plants. *Duganella*, an Oxalobacteraceae member, possesses antifungal properties toward pathogenic fungi, including Foc, through the synthesis of chitinase (Fu et al., 2020). Since the cell wall of fungi is composed mainly of chitin, exposure to chitinase will degrade the cell wall and kill the pathogenic fungi. Antifungal properties were also found in *Burkholderia* that produce volatile sulfurous compounds. These compounds rupture the fungal cell wall, alter hyphal morphologies and rupture the mycelia of pathogenic fungi, which leads to a reduction in Foc infection (Jamil et al., 2022; Xue et al., 2015). *Cupriavidus*, related to *Burkholderia*, was found to suppress *F. oxysporum* in vitro and in planta with other microbes. It increases plant susceptibility toward phytopathogens by lipopeptide siderophores (taiwachelin) that mediate iron competition (Fu et al, 2020).

Actinobacteriota

Actinomycetes are one of the prevalent beneficial soil microbes that reside in the top layer. Some of the most abundant genera in the banana microbiome include *Microbacterium*, *Mycobacterium*, and *Streptomyces*. These anaerobic saprophytes improve soil health through organic matter decomposition, enzyme secretion (chitinase, proteases), and agrobiochemical production. They are also found to suppress phytopathogens by secreting antimicrobial com-pounds or competing for available resources (Kaushal et al., 2020; Zolkhairi & Ismail, 2022). The extracts of siderophore-producing *Streptomyces* can serve as *Foc* TR4 suppressive agents as they were shown to inhibit mycelia growth, spore germination, and hyphae formation (Jamil et al., 2022).

Bacillota

Bacillus spp. is another well-studied biocontrol agent dominating the endosphere of bananas. These species were discovered to be negatively associated with both Fusarium wilt and bunchy top virus diseases (Köberl et al., 2015). These beneficial microbes colonize the plant root surfaces, forming a dense biofilm that restricts pathogen invasion (Shen et al., 2022). *Bacillus* spp. also promotes plant growth via phytohormone production and nutrient solubilization (Kaushal et al., 2020).

Bacteroidota

Similar to *Duganella*, chitin degradation and fungal cell wall lysis activities were found in chitinolytic enzyme-producing Chitinophagaceae and Cytophagales (Fu et al, 2020). The member of this phylum, *Chryseobacterium* was discovered to be involved in suppressing Fusarium wilt. Interactions of this bacteria with *Enterobacter*, *Pseudomonas*, *Bacillus*, and *Stenotrophomonas* enhance its antifungal activities against phytopathogens (Beltran-Garcia et al., 2021).

2.2.2 Fungi

The core fungal communities found across banana plants falls within the phylum Ascomycota and Mortierellomycota (Birt et al., 2023). Some other abundant phyla include Basidomycota, Glomeromycota, Zygomycota, and Chytridiomycota (Birt et al., 2023; Gomez-Lama Cabanas et al., 2021; Shen et al., 2022). Ascomycota represents the highest division among the core taxa, with members of Nectriaceae having the greatest relative abundance (Birt et al., 2023).

Ascomycota

Ascomycota is the most common fungi found in plants, including bananas. Members under Dothideomycetes, Eurotiomycetes, and Sordariomycetes were commonly found in banana plants. The most abundant family, Nectriacaeae of Sordariomycetes, hosts the causal agent of Fusarium wilt. Although many members of Fusarium present in this phylum, only one causes diseases in banana plants (Birt et al., 2023). Moreover, some non-pathogenic strains even showed a positive relationship toward suppressing Fusarium wilt in planta (Birt et al., 2023). *Fusicolla violacea*, also a member of Nectriacaeae, was found to have antagonistic effects against many phytopathogenic fungi (Wang et al., 2023).

Another abundant family in the Sordariomycetes phylum includes Hypocreaceae. *Trichoderma* is one of the most studied fungi as a biocontrol agent within this family, posing direct antifungal properties toward various pathogen invasions (Hong et al., 2022; Jamil et al., 2023; Zhou et al., 2019). Several Trichoderma strains were potent in suppressing Fusarium wilt in bananas. For instance, *T. asperellum* produces ion-binding siderophores that suppress *Foc* through iron competition, while *T. viride* which poses high root surface colonization ability that reduces the severity of Fusarium wilt symptoms (Mon et al., 2021). Another genus from the same family, *Acremonium*, could benefit banana as it alleviates plant biotic (insects) and abiotic (drought) stresses (Wang et al., 2023). *Nigrospora*, a genus in the Trichosphaeriaceae family, are leaf-degrading fungi found primarily on fallen banana leaves (Wang et al., 2023).

Other well-known fungi with antifungal properties are the *Penicillium*, un-der Eurotiomycetes (Jamil et al., 2023). *Penicillium* produces enzymatic anti-fungal compounds such as chitinase and β-1,3-glucanase, which destroy the cell walls of fungi and suppress the invasion of broad plant pathogens. Apart from fungal infection, *Penicillium* was also found to antagonize bacterial pathogens such as *Xanthomonas vasicola pv. musacearum* (causal agent of Xanthomonas wilt) by antibacterial metabolite production (Jamil et al., 2023).

Mortierellomycota

The *Mortierella* genus was shown to suppress the Fusarium wilt of bananas by producing antibiotic and antifungal compounds that antagonize pathogens (Birt et al., 2023; Hong et al., 2022; Zhou et al., 2019). However, the relative abundance of this microbe decreases in continuous monoculture fields (Birt et al., 2023). In other studies, members of *Mortierella,* such as *M. capitata* enrich and reshape the rhizosphere microbial population and improve plant growth by increasing available phosphorus in the soil for maize plants (Li et al., 2020; Wang et al., 2023).

Basidiomycota

The family Hydnodontaceae are the prolific producers of lignin and wood decomposition. They improve plant growth by balancing the carbon and nitrogen in the soil for plants. Although no evidence shows their ability to suppress *Foc*, they antagonize *F. oxysporium* species that cause root rot disease in ginseng plants (Wang et al., 2022).

2.2.3 Virus

Viruses are responsible for various diseases, but some might be beneficial. Bacteriophages are viruses that only infect bacteria. These viruses positively impact the plant microbiome, possibly through increasing the elements cycle in the soil to promote plant growth and microbiome attraction, maintaining a youthful bacterial community, releasing endoenzymes for complete mineralization of low molecular weight organic substances, and long-term carbon sequestration in soil (Kuzyakov & Mason-Jones, 2018). The banana plant bacteriophage community has been the subject of recent studies (Aghdam et al., 2023; Dion et al., 2020).

The most abundant family includes Ackermannviridae, Peduoviridae, Autographviridae, Persivirus, Naomviridae, Stephanstrimvirinae and Casjensviridae. Among them, *Ackermanniviridae* and *Rhizobium* phage are the most abundant taxa found in banana plants. Potential *Klebsiella* phages that targeted *Klebsiella* bacteria were also found (Aghdam et al., 2023). *Klebsiella* bacteria are beneficial plant colonizers that fix nitrogen, promote plant growth, and antagonize *Foc* TR4 (Nakkeeran et al., 2021). The divergence in *Klebsiella* phage-like sequence suggests a novel banana-specific *Klebsiella* bacteria strain present in the plant that might directly impact the management of Fusarium wilt (Aghdam et al., 2023).

In addition, phages targeting common plant pathogens that cause bacterial disease were found. For example, *Erwinia* phage targets pathogenic *Erwinia* spp., *Pectobacterium* phage targets bacteria causing soft rot disease, and *Ralstonia* phage targets bacteria causing banana Moko and blood disease (Aghdam et al., 2023).

2.3 *Factors Affecting the Microbial Diversity of Bananas*

2.3.1 Plant Compartments

Studies have found that plant compartments hold the strongest influence in determining microbial communities. The highest diversity is found in bulk soil, followed by ectorhizosphere, endorhizosphere, rhizome, pseudostem, and leaves. In the more proximal compartments, similar microbe compositions were discovered. This is due to some bacteria communities originating from adjoining compartments. The greater the distance between compartments, the lower the proportions of microbes derived from that compartment (Birt et al., 2022, 2023).

The rhizosphere is a niche where the soil microbes and plant roots interact. The ectorhizosphere consists of soil surrounding the root surface, while endorhizosphere consists of the root tissues (including endodermis and cortical cells) (Bashir et al, 2016). Soil microbes are thought to colonize the roots, and bacteria are the most prevalent microbes in the rhizosphere. Among them, Acidobacteriota were detected frequently in the ectorhizosphere, whereas some Actinobacteriota, such as *Mycobacterium* and *Streptomyces,* are more commonly found in the endorhizosphere of banana plants. As for Proteobacteriota, Xanthomonadales, *Pseudomonas,* and *Rhizobium* were more abundant in the rhizosphere compared to other compartments (Beltran-Garcia et al., 2021; Birt et al., 2022; Kaushal et al., 2022; Köberl et al., 2015).

On the other hand, the fungal community found in the rhizosphere mainly comprises species of Ascomycota, mainly *Fusarium*, *Colletotrichum* and *Pyrenchaetopsis* (Birt et al., 2022). Apart from bacteria and fungi, bacteriophages such as *Salmonella phage*, *Peduovirus*, *Quadragintavirus*, *Yersinia phage*, *Brunovirus*, *Eganvirus,* and *Klebsiella phage* are also abundant, especially in the endorhizospheres (Aghdam et al., 2023).

All above-ground parts of bananas, including the pseudostem and leaves, are collectively known as the phyllosphere (Mwajita et al., 2013). The microbial community of leaves is assumed to have a high proportion of airborne microorganisms, which explains the differences in microbiome richness between the ectorhizosphere and phyllosphere. Members of Bacilli can be found in higher abundance in the phyllosphere of bananas (Birt et al., 2022).

On the other hand, the dominant fungi occupying only the phyllosphere were *Phaeosphaeria oryzae* and *Cladosporium* spp. from the phylum Ascomycota (Birt et al., 2023). Members of Basidomycota were also commonly found in leaf compartments (Birt et al., 2022). As for the viral communities, *Vhulanivirus*, *Brucella phage* and *Noahvirus* were the members that are more abundant in banana leaves (Aghdam et al., 2023).

2.3.2 Biotic Stress

Diseased plants were shown to harbor different microbial communities compared to healthy or non-symptomatic plants. Symptomatic plants have higher diversity than asymptomatic plants (Jamil et al., 2022; Zhou et al., 2019). Diverse and abundant microbial communities are important in disease suppression either directly by attracting antimicrobial-secreting biocontrol agents or indirectly by changing the available nutrients in the soil and increasing beneficial microbe populations to maintain the pathogen below the minimum pathogenicity expression level (Köberl et al., 2017; Zhou et al., 2019). For instance, banana plants infected by *Foc* were dominated by bacterial species under Rhizobiales, Flavobacteriales, Enterobacteriaceae (Kaushal et al., 2020; Köberl et al., 2017). On the contrary, healthy plants have a higher abundance of members of Xanthomonadales, Pseudomonadales, Firmicutes and Burkholderiales, Streptomycetales (Jamil et al., 2022; Kaushal et al., 2020; Köberl et al., 2017). In addition, the presence of *Pseudomonas*, *Bacillus,* and *Stenotrophomonas* that are suppressive toward Fusarium wilt indicates healthy banana plants (Köberl et al., 2017). Healthy plants are also found with a lower abundance of Ascomycota and a higher abundance of *Penicillium* and *Trichoderma* (Jamil et al., 2023; Zhou et al., 2019).

2.3.3 Other Factors

Soil is the main exogenous factor influencing plant microbiome in the rhizosphere (Birt et al., 2022; Shakya et al., 2013). Soil composition, physiochemistry, structure, and pH changes cause differences in microbial communities apart from geographical areas. Soil pH was found to be negatively correlated with *Foc* abundance, thereby decreasing the disease incidence in a field (Hong et al, 2020). Despite different plant species harboring varied microbial communities, different genotypes within bananas have no substantial influence on microbial diversity (Birt et al., 2022, 2023; Pande et al., 2021). Other factors that can affect the microbial communities include seasonal variation, developmental stage, abiotic stress, chemical stimulant, and farming system.

2.4 Strategies in Manipulating Plant Microbiome for Fusarium Wilt Disease Management

The current strategies in managing plant disease by chemical application are considered effective. However, repeated use of chemicals poses a major threat to the environment and human health. These concerns demand more sustainable and cost-effective strategies. Thus, studies on disease management through the manipulation of soil microbiome have attracted much attention in recent years, especially in managing Fusarium wilt of bananas. The soil microbiome was the focus of many studies as the majority of microbes that hold critical roles in soil-borne disease suppression reside in the rhizosphere and roots (Lv et al., 2023).

2.4.1 Direct Introduction of Biocontrol Agents

Although pathogen resistance is governed by microbial consortia instead of a single taxon, increasing the abundance and survival of specific keystone microbes in the plant system or soil is advantageous as they are vital in inducing disease-suppressive ability (Nakkeeran et al., 2021; Zhou et al., 2019). Introducing biocontrol strains can be considered a practical approach as it is environmentally sustainable, safe, lower cost, and has a long-term positive effect in controlling plant diseases (Zolkhairi & Ismail, 2022). Several strategies are available to introduce these biocontrol strains into the plant system, either through organic fertilizer as a carrier or direct inoculation into the plant.

2.4.2 Biofertilizer Application

Organic fertilizers are preferred in farming as they were shown to increase plant biomass and nutrient uptake, improve soil available nutrients, and prime microbial activities (Zhang et al., 2019). However, their ability to suppress soil-borne disease is minute. Biofertilizers are organic fertilizers enriched with isolates of beneficial microorganisms or biocontrol agents. The most studied biofertilizer against Fusarium wilt in banana plants comprises the members of Bacillus (*B. amyloliquefaciens* NJN-6 and W19) (Fu et al., 2017; Tao et al., 2020; Xue et al., 2015). The biofertilizer improves plant growth by enhancing root system development stimulated by IAA production. It also increases available nutrients in the soil for root uptake, resulting in higher production of root exudate that will attract and enrich the microbial community. High bacterial diversity and richness create a competitive environment for nutrients and resources that will limit the abundance and invasion of pathogens. These *Bacillus* strains also have direct antagonistic effects toward *Foc* and can attract more beneficial microorganisms. For instance, *B. amyloliquefaciens* W19 attracts more antagonistic *Pseudomonas* toward *Foc*. *Pseudomonas* improves biofilm production and root colonization in the presence of *B. amyloliquefaciens* W19, shielding the root surface from pathogen invasion (Tao et al., 2020). In short, the combination of direct effects on the pathogen and the indirect effects from reshaping the soil microbial community by biofertilizers contribute to disease suppression.

2.4.3 Direct Microbiome Inoculation

Direct inoculation of beneficial microbes is advantageous as it can be applied early, for example, at the hardening phase of tissue-cultured plants, before introducing the plantlets to potential disease-conducive soil. A combined application of different biocontrol agents either provides additional functions or enhances the ability of existing strains to suppress Fusarium wilt in banana plants. For instance, soil pre-treated with *Piriformospore indica* serves a preventive role by reducing the chances of pathogen infection through iron chelation, as iron is an essential nutrient for pathogen-plant interaction (Zhu et al., 2023). Inoculation of *Streptomyces morookaensis* to banana plantlets before planting was shown to restrict *Foc* growth in plant endophytic compartments (Zhu et al., 2023). Other studies that introduced several strains of *Pseudomonas* or different beneficial fungi also showed enhanced efficiency of Fusarium wilt suppression. Otherwise, a single treatment of the same microbes was shown to negatively affect plant growth or have low suppressive ability (Cheng et al., 2022; Lv et al., 2023). In addition, incorporating chemotactic substances such as glycine enhances the biocontrol efficacies of certain strains toward Fusarium wilt (Yang et al., 2022c). Interestingly, introducing beneficial microbes from non-host plants can also induce the immune system of bananas against *Foc* (Nakkeeran et al., 2021).

2.4.4 Intercropping and Crop Rotation

Banana crops are mainly grown in large monoculture plantations. However, monocropping decreases soil microflora diversity and some key soil functions, which drives the abundance of pathogenic *Fusarium* in the field (Jamil et al., 2023; Wang et al., 2023). Although introducing beneficial strains were efficient in decreasing disease incidence, containment of *Fusarium* abundance for a short period does not improve the imbalance of the microbial community fundamentally. The exudates of banana roots will eventually attract the pathogens again after a period of planting. Thus, intercropping or crop rotation is introduced to gradually restore the ecological balance of the plantation field (Yang et al., 2022a). Green manure intercropping using Sirato (legume), *Vicia villosa,* and *Trifolium repens* positively affected the microbiome diversity. These cover crops secrete bioactive compounds that change the microflora, increasing the abundance of antagonistic microbes for disease suppression (Wang et al., 2023; Yang et al., 2022a). Intercropping plants other than cover crops is a challenge since the shape of banana leaves creates shade in the field. Therefore, crop rotation is more applicable as one of the strategies to reduce disease incidence in bananas. Studies have found chili pepper or pineapple rotation in banana fields suppresses Fusarium wilt disease by improving soil quality and enhancing beneficial microbiome populations to inhibit the exponential growth of *Foc* (Hong et al., 2020; Yang et al., 2022b). Litter decomposition and root exudates of rotated crops in the field improve soil physiochemistry and increases the soil's pH, which decreases *Foc* abundance in a field. However, the disadvantage of using this system is the long period of soil remediation.

2.4.5 Plant Residue Manipulation

Similar to agroforestry, plant residue application improves soil's biological and physiochemical properties, promotes nutrient cycling, and reduces pathogen pressure. The decayed plant residues increase soil carbon and induce colonization of residual host-specific microorganisms that can inhibit certain pathogens (Hong et al., 2022; Yuan et al., 2021). In banana plantations, the effect of chili and pineapple plant residue on Fusarium wilt has been established. The chili plant residue extract was found to inhibit the growth of *Foc* mycelium directly. In addition, this residue indirectly attracts *Bacillus* population by increasing the soil's alkali-hydrolyzable nitrogen content (Hong et al., 2022). Although pineapple plant residue extract does not have a direct effect on the growth of *Foc,* it attracts antagonists such as *Aspergillus fumigatus* and *F. solani* that either secrete antimicrobial compounds or compete for resources to reduce *Foc* abundance (Yuan et al., 2021).

2.4.6 Soil Fumigation and Disinfection

Biofertilizers and agroforestry have been found to reduce the abundance of *Foc* in soil to control Fusarium wilt in bananas. However, these strategies had low efficacy, especially in highly infested fields, as the number of pathogens is enormous and long treatment periods are needed to suppress the disease. Fumigation is a promising method to reduce or eliminate pathogens from a field (Shen et al., 2019). Ammonia rather than traditional methyl bromide as a fumigation agent was suggested due to several advantages: it improves banana plant growth by introducing nitrogen into the soil; it reduces *Foc* abundance by mycelia destruction and spore germination suppression; it is environmental- and cost-friendly (Shen et al., 2018, 2019). Nevertheless, ammonia can still negatively impact the ecosystem and biodiversity despite being more eco-friendly compared to methyl bromide. Lime treatment with ammonium bicarbonate fumigation effectively reduces *Foc* abundance (Shen et al., 2018), as higher pH converts ammonium to ammonia, and pH is known to directly affect *Foc* abundance. Fumigation will result in a reduction in the microbial community, especially ammonium-sensitive species. However, the beneficial microbes in the soil can be recovered by adding biofertilizers after the fumigation. This manipulation creates a unique soil microbiome, further reducing the pathogenic strains in the field (Shen et al., 2018, 2019).

2.4.7 Genetic Engineering

Manipulating naturally occurring beneficial microorganisms to control Fusarium wilt is challenging, and complete replacement of chemical application might require engineered microorganisms with enhanced potential directly against pathogens. Efforts have been made to engineer 1-aminocyclopropane-1-carboxylate (ACC) deaminase that prevents ethylene production. Ethylene produced by plants promotes the formation of vascular system-obstructing gels in *Foc* during infection, causing wilting symptoms in banana plants (VanderMolen et al., 1983). Introducing plant growth-promoting strains engineered with ACC deaminase that cleaves ACC (ethylene synthesis precursor) might prevent the formation of gels blocking the vascular system, thereby decreasing the disease incidence of Fusarium wilt (Liu et al., 2019). In addition, engineering bacteria with chitinase expression also inhibited hyphae penetrations of *F. oxysporum* (García-Fraga et al., 2015; Kaur et al., 2021). However, host-specific promoters, terminators and operons limit the potential of recombinant bacteria to express the desired trait. Thus, building host-specific libraries for these biological parts and regulatory sequences is important to apply this strategy in managing plant diseases (Kaur et al., 2021).

References

Aghdam, S. A., Lahowetz, R. M., & Brown, A. (2023). Divergent endophytic viromes and phage genome repertoires among banana (*Musa*) species. *Frontiers in Microbiology, 14*, 1127606.

Bashir, O., Khan, K., Hakeem, K. R., Mir, N. A., Rather, G. H., & Mohiuddin, R. (2016). Soil microbe diversity and root exudates as important aspects of rhizosphere ecosystem. *Plant, Soil and Microbes: Mechanisms and Molecular Interactions, 2*, 337–357.

Beibei, W. A. N. G., Zongzhuan, S. H. E. N., Zhang, F., Waseem, R. A. Z. A., Jim, Y. U. A. N., Huang, R., et al. (2016). *Bacillus amyloliquefaciens* strain W19 can promote growth and yield and suppress Fusarium wilt in banana under greenhouse and field conditions. *Pedosphere, 26*(5), 733–744.

Beltran-Garcia, M. J., Martinez-Rodriguez, A., Olmos-Arriaga, I., Valdez-Salas, B., Chavez-Castrillon, Y. Y., Di Mascio, P., & White, J. F. (2021). Probiotic endophytes for more sustainable banana production. *Microorganisms, 9*(9), 1805.

Berg, G., Rybakova, D., Fischer, D., Cernava, T., Vergès, M. C. C., Charles, T., et al. (2020). Microbiome definition re-visited: Old concepts and new challenges. *Microbiome, 8*(1), 1–22.

Birt, H. W., Pattison, A. B., Skarshewski, A., Daniells, J., Raghavendra, A., & Dennis, P. G. (2022). The core bacterial microbiome of banana (*Musa* spp.). *Environmental Microbiome, 17*(1), 1–19.

Birt, H. W., Pattison, A. B., Skarshewski, A., Daniells, J., Raghavendra, A., & Dennis, P. G. (2023). The core fungal microbiome of banana (*Musa* spp.). *Frontiers in Microbiology, 14*, 1127779.

Caro, L. P. (2020). *Wages and working conditions in the banana sector: The case of Costa Rica, Ethiopia, India, Indonesia, and Vietnam: Background note.* International Labour Organization.

Cheng, C., Liu, F., Wang, B., Qu, P., Liu, J., Zhang, Y., et al. (2022). Influences of *Serendipita indica* and *Dictyophorae echinovolvata* on the growth and Fusarium wilt disease resistance of banana. *Biology, 11*(3), 393.

Custer, G. F., Gans, M., van Diepen, L. T., Dini-Andreote, F., & Buerkle, C. A. (2023). Comparative analysis of core microbiome assignments: Implications for ecological synthesis. *Msystems, 8*(1), e01066-22.

Dion, M. B., Oechslin, F., & Moineau, S. (2020). Phage diversity, genomics and phylogeny. *Nature Reviews Microbiology, 18*(3), 125–138.

Dita, M., Barquero, M., Heck, D., Mizubuti, E. S., & Staver, C. P. (2018). Fusarium wilt of banana: Current knowledge on epidemiology and research needs toward sustainable disease management. *Frontiers in Plant Science, 9*, 1468.

FAOSTAT. (2023). *Crops and livestock products trade statistics*. www.fao.org/faostat/en/#data. Accessed 7 July 2023.

Fu, L., Penton, C. R., Ruan, Y., Shen, Z., Xue, C., Li, R., & Shen, Q. (2017). Inducing the rhizosphere microbiome by biofertilizer application to suppress banana Fusarium wilt disease. *Soil Biology and Biochemistry, 104*, 39–48.

Fu, L., Xiong, W., Dini-Andreote, F., Wang, B., Tao, C., Ruan, Y., et al. (2020). Changes in bulk soil affect the disease-suppressive rhizosphere microbiome against Fusarium wilt disease. *Frontiers of Agricultural Science and Engineering, 7*(3), 307–316.

García-Fraga, B., Da Silva, A. F., López-Seijas, J., & Sieiro, C. (2015). A novel family 19 chitinase from the marine-derived *Pseudoalteromonas tunicata* CCUG 44952T: Heterologous expression, characterization and antifungal activity. *Biochemical Engineering Journal, 93*, 84–93.

Gomez-Lama Cabanas, C., Fernández-González, A. J., Cardoni, M., Valverde-Corredor, A., López-Cepero, J., Fernández-López, M., & Mercado-Blanco, J. (2021). The banana root endophytome: differences between mother plants and suckers and evaluation of selected bacteria to control *F. oxysporum* f. sp. *cubense*. *Journal of Fungi, 7*(3), 194.

Hong, S., Jv, H., Lu, M., Wang, B., Zhao, Y., & Ruan, Y. (2020). Significant decline in banana Fusarium wilt disease is associated with soil microbiome reconstruction under chilli pepper-banana rotation. *European Journal of Soil Biology, 97*, 103154.

Hong, S., Jv, H., Yuan, X., Geng, J., Wang, B., Zhao, Y., et al. (2022). Soil organic nitrogen indirectly enhances pepper-residue-mediated soil disease suppression through manipulation of soil microbiome. *Agronomy, 12*(9), 2077.

Jamil, F. N., Hashim, A. M., Yusof, M. T., & Saidi, N. B. (2022). Analysis of soil bacterial communities and physicochemical properties associated with Fusarium wilt disease of banana in Malaysia. *Scientific Reports, 12*(1), 999.

Jamil, F. N., Hashim, A. M., Yusof, M. T., & Saidi, N. B. (2023). Association of soil fungal community composition with incidence of Fusarium wilt of banana in Malaysia. *Mycologia, 115*(2), 178–186.

Jiao, S., Chen, W., Wang, J., Du, N., Li, Q., & Wei, G. (2018). Soil microbiomes with distinct assemblies through vertical soil profiles drive the cycling of multiple nutrients in reforested ecosystems. *Microbiome, 6*, 1–13.

Kaur, R., Kaur, S., & Kaur, G. (2021). Molecular and physiological manipulations in rhizospheric bacteria. *Acta Physiologiae Plantarum, 43*(5), 77.

Kaushal, M., Kolombia, Y., Alakonya, A. E., Kuate, A. F., Ortega-Beltran, A., Amah, D., & Masso, C. (2022). Subterranean microbiome affiliations of plantain (*Musa* spp.) under diverse agroecologies of western and central Africa. *Microbial Ecology, 12*, 1–14.

Kaushal, M., Mahuku, G., & Swennen, R. (2020). Metagenomic insights of the root colonizing microbiome associated with symptomatic and non-symptomatic bananas in Fusarium wilt infected fields. *Plants, 9*(2), 263.

Köberl, M., Dita, M., Martinuz, A., Staver, C., & Berg, G. (2015). Agroforestry leads to shifts within the gammaproteobacterial microbiome of banana plants cultivated in Central America. *Frontiers in Microbiology, 6*, 91.

Köberl, M., Dita, M., Martinuz, A., Staver, C., & Berg, G. (2017). Members of Gammaproteobacteria as indicator species of healthy banana plants on Fusarium wilt-infested fields in Central America. *Scientific Reports, 7*(1), 45318.

Kuzyakov, Y., & Mason-Jones, K. (2018). Viruses in soil: Nano-scale undead drivers of microbial life, biogeochemical turnover and ecosystem functions. *Soil Biology and Biochemistry, 127*, 305–317.

Lederberg, J., & McCray, A. T. (2001). Ome SweetOmics: A genealogical treasury of words. *The Scientist, 15*(7), 8–8.

Li, F., Zhang, S., Wang, Y., Li, Y., Li, P., Chen, L., et al. (2020). Rare fungus, *Mortierella capitata*, promotes crop growth by stimulating primary metabolisms related genes and reshaping rhizosphere bacterial community. *Soil Biology and Biochemistry, 151*, 108017.

Liu, Y., Zhu, A., Tan, H., Cao, L., & Zhang, R. (2019). Engineering banana endosphere microbiome to improve Fusarium wilt resistance in banana. *Microbiome, 7*(1), 1–15.

Lv, N., Tao, C., Ou, Y., Wang, J., Deng, X., Liu, H., et al. (2023). Root-associated antagonistic *Pseudomonas* spp. contribute to soil suppressiveness against banana Fusarium wilt disease of banana. *Microbiology Spectrum, 11*(2), e03525-22.

Mon, Y. Y., Bidabadi, S. S., Oo, K. S., & Zheng, S. J. (2021). The antagonistic mechanism of rhizosphere microbes and endophytes on the interaction between banana and *F. oxysporum* f. sp. *cubense*. *Physiological and Molecular Plant Pathology, 116*, 101733.

Mwajita, M. R., Murage, H., Tani, A., & Kahangi, E. M. (2013). Evaluation of rhizosphere, rhizoplane and phyllosphere bacteria and fungi isolated from rice in Kenya for plant growth promoters. *Springerplus, 2*(1), 1–9.

Nakkeeran, S., Rajamanickam, S., Saravanan, R., Vanthana, M., & Soorianathasundaram, K. (2021). Bacterial endophytome-mediated resistance in banana for the management of Fusarium wilt. *3 Biotech, 11*(6), 267.

Orozco-Mosqueda, M. D. C., Rocha-Granados, M. D. C., Glick, B. R., & Santoyo, G. (2018). Microbiome engineering to improve biocontrol and plant growth-promoting mechanisms. *Microbiological Research, 208*, 25–31.

Pande, A., Mun, B. G., Lee, D. S., Khan, M., Lee, G. M., Hussain, A., & Yun, B. W. (2021). No network for plant–microbe communication underground: A review. *Frontiers in Plant Science, 12*, 658679.

Pérez-Vicente, L., Dita, M. A., & Martínez-de la Parte, E. (2014). Technical manual prevention and diagnostic of Fusarium Wilt (Panama disease) of banana caused by *F. oxysporum* f. sp. *cubense* Tropical Race 4 (TR4). *Food and Agriculture Organization of The United Nations. Technical manual.*

Shakya, M., Gottel, N., Castro, H., Yang, Z. K., Gunter, L., Labbé, J., et al. (2013). A multifactor analysis of fungal and bacterial community structure in the root microbiome of mature *Populus deltoides* trees. *PLoS ONE, 8*(10), e76382.

Shen, Z., Thomashow, L. S., Ou, Y., Tao, C., Wang, J., Xiong, W., et al. (2022). Shared core microbiome and functionality of key taxa suppressive to banana Fusarium wilt. *Research, 2022.*

Shen, Z., Xue, C., Penton, C. R., Thomashow, L. S., Zhang, N., Wang, B., et al. (2019). Suppression of banana Panama disease induced by soil microbiome reconstruction through an integrated agricultural strategy. *Soil Biology and Biochemistry, 128*, 164–174.

Shen, Z., Xue, C., Taylor, P. W., Ou, Y., Wang, B., Zhao, Y., et al. (2018). Soil pre-fumigation could effectively improve the disease suppressiveness of biofertilizer to banana Fusarium wilt disease by reshaping the soil microbiome. *Biology and Fertility of Soils, 54*, 793–806.

Soorianathasundaram, K., Narayana, C. K., & Paliyath, G. (2016). Banana and plantains. In B. Caballero, P. M. Finglas, & F. Toldrá (Eds.), *Encyclopedia of Food and Health* (pp. 320–327). Elsevier.

Tao, C., Li, R., Xiong, W., Shen, Z., Liu, S., Wang, B., et al. (2020). Bio-organic fertilizers stimulate indigenous soil *Pseudomonas* populations to enhance plant disease suppression. *Microbiome, 8*, 1–14.

Thakur, M. P., & Geisen, S. (2019). Trophic regulations of the soil microbiome. *Trends in Microbiology, 27*(9), 771–780.

Thomas, P., & Sekhar, A. C. (2017). Cultivation versus molecular analysis of banana (*Musa* sp.) shoot-tip tissue reveals enormous diversity of normally uncultivable endophytic bacteria. *Microbial Ecology, 73*(4), 885–899.

Turner, T. R., James, E. K., & Poole, P. S. (2013). The plant microbiome. *Genome Biology, 14*(6), 1–10.

VanderMolen, G. E., Labavitch, J. M., Strand, L. L., & DeVay, J. E. (1983). Pathogen-induced vascular gels: Ethylene as a host intermediate. *Physiologia Plantarum, 59*(4), 573–580.

Wang, L., Zhang, Y., Wang, Y., Suo, M., Wu, H., Zhao, M., & Yang, H. (2022). Inoculation with *Penicillium citrinum* aids ginseng in resisting *F. oxysporum* by regulating the root and rhizosphere microbial communities. *Rhizosphere, 22*, 100535.

Wang, Y., Zhang, W., Goodwin, P. H., Zheng, S. J., Li, X., & Xu, S. (2023). Effect of natural weed and Siratro cover crop on soil fungal diversity in a banana cropping system in southwestern China. *Frontiers in Microbiology, 14*, 1138580.

Were, E., Viljoen, A., & Rasche, F. (2022). Back to the roots: Understanding banana below ground interactions is crucial for effective management of Fusarium wilt. *Plant Pathology, 10*, 1–20.

Xue, C., Ryan Penton, C., Shen, Z., Zhang, R., Huang, Q., Li, R., et al. (2015). Manipulating the banana rhizosphere microbiome for biological control of Panama disease. *Scientific Reports, 5*(1), 11124.

Yang, J., Duan, Y., Liu, X., Sun, M., Wang, Y., Liu, M., et al. (2022a). Reduction of banana fusarium wilt associated with soil microbiome reconstruction through green manure intercropping. *Agriculture, Ecosystems and Environment, 337*, 108065.

Yang, J., Ren, X., Liu, M., Fan, P., Ruan, Y., Zhao, Y., & Wang, B. (2022b). Suppressing soil-borne Fusarium pathogens of bananas by planting different cultivars of pineapples, with comparisons of the resulting bacterial and fungal communities. *Applied Soil Ecology, 169*, 104211.

Yang, L., Zhou, Y., Guo, L., Yang, L., Wang, J., Liang, C., & Huang, J. (2022c). The effect of banana rhizosphere chemotaxis and chemoattractants on *Bacillus velezensis* LG14-3 root colonization and suppression of banana Fusarium wilt disease. *Sustainability, 15*(1), 351.

Yuan, X., Hong, S., Xiong, W., Raza, W., Shen, Z., Wang, B., et al. (2021). Development of fungal-mediated soil suppressiveness against Fusarium wilt disease via plant residue manipulation. *Microbiome, 9*(1), 1–15.

Zhang, J., Bei, S., Li, B., Zhang, J., Christie, P., & Li, X. (2019). Organic fertilizer, but not heavy liming, enhances banana biomass, increases soil organic carbon and modifies soil microbiota. *Applied Soil Ecology, 136*, 67–79.

Zhou, D., Jing, T., Chen, Y., Wang, F., Qi, D., Feng, R., et al. (2019). Deciphering microbial diversity associated with Fusarium wilt-diseased and disease-free banana rhizosphere soil. *BMC Microbiology, 19*, 1–13.

Zhu, Z., Wu, G., Deng, R., Hu, X., Tan, H., Chen, Y., et al. (2023). Spatiotemporal biocontrol and rhizosphere microbiome analysis of Fusarium wilt of banana. *Communications Biology, 6*(1), 27.

Zolkhairi, N. A. M., & Ismail, I. N. A. (2022). Soil microbiome and banana plant diseases: A review. *Malaysian Journal of Science Health and Technology, 8*(1), 85–91.

Utilisation of Plant-Based Product in Post-harvest Disease Management of Fruits

Nur Ain Izzati Mohd Zainudin, Nur Baiti Abd Murad, and Fatin Nabila Shaari

Abstract Fungal disease has become a significant threat to plants, animals, and the public. Fungal plant pathogens affect the profitability, quality, and quantity of crop yields through various mechanisms of pathogenesis that compromise the immune system of the plants, such as developing ways to attack the plants, seeking entry via open pores or exposed surfaces or wounds, as well as sourcing nutrients forcefully for their growth and development. These mechanisms will negatively affect the plant's health, homeostasis, and physiology while causing systemic damage, in some cases at severe levels. Synthetic chemical fungicide has been commercially used to control fungal diseases in crops. Unfortunately, the unregulated and excessive use of fungicides can threaten public health due to the toxicological safety of fresh produce, which encourages environmental pollution associated with these chemicals and enhances the development of pathogen resistance to the fungicide. The increasing global population today requires sufficient, safe and synthetic chemical free fresh produce. Therefore, finding alternative ways to improve productivity, quality, and safety of crop yields is of utmost importance.

1 Introduction

The post-harvest losses are estimated to range from 10 to 50% of the global production of fresh fruits and vegetables per year, with approximately 20% of post-harvest losses reported for fruits and vegetables in Malaysia, even with the expenditure of modern storage facilities and techniques. Therefore, fruit and vegetable loss and waste need to be reduced to provide sustainable feeding for the world's population in the future (Iordăchescu et al., 2019; Mahmud, 2017; Mohajer et al., 2015; Singh et al., 2017).

N. A. I. Mohd Zainudin (✉) · N. B. Abd Murad · F. N. Shaari
Department of Biology, Faculty of Science, Universiti Putra Malaysia, 43400 Serdang, Selangor, Malaysia
e-mail: ainizzati@upm.edu.my

N. A. I. Mohd Zainudin
Laboratory of Sustainable Agronomy and Crop Protection, Institute of Plantation Studies, Universiti Putra Malaysia, 43400 Serdang, Selangor, Malaysia

M. Wong (ed.), *Advances in Tropical Crop Protection*,
https://doi.org/10.1007/978-3-031-59268-3_9

Reduction in fruit quantity and quality should be expected when estimating post-harvest disease losses as certain diseases may not reduce the unsalable product but can still reduce their value in terms of physical appearance, shelf life and nutritional contents (Brauer et al., 2019).

In the year 2050, the world population is expected to exceed about 9 billion; hence, food availability needs to be increased urgently in a sustainable manner by reducing post-harvest losses (Toivonen et al., 2014). Henceforth, post-harvest losses should be reduced at field, storage, markets, and consumer levels to increase the amount of food available for human consumption. Moreover, Malaysia has currently identified post-harvest as one of the sectors that need to be exploited to ensure the country's food security and safety (Mahmud, 2017).

Natural plant extracts are usually known as botanical extracts that include an extensive diversity of constituents with different properties and biological activities. The main characteristics of the biological alternatives are easy to extract, eco-friendly, biodegradable, possess low toxicity against living things, are cheap and are effective against extensive pests (Zaker, 2016). The outcomes from several previous studies revealed that some of the plant extracts could control plant pathogenic pests or at least be used as a model for constructing new anti-fungal compounds (Amini et al., 2012).

2 Post-harvest Problems Caused by Fungal Phytopathogens

Post-harvest diseases can occur during various stages of processing, such as harvesting, handling, storage, packaging and transportation, or in the hands of the consumer (Agrios, 2005). Fresh fruit and vegetable rot in post-harvest processes are mainly caused by fungal pathogens (Gatto et al., 2011). According to Tripathi and Dubey (2004), the majority of post-harvest diseases are caused by more than 100 species of fungi, where post-harvest diseases can destroy about 10–30% of crop yield (Agrios, 2005). For the record, loss of fresh commodities can be as high as 50% in developing countries and tropical regions (Tripathi & Dubey, 2004).

The occurrence of decay, decomposition, or disintegration of plant tissue is known as rot. The rot condition can be either hard, dry decay, or soft and squashy decay. The rot phenomenon is capable of affecting the root or rhizome, stem, tree trunk, blossom, and fruit, while some rots can even affect the leaves of a plant, which can be seen in the form of leaf spots or blights (Horst, 2001). Fungal phytopathogens affect crops via various mechanisms of infection. The pathogens enter the host via unprotected wounds; thus, precautionary steps should be taken right away, as it is typically too late to do anything about the disease when the infection signs appear (Horst, 2001).

Worldwide fruit rot losses have been estimated at around 50% due to fungal and bacterial infection. Fruits and vegetables are easily and always at high risk of being infected by fungal spoilage during the pre- and post-harvest stages. The significant fungal pathogens' genera include *Pythium, Phytophthora, Fusarium, Penicillium, Alternaria, Botrytis, Geotrichum, Sclerotinia,* and *Rhizoctonia* spp. The fungal growth of fresh fruits and vegetables will lead to food spoilage and numerous plant diseases. The growth also depends on some abiotic factors such as pH, water activity (aw), solute concentration, temperature, atmosphere and time. However, the main variables to determine the development of fungal pathogens are the surrounding temperature and water activity of the host (da Cruz Cabral et al., 2013). Some important fungal phytopathogens and their associated agricultural fruit and vegetable crop diseases are listed in Table 9.1.

Undoubtedly, post-harvest care for fresh fruits and vegetables is difficult and complicated in that it is impacted by the location of the different producers' economic backgrounds and the projected market, either local, regional, national or international (Toivonen et al., 2014). Besides, the risk of using fungicides during the production of fruits and vegetables arises as a common issue, including the use of unregistered ingredients, application at higher than permitted dosage, insufficient safety clothing, and training of workers and applications close to harvest season and consumption. Consumption of raw fruits and vegetables increases the chances that contaminated yields may cause illness to the consumers. Numerous recommended strategies for decreasing the risk of fresh yield contamination with human pathogens and chemically synthetic fungicides are particularly challenging to practice, especially in developing countries (Toivonen et al., 2014).

In current years, several alternative approaches, such as the use of plant extract known to produce an extensive variety of secondary metabolites, are starting to be acknowledged. Moreover, the plant fungal pathogens use different strategies to attack and enter their host. Some of the pathogens can come into their host using mechanical pressure and chemical action, while some require an opening site, such as through wounds and stomata, to enter the host tissue. In addition, plants have also developed defence mechanisms against fungal pathogens through evolution, which indicates a good hint to investigate the presence of anti-fungal compounds in plants (Eloff & McGaw, 2014).

3 Drawbacks of Using Synthetic Fungicides

Recently, farmers have applied synthetic chemical fungicides to control post-harvest diseases by spraying or soaking fruits and vegetables. The use of chemical fungicides such as benomyl, thiabendazole, prochloraz, and carbendazim to control post-harvest diseases in fruits is typically very effective. However, the use of chemical fungicides such as carbendazim to control the post-harvest disease of fruits could end up with chemical deposition in the fruits, which can later affect human health and cause pollution in the environment even at very low doses of application. The introduction of benomyl, also known as [methyl1-(butylcarbomyl)-2-benzimidazolecarbanate] and its major active metabolite, carbendazim (methyl2-benzimidazolecarbamate),

Table 9.1 Fungal pathogens associated with fruit and vegetable crop diseases

Fungal pathogens	Fruit /vegetable crops	Diseases	References
Acanthorhynchus vaccinii	Cranberry	Blotch rot	Horst (2001)
Alternaria spp.	Pome fruit	Alternaria rot	Singh et al. (2017)
Alternaria alternata	Stone fruit, mango, tomato, potato, pomegranate	Alternaria rot, black heart	Singh et al. (2017)
Alternaria citri	Citrus fruit, grapefruit	Black centre rot	Singh et al. (2017)
Alternaria mali	Apple	Calyx-end rot	Horst (2001)
Alternaria radicina	Carrot	Black rot	Horst (2001)
Alternaria solani	Tomato	Collar rot	Horst (2001)
Aspergillus fumigatus	Beet	Wound rot	Horst (2001)
Aspergillus niger	Dates, grapes, pomegranate, peach, peanut, apple, potato, onion	Calyx-end rot, bunch mould, black mould, crown rot, storage rot	Horst (2001)
Aspergillus niger	Mango	Black mould	Singh et al. (2017)
Botrytis cinerea	Strawberry, pome fruit, stone fruit, grape, berries, mango, peach, cherry, plum, pomegranate, quince, pear, grape, pepper, tomato, eggplant	Grey mould	Brauer et al. (2019), Horst (2001)
Botryosphaeria dothidea	Peach, grape	Fruit rot	Horst (2001)
Botryosphaeria obtusa	Peach	Fruit rot	Horst (2001)
Botryosphaeria rodina	Peach	Fruit rot	Horst (2001)
Botryosphaeria ribis	Avocado, citrus, apple, pear, Tung oil	Dothiorella rot, black fruit rot, nut rot	Horst (2001)
Brachysporium tomato	Tomato	Fruit rot	Horst (2001)
Catelunaria fuliginea	Dates	Fruit rot	Horst (2001)
Cladosporium spp.	Berries	Cladosporium rot	Singh et al. (2017)
Cladosporium carpophilum	Peach	Fruit scab	Hartman (2007)
Colletotrichum spp.	Mango, banana, papaya	Anthracnose	Brauer et al. (2019)
Colletotrichum acutatum	Avocado	Anthracnose	Singh et al. (2017)

(continued)

Table 9.1 (continued)

Fungal pathogens	Fruit /vegetable crops	Diseases	References
Colletotrichum capsici	Chilli	Fruit rot	Aini et al. (2005)
Colletrotrichum gloeosporioides	Pome fruit, avocado, mango, papaya	Bitter rot, fruit surface rot	Singh et al. (2017)
Colletotrichum orbiculare	Watermelon, muskmelon	Anthracnose	Ziv and Zitter (1992)
Colletrotrichum musae	Banana, papaya	Anthracnose, crown rot, fruit surface rot	Singh et al. (2017), Alvindia and Hirooka (2011)
Choanephora cucurbitarum	Summer squash	Choanephora fruit rot	Seebold (2010)
Diaporthe phaseolorum	Tomato	Fruit rot	Horst (2001)
Didymella bryoniae	Squash, pumpkin, gourd	Black rot	Ziv and Zitter (1992)
Dothiorella spp.	Avocado	Stem end rot	Singh et al. (2017)
Diplodina persicae	Peach	Fruit rot	Horst (2001)
Fusarium spp.	Banana	Crown rot	Singh et al. (2017)
Fusarium acuminatum	Squash, pumpkin	Fruit rot	Horst (2001)
Fusarium culmorum	Squash, pumpkin	Fruit rot	Horst (2001)
Fusarium equiseti	Pumpkin, squash, tomato,	Fruit rot	Abu Bakar et al. (2013), Horst (2001)
Fusarium oxysporum	Tomato, banana, guava	Fruit rot	Abu Bakar et al. (2013), Zakaria et al. (2012)
Fusarium semitectum	Tomato, banana, papaya, guava	Fruit rot	Abu Bakar et al. (2013), Zakaria et al. (2012)
Fusarium solani	Tomato, banana, papaya	Fruit rot	Abu Bakar et al. (2013), Zakaria et al. (2012)
Fusarium subglutinans	Tomato	Fruit rot	Abu Bakar et al. (2013)
Fusarium verticillioides	Banana	Crown rot	Alvindia and Hirooka (2011)
Lasiodiplodia theobromae	Citrus, peach, mango, avocado, banana	Fruit rot, crown rot	Horst (2001), Alvindia and Hirooka (2011)
Geotrichum candidum	Tomato	Fruit rot	Horst (2001)
Glomerella cingulate	Peach, apple, pear	Fruit rot, bitter rot	Horst (2001)
Guignardia bidwellii	Grape	Black rot	Horst (2001)

(continued)

Table 9.1 (continued)

Fungal pathogens	Fruit /vegetable crops	Diseases	References
Gliocladium roseum	Citrus, lemon	Fruit rot	Horst (2001)
Monilia spp.	Stone fruit	Brown rot	Singh et al. (2017)
Mucor mucedo	Tomato	Post-harvest rot	Horst (2001)
Mucor piriformis	Tomato, cherry	Fruit rot	Horst (2001)
Mucor racemosus	Citrus	Fruit rot	Horst (2001)
Monilinia fructicola	Peach	Brown rot	Hartman (2007)
Monilinia laxa	Peach	Brown rot	Hartman (2007)
Monilinia fructigena	Stone and pome fruits	Brown rot	Hartman (2007)
Musicillium theobromae	Banana	Finger rot	Twumasi et al. (2014)
Nigrospora spharica	Banana	Black end	Singh et al. (2017)
Penicillium spp.	Pome fruit, stone fruit, grape, berries, pomegranate	Blue mould	Singh et al. (2017)
Penicillium digitatum	Citrus fruit	Green mould	Singh et al. (2017)
Penicillium expansum	Mango, apple, pear, avocado, pomegranate, Japanese persimmon, quince, feijoa	Blue mould, soft rot	Singh et al. (2017), Horst (2001)
Penicillium funiculosum	Pineapple	Fruit let core rot	Singh et al. (2017)
Penicillium italicum	Citrus fruit	Blue mould	Singh et al. (2017)
Phoma caricae-papayae	Papaya	Black rot	Singh et al. (2017)
Phoma cucurbitacearum	Squash, pumpkin, gourd	Black rot	Ziv and Zitter (1992)
Phomopsis sp.	Peach	Fruit rot	Horst (2001)
Phomopsis caricae-papayae	Papaya	Phomopsis rot	Singh et al. (2017)
Phomopsis citri	Citrus fruit	Stem end rot	Singh et al. (2017)
Phomopsis mali	Apple	Fruit rot	Horst (2001)
Phomopsis mangifera	Mango	Stem end rot	Singh et al. (2017)
Phomopsis vaccinii	Blueberry	Fruit rot	Horst (2001)
Physalospora vaccini	Cranberry	Blotch rot	Horst (2001)
Phytium sp.	Cucumber, squash	Cottony leak	Seebold (2010)
Phytium ultimum	Muskmelon	Fruit rot	Horst (2001)
Phytophthora cactorum	Strawberry	Leather rot	Hartman and Kaiser (2008)

(continued)

Table 9.1 (continued)

Fungal pathogens	Fruit /vegetable crops	Diseases	References
Phytophthora capsici	Cucumber, gourd, honeydew, watermelon, eggplants, peppers, tomatoes	Blight and fruit rot	Babadoost (2005)
Rhizoctonia solani	Potato, cucumber	Stem canker, black scurf, belly rot	Brauer et al. (2019), Seebold (2010)
Saccharomyces spp.	Pineapple	Yeasty rot	Singh et al. (2017)
Sclerotinia sclerotiorum	Squash, pumpkin	White mould	Ziv and Zitter (1992)
Thielaviopsis paradoxa	Banana, pineapple	Ceratocystis fruit rot, water blister	Singh et al. (2017)
Mycosphaerella sp.	Papaya	Dry rot	Alvarez and Nishijima (1987)
Rhizopus spp.	Jackfruit	Rhizopus rot	Nelson (2005)

are benzimidazole fungicides that have been widely used in the agriculture sector since the 1970s. Benomyl and carbendazim have been found to be able to suppress fungal growth by disrupting tubulin polymerisation (Zhou et al., 2015).

The US Environmental Protection Agency has cancelled the registration of the fungicides for over a decade due to their accumulating toxicological evidence that might threaten human health in the long term, even though European countries still predominantly apply benomyl and carbendazim fungicides in their agriculture sector; in fact, they are also widely used in some developing countries like China (Zhou et al., 2015). Benomyl could quickly degrade into its metabolite, known as carbendazim, due to its short half-life, which is only within a few hours in environmental units, including soils, water and plants. This differs from carbendazim, where its half-life period is longer, from several days to a few months in water and soils. In addition, carbendazim has been reported as among the most common pesticides detected in fruits and vegetables as its residual half-life is nearly 6.5 to 7 days, correspondingly from pre- and post-harvest treatment (Bhattacherjee et al., 2009). Plus, the concentration can be at a level of up to 1 mg/L (5.5 M) in the fruits, as reported from studies in China and Italy (Liu et al., 2009). Hence, the deposition of benomyl and carbendazim residuals in the environment and the potential to be exposed to them through diet have become ongoing concerns regarding the effects of these fungicides towards human health.

There was much evidence reported in the laboratory of mammals supporting the lethal effects of benomyl and carbendazim on the male reproduction system. For instance, the direction of benomyl to male rats and mice can cause a decrease in testicular, epidydimal weights, epidydimal sperm counts and fertility rate, which will lead to reproductive damage (Zhou et al., 2015). Meanwhile, carbendazim can cause morphological and functional alteration of Sertoli cells, increase Leydig cellular oxidative stress, apoptosis in germ cells of testes in the rats and toxic effects of both fungicides on the female reproduction system and foetus development. In addition, the benomyl had teratogenic effects, which can be observed in the decreasing morphological score, somite number and optic development on cultured rat embryos, while the exposure of the pregnant mice to carbendazim had induced maternal and developmental toxicity, as well as increasing post-implantation loss (Farag et al., 2011).

4 Plant Extracts as an Alternative in Post-harvest Disease Control

There are many ways to recognise plant extracts as an alternative control for post-harvest diseases. One of the ways, as described by Choudhury et al. (2018), is by practising the combined application of the plant extracts. The strategy involves the combination of the plant extract with other biocontrol agents to improve its bioactivity mechanism of action against plant diseases. For example, the application of the combined botanical extracts and biocontrol agents was found to be effective in reducing the Fusarium wilt of bananas in India. In addition, a synergistic effect can also be observed when a mixture of aromatic flower bud extract and piper beetle leaf extract gave significant inhibition against *Fusarium oxysporum* f. sp. *vanilla*, thus causing stem rot disease in vitro with 92% inhibition percentage of fungal growth over the control; this formulation was also reported to successfully suppress the growth population of the pathogen in the soil.

Another example is the combination of *Corni fructus,* cinnamon and Chinese chive extracts, which were found to be effective against some of the common food-borne microorganisms in terms of mixture stability under the heat treatment and showed excellent inhibitory effects against the wide spectrum microorganism. The next similar strategy has also been suggested by Mehta and Sharma (2016), who proposed bioformulation. Since the plants are enriched by a variety of secondary metabolites, such as volatile oils, glycosides, steroids, phenols, flavonoids, alkaloids, terpenoids, and tannins that have been reported to possess anti-microbial properties, further investigation on the phytochemical constituents in the plants is necessary to develop bioformulation that can be used in managing, treating, and controlling the plant disease efficiently and safely. Scientists should take opportunities to use plants to treat plant diseases since this approach can potentially increase food security and safety for rural farmers and can gain commercial rewards on the application of the extract mixtures in greenhouse or field experiments. Even if the extracts are found with a certain degree of toxicity to humans and animals, they may still be used in the floriculture industry (Shuping & Elof, 2017).

Almost all plants have medicinal value. Plants contain other phytochemical constituents known as secondary metabolites, which are produced as by-products and are indirectly useful to humans and animals. The medicinal value of the plants has been contributed by the secondary metabolites present in the plants, which are made up of alkaloids, tannins, saponins, flavonoids, anthraquinones, glycosides, volatile oils, terpenes, essential oils, and resin. Therefore, a medicinal plant can be described as each of its organs, such as leaf, bark, petal, flower, stem, and root, containing substances that can be used for therapeutic purposes. The interest in the investigation of different extracts obtained from traditional medicinal plants has increased over the past two decades to seek potential sources of new anti-microbial agents. In current times, the use of herbal products has deliberately increased in developed countries and several other countries. An estimation made by the World Health Organisation showed that 80% of the world's population presently uses herbal medicine for some aspect of primary health-care maintenance (Mazid et al., 2012).

Many plant species have been used in traditional medicines for treating many diseases (Bussmann et al., 2010). Natural products are valuable sources of various structures of chemical compounds isolated from different parts of the plants and applied to clinical medical practices. Fruits and vegetables are among the most widely studied natural resources to discover new alternatives for anti-microbial, anticancer, antioxidants, anti-inflammatory, and other biological functional activities. Recently, some underutilised fruits have strained the attention of many researchers as a natural source for alternative treatment to cure various diseases (Mokhtar & Abd Aziz, 2016). There are many anti-microbial activities obtained from some plant extracts that have been reported by many scientists previously, as recorded in Table 9.2.

4.1 Averrhoa bilimbi *L.*

Averrhoa bilimbi Linn. is a fruit plant under the family Oxalidaceae, which is commonly known as bilimbi. It is a common plant in Asia that grows up to 15 m tall and 30 cm in diameter. The fruits have cylindrical shapes with five broads, rounded longitudinal lobes and are formed in clusters. The weight and the size of the fruits can be increased to maximum value during the maturity stage, and their green colour peel will slowly turn to light yellow. Bilimbi fruits have a sour taste and are commonly used to produce vinegar, wine, and pickles, as well as in the preparation of traditional dishes. Oxalic acid and vitamin C have been identified as dominant chemical constituents of the *A. bilimbi* fruit. It is extensively used to cure cough, mumps, rheumatism, and pimples and can be used as an antioxidant in traditional medicine (Rimal Isaac et al., 2013).

A previous study reported by Alhassan and Ahmed (2016) and Aziz (2016) revealed the positive anti-microbial activity of aqueous extracts of fruits and leaves of *A. bilimbi* against several Gram-positive bacteria, such as *Staphylococcus aureus, S. epidermis, Bacillus cereus, Corneybacterium diphtheria* and *Kochuria rhizophila,* as well as against several Gram-negative bacteria including *Salmonella typhi, Citrobacter fuendii,* and *Aeromonas hydrophila.* Besides, a study conducted by Aziz et al. (2014) showed that the aqueous extract of *A. bilimbi* fruit produced maximum efficacy against most of the bacteria compared to the ethanol and methanol extracts.

Table 9.2 List of the plant species used as anti-fungal agent

Scientific name (common name) of plant	Fungal species tested	References
Adenocallima alliaceum (Wild garlic)	*Alternaria alternata* and *Fusarium oxysporum*	Rai et al. (2000)
Aframomum melegueta (Malegueta pepper)	*Aspergillus niger, Botryodiplodia theobromae, Fusarium solani* and *Penicillium oxalicum*	Okigbo et al. (2009)
Agapanthus africanus (Cape agapanthus)	*Pythium ultimum, Fusarium oxysporum, Alternaria alternata, Mycosphaerella pinodes* and *Ascochyta* sp.	Tegegne et al. (2008)
Aloe barbadensis (Aloe vera)	*Aspergillus niger*	Anjum et al. (2016)
Allium sativum (Garlic)	*Aspergillus niger*	Anjum et al. (2016)
Annona squamosa Linn (Sugar apple)	Anthracnose of pepper	Nduagu et al. (2008)
Anthemis nobilis L. (Chamomile)	*Aspergillus candidus, A. niger, Penicillium* sp. and *Fusarium culmorum*	Magro et al. (2006)
Artemisia dubia	*Alternaria brassicicola*	Parajuli et al. (2005)
Artemisia gmelinii (Gmelin's wormwood)	*Alternaria brassicicola*	Parajuli et al. (2005)
Azadirachta indica A. Juss (Neem)	*Alternaria solani, Fusarium oxysporium, Aspergillus niger, Botryodiplodia theobromae, F. solani, Penicillium oxalicum* and *P. expansum*	Siddiqui et al. (2004)
Calotropis procera (Thorn apple)	*Curvularia lunata*	Manoharachary and Gourinath (1988)
Cassia sp. (Cassia)	*Alternaria alternate*	Feng et al. (2008)
Chenopodium ambrosioides (American wormseed or Mexican tea)	*Aspergillus flavus, A. fumigatus, Botryodiplodia theobromae, Fusarium oxysporum, Phythium debaryanum* and *Sclerotium rolfsii*	Kumar et al. (2007)
Chenopodium quinoa (Quinoa or Quinua)	*Botrytis cinerea*	Stuardo and San Martin (2008)
Cryptomeria japonica (Japanese cedar or sugi)	*Rhizoctonia solani, Collectotrichum gloeosporioides, Fusarium solani* and *Ganoderma australe*	Cheng et al. (2005)

(continued)

Table 9.2 (continued)

Scientific name (common name) of plant	Fungal species tested	References
Cymbopogon nardus (Citronella grass)	*Fusarium oxysporum, F. verticillioides, F. proliferatum, Penicillium expansum, P. brevicompactum, Aspergillus flavus* and *A. fumigatus and Colletotrichum musae*	Anthony et al. (2004), Zabka et al. (2009)
Cuminum cyminum (Cumin)	*Fusarium solani, F. oxysporum, F. verticillioides, F. poae, F. equiseti, Rhizoctonia solani, F. oxysporum* f. sp. *tulipae, Botrytis cinerea and Alternaria citri*	Naeini et al. (2010), Boyraz and Özcan (2005)
Curcuma longa Linn (Turmeric)	*Phytopthora infestans, Fusarium solani and Pyricularia oryzae*	Bandara et al. (1989)
Datura metel L (Metel)	*Alternaria solani*	Latha et al. (2009)
Drimys winteri (Winter's bark)	*Gaeumannomyces graminis* var *tritici*	Monsálvez et al. (2010)
Echinophora tenuifolia (Pickling herb)	*Rhizoctonia solani, Fusarium oxysporum* f. sp. *tulipae, Botrytis cinerea* and *Alternaria citri*	Boyraz and Özcan (2005)
Eucalyptus globulus (Eucalyptus)	*Aspergillus flavus* and *A. parasiticus*	Vilela et al. (2009)
Eugenia aromatica (Clove)	*Alternaria solani*	Khallil (2001)
Flourensia cernua (Tarbush)	*Alternaria* sp., *Rhizoctonia solani* and *Fusarium oxysporum*	Jasso de Rodriguez et al. (2007)
Flourensia microphylla	*Alternaria* sp., *Rhizoctonia solani* and *Fusarium oxysporum*	Jasso de Rodriguez et al. (2007)
Flourensia retinophylla	*Alternaria* sp., *Rhizoctonia solani* and *Fusarium oxysporum*	Jasso de Rodriguez et al. (2007)
Jatropha curcas (Purging nut or Barbados nut)	*Aspergillus niger*	Anjum et al. (2016)
Juniperus recurve (Drooping juniper)	*Alternaria brassicicola*	Parajuli et al. (2005)
Lantana camara Linn. (Spanish flag or Shrub verbena)	*Alternaria solani, Botrytis cinerea, B. ricini Pythium ultimum, Rhizoctonia solani* and *Verticillium dahlia*	Batthiprolu and Bhattiprolu (2006)
Laurus nobilis (Sweet bay)	*Botrytis cinerea, Monilinia laxa* and *Penicillium digitatum*	Corato et al. (2010)

(continued)

Table 9.2 (continued)

Scientific name (common name) of plant	Fungal species tested	References
Lowsonia inermis Linn (Hena)	*Dreshslera oryzae*	Natarajan and Lalithakumari (1987)
Malva sylvestris L. (Malva)	*Aspergillus candidus, A. niger, Penicillium* sp. and *Fusarium culmorum*	Magro et al. (2006)
Melia azadrach (Chinaberry)	*Alternaria solani* and *Fusarium oxysporium*	Siddiqui et al. (2004)
Mentha spicata Linn. (Spearmint)	*Rhizoctonia solani* and *Sclerotium sclerotiorum*	Yegen et al., 1992)
Metasequoia glyptostroboides (Dawn redwood)	*Fusarium oxysporum, F. solani, Sclerotonia sclerotiorum, Rhizoctonia solani, Colletotricum capsici, Botrytis cinerea* and *Phytophthora capsici*	Bajpai et al. (2007)
Nardostachys grandiflora (Spikenard)	*Alternaria brassicicola*	Parajuli et al. (2005)
Nigelia sativa Linn (Black cumin)	*Fusarium oxysporum, A. niger and A. flavus*	Sitara et al. (2008)
Ocimum basilicum (Sweet basil)	*Colletotrichum musae* and *Fusarium proliferatum*	Anthony et al. (2004)
Ocimum sanctum L (Holy basil)	*Alternaria solani*	Latha et al. (2009)
Origanum hercleoticum (Oregano)	*Fusarium oxysporum and Phoma tracheiphila*	Salomone et al. (2008)
Pelargonium roseum (Geranium)	*Fusarium oxysporum, F. verticillioides, Penicillium expansum, P. brevicompactum, Aspergillus flavus* and *A. fumigatus*	Zabka et al. (2009)
Pimenta dioica (Jamaica pepper)	*Fusarium oxysporum, F. verticillioides, Penicillium expansum, P. brevicompactum, Aspergillus flavus* and *A. fumigatus*	Zabka et al. (2009)
Piper longum (Bengal pepper)	*Pyricularia oryzae, Rhizoctonia solani, Botrytis cineria, Phytophthora infestans, Puccinia recondite* and *Erysiphe graminis*	Lee et al. (2001)
Peumus boldus (Boldo)	*Aspergillus niger, A. flavus* and *Fusarium* spp.	Souza et al. (2005)
Portulaca olaracea Linn (Perslane)	*Helminthosporium maydis*	Noriel and Robles (1990)

(continued)

Table 9.2 (continued)

Scientific name (common name) of plant	Fungal species tested	References
Robinia pseudoacacia (Black locust)	*Sphaerotheca fuliginea*	Zhang et al. (2008)
Rumex crispus (Curly dock)	*Blumeria graminis* f. sp. *hordei*	Choi et al. (2004)
Salvia fruticosa Mill (Greek Sage)	*Rhizoctonia solani* and *Sclerotium sclerotiorum*	Yegen et al. (1992)
Salvia officinalis Sage	*Penicillium* sp., *Aspergillus* sp., *Cladosporium* sp. and *Fusarium* sp.	Pinto et al. (2007)
Satureja hortensis Sater	*Rhizoctonia solani, Fusarium oxysporum* f. sp. *tulipae, Botrytis cinerea* and *Alternaria citri*	Boyraz and Özcan (2005)
Thevetia peruviana (Yellow oleander)	*Aspergillus niger* and *Penicillium* spp.	Ravikumar Patil et al. (2007)
Thymus linearis (Himalayan thyme)	*Alternaria brassicicola*	Parajuli et al. (2005)
Thymus vulgaris (Thyme)	*Fusarium oxysporum, F. verticillioides, Penicillium expansum, P. brevicompactum, Aspergillus flavus* and *A. fumigatus*	Zabka et al. (2009)
Vinca rosea L. (Madagascar periwinkle)	*Alternaria solani*	Latha et al. (2009)
Zanthoxylum armatum (Winged prickly ash)	*Alternaria brassicicola*	Parajuli et al. (2005)
Zataria multiflora (Zataria)	*Fusarium solani, F. oxysporum, F. verticillioides, F. poae* and *F. equiseti*	Naeini et al. (2010)
Zingiber officinale Rosc. (Ginger)	*Phytopthora infestans, Fusarium solani and Pyricularia oryzae*	Bandara et al. (1989)

However, there is a lack of reports on the efficacy of *A. bilimbi* fruit extract against the growth of fungal species; yet, a finding from a study reported by Rana et al. (2014) showed a positive effect of *A. bilimbi* fruit methanol extract on growth of *Candida albicans* and *Aspergillus niger*. It has been reported that the anti-microbial activity of *A. bilimbi* fruit was mainly attributed to its highly acidic pH (De Lima et al., 2001), as the pH of *A. bilimbi* extracts ranges from pH 0.9–1.5. *A. bilimbi* fruit has a high level of oxalic acid ranging from 8.57 to 10.32 mg/100 g and ascorbic acid ranging from 32.23 to 60.95 mg/100 g. The anti-microbial properties of the fruits have been attributed to the high content of bioactive components in the first stage of fruits. *A. bilimbi* contains oxalic acid, a strong acid that might have an action on bacterial growth and its potential to become a potent source of antioxidant and anti-microbial against *S. aureus* (Mokhtar & Abd Aziz, 2016).

Furthermore, *A. bilimbi* fruit extract possesses one of the crucial pharmacological profiles, which is anti-microbial, owing to the presence of some of the bioactive compounds such as alkaloid, tannins, saponins, flavonoids, cardiac glycoside, triterpenes, phenolic, triterpenoid, oxalic acids, D-3-hexanal, citric acid, and cyaniding-3-O-β-D-glucoside, contributing to its anti-fungal, anti-microbial, antilipidemic and antioxidant properties (Nakyinsige et al., 2016; Vince Carlo et al., 2017; Zulhaimi et al., 2017).

4.2 Garcinia mangostana *L.*

Garcinia mangostana L., known as mangosteen as its common name, is a tropical queen of fruits that belongs to the family of Clusiaceae, which is normally found in India and Southeast Asia, such as Indonesia, Malaysia, Myanmar, Philippines, Sri Lanka, and Thailand (Ji et al., 2007). Mangosteen has been planted as one of the export commodities of fruits in Indonesia due to its sweet, sour and pleasant taste (Muchtaridi et al., 2017). Mangosteen is planted as a source of its highly palatable fruit comprising the white colour of pulp contained in a dark purple colour of the pericarp. This fruit is typically consumed freshly, thus producing a high number of waste due to its unconsumable pericarps and seeds. Therefore, some scientists are interested in gaining benefits from the unused part (the pericarps and seeds) of mangosteen fruit to obtain some phytochemical components via extraction processes to be used for various biological activities (Ovalle-Magallanes et al., 2017; Rohman et al., 2019).

There were several biological activities have been reported for the *G. mangostana* pericarp extract, which include antioxidant, anti-microbial, anti-diabetic, anti-proliferative and anti-tumour. Previous studies of phytochemical analysis of *G. mangostana* pericarp discovered that the pericarp is rich in α-mangostin, other phenolic compounds such as ferulic acid, p-coumaric acid, veratric acid, t-cinnamic acid, vanillic acid, cinnamic acid, caffeic acid, mandelic acid, gentisic acid, and sinapic acid, as well as some flavonoids such as epicatechin and quercetin. The previous studies also reported that α-mangostin has contributed to most of the biological activities of *G. mangostana.* The α-mangostin isolated from the extract of dried *G. mangostana* pericarp exhibited antioxidant, anticancer, and cytotoxicity activities (Ghasemzadeh et al., 2018).

Besides, the mangosteen pericarps also contain xanthone, a secondary metabolite that is higher than the other parts of mangosteen. Xanthone is a polyphenol compound with a chemical structure containing aromatic tricycles (three consecutive aromatic rings differentiated with side chains). Moreover, several types of xanthones obtained from the mangosteen pericarp had been proven to have strong antioxidant activity, including α-mangostin. Besides containing antioxidant properties, the mangosteen pericarp also exhibits other biological activities such as anti-inflammatory, anti-allergy, antibacterial, antiviral, and anti-fungal activities. To date, at least 70 xanthones have been characterised (Melia et al., 2019).

Interestingly, the bioactivity of xanthone can be influenced by modification of the side chains. Several types of xanthones are usually isolated from the mangosteen pericarp, including α-mangostin, β-mangostin, γ-mangostin, gartanin, 8-deoxygartanin and garcinone E. Other components, including anthocyanins such as cyanidin-3-sophoroside and cyanidin-3-glucoside, can also be found lavishly in the fruit pericarp (Aizat et al., 2019). The mangosteen pericarp also contains other functional bioactive compounds such as terpenes, anthocyanins, tannins, flavonoids, and polyphenols (Narasimhan et al., 2017).

4.3 Tamarindus indica *L.*

Tamarind is known scientifically as *Tamarindus indica,* grouped under the family Fabaceae, a tropical plant normally found in Asia and Africa. It is highly valued for its pulp, which has a sweet, acidic taste owing to a combination of high content of tartaric acid and reducing sugars, which are commonly used as food seasoning and preparation of juice and other beverages, while its leaves and flowers can be eaten as vegetables and used in preparation of variety dishes (Abubakar et al., 2010). *T. indica* is a species commonly used worldwide, as its distribution covers almost all tropical regions. The chemical components that can be found in each part of the tree make the tree unique, whereas the beneficial bioactive metabolites, including tannin, can be extracted from the pulp (fruit), pod, leaves, bark, and stem of the tamarind tree possessing an anti-fungal property. Apart from this, *T. indica* has been traditionally used to cure certain chronic and acute diseases naturally. For example, *T. indica* can cure snakebite, diarrhoea, abdominal pain, and inflammations. Moreover, the fruit, leaf, and seed of the tamarind tree are rich in antioxidant properties due to the presence of phenolic compounds, which enhance the natural defence mechanisms against cardiovascular disease and are effective in reducing chronic inflammations of a certain type of cancer. Additionally, tamarind leaf is also enriched in flavonoids, vitamins, lipids, and fatty acids (Sandra, 2016).

Based on the research conducted by Soradech et al. (2016), tamarind seed coat also contains active antioxidants such as tannin. In addition, methanolic leaf extract can help in the inhibitory process of certain bacteria like *Bacillus subtilis* and has some laxative effects. Besides having medicinal purposes, fruits of *T. indica* are commonly used in the food manufacturing process and have been used widely as among the sources of food. For instance, mature ripe fruit can be eaten as a snack and used in the preparation of porridge, while the husk of fruit can be used to make beverages. Tamarind is a species that is beneficial in many ways since each part of the tree can be used in numerous ways. For example, the pulp of the tamarind tree can be used as a laxative in beverages and porridge. Medically, the pulp of tamarind trees can be utilised to treat malaria disease.

Studies on the anti-fungal properties of tamarind, as reported by Garcia (2011), showed that the ripe fruit exhibited bioactivity against bacteria, yeast and fungi. The fruits have been stated to have anti-fungal and antibacterial properties when extracted with benzene and dissolved with petrol, producing a liquid with a brown colour and odourless known as "tamarindienal" that has been identified as 5-hydroxy 2-oxo-hexa-3,5-dienal. This liquid was tested for a strong fungicidal agent against *Aspergillus niger* and *Candida albicans* and exhibited antibacterial activity against *Salmonella* sp., *Bacillus subtilis, Staphylococcus aureus, Escherichia coli,* and *Pseudomonas aeruginosa.* Besides, the liquid also displayed molluscicidal activity against *Bulinus transcatus* snail, which was contributed by the presence of 33 saponins in the fruit.

4.4 Pilea microphylla *(L.) Liebm.*

Pilea microphylla is a weed type under the family Urticaceae, which has been used as folk medicine to treat several allergies and wounds in Peninsular Malaysia, especially Penang Island. Previously, it was reported to have antibacterial activity, moderate antioxidant activity and containing a total number of phenolic contents. It has also been used to treat infertility, inflammations and womb cleansers. This weed displayed a variety of anti-microbial activities against several pathogenic microorganisms. Most of the weed crude extracts, which inhibited the growth of bacteria, were found to be active against Gram-positive bacteria, such as *Bacillus cereus, B. subtilis, Methicillin resistant,* and *Staphylococcus aureus* (Bansal et al., 2011; Chahardehi et al., 2010).

4.5 Peperomia pellucida *Kunth*

Peperomia pellucida (L.) is a traditional medicinal plant belonging to the family Piperaceae. It is also known as a common annual weed native in tropical North and South America and is now spread as pantropic and found abundantly in Malaysia. The weed is well-known by some other local's names in Malaysia, such as "sirih cina", "ketumpang air", or "tumpang angin", and is famous in other countries with various names such as "pak-krasang" in Thailand, "càng cua" in Vietnam, "pansit-pansitan" or "ulasimang bato" in Philippines, "silverbush", "man-to-man", "pepper elder" or "rat ear" in North America, "kaca-kaca" or "surukan" in Indonesia, "usuba sunakosho" in Japan, "alumbre" or "erva-de-vidro" in Spain, "mashitandu chedi", "pononoa", "toyakandha" or "varshabhoo" in India, "pépéromie" or "herbe à couleuvre" in France and "luchi pata" in Bangladesh (Der-Jiun et al., 2012).

Moreover, the weed has been traditionally used to treat a variety of inflammatory disorders. The herbal extract preparation of the weed uses its leaves and stalks through infusion and decoction methods and is used to treat gout and arthritis, besides aiding in old wounds healing. The plant is also used as a topical medicine to treat some skin-related problems such as acne, pimples, and abnormal complexion. In other Asian countries such as Bangladesh, the aerial parts, including the leaves of the weed, have also been used by local traditional healers to treat excited mental disorders (Khan et al., 2008). Despite this, several scientific studies have also reported the antibacterial and analgesic activities of *P. pellucida,* while some other studies stated the isolation of some bioactive phytoconstituents with strong anti-fungal and anticancer properties from the weed (Mutee et al., 2010). In addition, *P. pellucida* leaf was reported to exhibit a broad spectrum of bactericidal and fungicidal activity against *Helminthosporium oryzae* and *Trichophyton mentagrophytes* (Der-Jiun et al., 2012).

In a past study, the discovery of phytochemical constituents of *P. pellucida* revealed the presence of flavonoids, apiols, phytosterols, substituted styrenes, secolignans, tetrahydrofuran lignans, arylpropanoids, sesamin, isoswertisin, and xanthone glycoside, as well as peperomins A, B, C, and E. The arylpropanoids, peperomins and patuloside A, which is a type of xanthone glycoside, exhibited anti-fungal, anticancer and antibacterial activities, respectively (Der-Jiun et al., 2012). Besides, the major compound identified in methanol extracts of *P. pellucida* is phytol, which is one of the most important diterpenes that exhibit both anti-microbial and anticancer activities. The other detected constituents, including hexadecanoic acid, methyl ester and 9, 12-octadecadienoic acid, have displayed antioxidant and anticancer properties. Hence, further work needs to be done in the future to correlate the specific compound with its biological properties (Wei et al., 2011).

4.6 Persicaria odorata *L.*

Persicaria odorata (L) is a herbal plant under the family Polygonaceae, which is a native plant in Southeast Asia known as "Vietnamese coriander" and is distributed throughout Asia. It is known by several common names in different countries, like Vietnamese cilantro, Vietnamese mint and Vietnamese coriander. In Malaysia, Brunei and Singapore, it is known as "Daun Kesum", "phak phai" in Thailand and "rau ram" in Vietnam. *P. odorata* is a perennial herb with a 30–35 cm height. The leaves are globous, 6–15 cm in length, with dark purple markings in the centre. It has filiform inflorescence and a strong odour. *P. odorata* grows well in tropical and subtropical warm and wet environments with rich and moist soil, as well as in semi-shade areas. It is planted and consumed as a vegetable and used in several cooking preparations (Kantachot et al., 2010; Ridzuan et al., 2013).

Persicaria odorata exhibits an assortment of beneficial biological properties, such as anti-microbial, anti-inflammatory, antitumor-promoting, and anti-feedant activities, as well as antioxidant due to the presence of some potential source of natural aliphatic components like aldehydes and terpenes that can be isolated from its essential oil. Many other organic compounds such as (Z)-3-hexenal, (Z)-3-hexenol and some aldehydes like decanal, undecanal, and dodecanal, can be extracted from the aerial part of the plant, which is the highly aromatic part. Moreover, its extract possesses high phenolic content due to the presence of flavonoids like rutin, catechin, quercetin, kaempferol, and isorhamnetin, which has been reported to show strong antioxidant and antibacterial activities against *Bacillus cereus, Listeria monocytogenes,* and *Staphylococcus aureus* (Yanpirat & Vajrodaya, 2015).

Persicaria odorata leaf is one of the natural sources that has been traditionally used worldwide in several sectors, including medicine and culinary, to boost the flavour and aroma of food and used as a preservative in the pharmaceuticals and cosmetics industries. Previously, some studies showed that different methods of extraction may influence the chemical and constituent properties of the plant (Hashemi et al., 2008). The plant is also reported to contain several compounds like caryophyllene, alpha-caryophyllene, drimenol, and decanal6, 3-sulfanylhexanal, and 3-sulfanyl-hexanol (Dash & Zakaria, 2016).

4.7 Cymbopogon citratus *(DC.) Stapf*

Cymbopogon citratus from the Poaceae family is generally known as lemongrass in its common name. It is a perennial tropical grass with thin, long leaves and is used as one of the main medicinal and aromatic plants. It is planted in tropical and subtropical regions of Asia, South America, and Africa. It is famous on demand for its essential oil (EO) worldwide (Boukhatem et al., 2014). The plant has lemon-scented leaves and is widely used in culinary, ornamental purposes, and as traditional medicine. Lemongrass has been found to contain bioactive compounds such as tannins, flavonoids, and alkaloids (Lawal et al., 2017).

In a phytochemical analysis by Falah et al. (2015), it was found that *Cymbopogon citratus* leaf contained various compounds such as alkaloids, flavonoids, phenols, saponins, steroids, and tannins. Gas chromatography analysis by Gazwi (2019) presented 38 compounds in the *Cymbopogon citratus* ethanolic leaf extract. The finding revealed that 4-Methyl-2,6-di-tert-butylphenol (29.24%), also known as butylated hydroxytoluene, was the most abundant compound. Lemongrass leaf extract has been shown to exhibit some inhibitory activities against the growth of fungi such as *Aspergillus flavus, Aspergillus niger, Colletotrichum* sp., and *Fusarium verticillioides* (Gyasi et al., 2022).

A strong anti-fungal activity was observed against *Fusarium oxysporum* on lentils using an aqueous extract of lemongrass (Aourach et al., 2021). The inhibitory activity of lemongrass extract was also observed on fungal strains *Botryodiplodia theobromae, Corticium salmonicolor, Rhizoctonia* spp., and *Penicillium* spp. (Okigbo et al., 2018). In a study conducted by Zulfa et al. (2016), the extract of *C. citratus* was reported to have anti-microbial activity against *Bacillus cereus, Escherichia coli, Klebsiella pneumoniae, Staphylococcus aureus,* and against yeast *Candida albicans.*

Numerous studies have also been conducted to test the anti-fungal activity of essential oils derived from lemongrass. The essential oil from lemongrass has been found to exhibit high antioxidant activity. In a recently published paper by Sawadogo et al. (2022), the essential oil was seen as effective in inhibiting the mycelial growth of *Aspergillus flavus* and *Aspergillus parasiticus*. The inhibitory effect was amplified when used in combination with essential oils from *Cymbopogon nardus* (citronella grass) and *Cymbopogon schoenanthus* (camel grass). The composition of the essential oil was also analysed. Neral, geranial, and oxygenated monoterpenes were found in the lemongrass essential oil. Similar studies by Rhimi et al. (2022) also showed anti-fungal activity on *Candida* spp. and *Malassezia furfur*.

4.8 Piper sarmentosum *Roxb. ex W. Hunter*

Piper sarmentosum, commonly known as wild pepper, is part of the Piperaceae family. This woody perennial creeping plant is native to countries like Malaysia, Indonesia, and India. The plant is widely utilised in traditional medicine and culinary (Rahman et al., 2016). The whole plant, from root to leaves, has been used in traditional medicines to treat various ailments as they exhibit therapeutic values. Traditional uses of the plant include the treatment of fever, stomachache, diabetes, and toothache. Compounds such as glutathione, uric acid, bilirubin, superoxide dismutase, catalase and glutathione peroxidase have also been identified as antioxidants in plant extract (Ismail et al., 2018). According to Sun et al. (2020), there are over 140 bioactive compounds identified in *Piper sarmentosum* so far, which include around 89 essential oils, 26 alkaloids, 13 flavonoids, and various other compounds such as lignans and steroids. Among these, the major constituents of the *Piper sarmentosum* extract are essential oils and alkaloids.

Among all the essential oils found in *Piper sarmentosum*, 40 of them are sesquiterpenes and 17 are phenylpropanoids. In addition, monoterpenes, alkanes and diterpenoids have been identified (Sun et al., 2020). According to research by Qin et al. (2010), myristicin (65.22%) and trans-caryophyllene (13.89%) are the primary components of *Piper sarmentosum* essential oil. The essential oil showed anti-feedant and toxic activity, which negatively affected organisms such as insects that feed on the leaves. It was suggested that various enzymes in the larvae of *Brontispa longissima* (Hispid beetles) are inhibited by essential oil from *Piper sarmentosum*, which in turn interfered with the ability of the insect to detoxify the plants and poisoning of the nervous system.

In a phytochemical analysis study by Sakilan et al. (2019), it was revealed that ethanolic extract of *Piper sarmentosum* leaf contains several bioactive compounds such as alkaloids, flavonoids, saponins, steroids, and tannins. These compounds have been suggested to confer anti-microbial activity of the leaf extract. *Piper sarmentosum* leaf ethanolic extract contained the presence of bioactive compounds isocaryophyllene, myristicin, α -asarone, ß-asarone, apiol, 1,1 –dichlorocyclopentane, and 14-chloro-1 -tetradecanol at different concentrations (Ibrahim & Emlee,

2020). Among these, ß-asarone (37.37%), α-asarone (30.39%) and myristicin (24.26%) are the primary constituents of *Piper sarmentosum* leaf extract.

Ethanol extract of *P. sarmentosum* at 100% concentration exhibited very strong inhibitory activity against *Microsporum gypseum* and strong anti-fungal effectiveness against *Candida albicans* (Khusnul et al., 2019). A recent study conducted by Ibrahim and Emlee (2020) demonstrated that aqueous extract of *P. sarmentosum* at a concentration of 100 mg/ml was able to inhibit the growth of pathogenic fungi *Fusarium verticilloides*. In another study by Shi et al. (2017), amides isolated from the leaf of *P. sarmentosum* displayed anti-fungal activity against *Cryptococcus neoformans*.

4.9 Psidium guajava *L.*

Guava trees *(Psidium guajava)* belong to the Myrtaceae family and are cultivated in tropical and subtropical regions. The leaves from guava trees have also been used in traditional medicine and can be administered topically or orally. Analysis of phytochemical constituents of *Psidium guajava* leaf ethanolic extract by Manikandan et al. (2016) showed the presence of numerous compounds, including alkaloids, flavonoids, phenols, quinones, saponins, tannins, and terpenoids.

Gas chromatography–mass spectrometry (GC–MS) analysis by El-Sesy and Mahran (2020) showed the presence of 25 compounds in the ethanol extract of *Psidium guajava* leaf. It was found that Trans-Nerolidol (20.97%), Caryophyllene (18.76%), and Beta-Caryophyllene oxide (9.55%) were the most abundant in the leaf extract. In a study conducted by Neela et al. (2014), both ethanol and acetone extract of guava leaves significantly inhibited the mycelial growth of *Fusarium oxysporum*, with 100% inhibition reported at 15% concentration of guava leaves ethanol extract. Another research showed that tannin extracted from guava leaves was not only inhibitive towards fungi *Aspergillus niger* and *Candida albicans* but also displayed anti-microbial activity against several bacteria species such as *Escherichia coli, Pseudomonas aeruginosa,* and *Staphylococcus aureus* (Mailoa et al., 2013).

This finding is consistent with the results obtained from a study by Bezerra et al. (2018). The extract of guava leaves tested against fungi *Candida albicans, Candida tropicalis,* and *Candida krusei* showed inhibitory action. The flavonoid and tannin fractions were also analysed to identify the chemical constituents. The study later concluded that the most effective anti-fungal activity of *P. guajava* leaf was attributed to tannin.

Research conducted by Das and Goswami (2019) also displayed similar anti-microbial activity of guava leaf extract on *Bacillus subtilis, Escherichia coli, Saccharomyces cerevisiae,* and *Aspergillus niger*. Stronger inhibition of microbial growth was observed with gram-positive bacteria (*Bacillus subtilis*) and fungal strain tested, compared to gram-negative bacteria (*Escherichia coli*). Analysis of the leaf extracts revealed a high content of flavonoids, phenols and tannin, which were purported to be responsible for the inhibitory activity. The guava leaf extract also contains alkaloids, flavonoids, saponins, and triterpenes. A recent study by Yadav and Mohite (2020) also reported mycelial growth inhibition of *Aspergillus niger* and *Candida albicans* using guava leaf extract. In addition, volatile oils from the guava leaves showed potential as an anti-microbial agent as they inhibited the growth of *Bacillus subtilis, Escherichia coli, Staphylococcus aureus, Streptococcus faecalis,* and *Pseudomonas aeruginosa.* The volatile oils were found to contain limonene as the primary hydrocarbon, with 1, 8-cineole as the major oxygenated monoterpenoid. In vitro assay on *Candida albicans* also displayed moderate anti-fungal activity (Soliman et al., 2016).

5 Extraction Techniques of Bioactive Compounds from Plant Extracts

Many solvents can be used to extract anti-fungal compounds. Solvent selection is vital for solvent extraction. Many factors need to be considered during the solvent's selection, including selectivity, solubility, cost and safety. Methanol is commonly used as an extraction solvent for phytochemical investigation to extract a wide range of mostly polar compounds from living samples, including plant materials that are responsible for many biological activities. Some non-polar compounds are also soluble in this solvent. It has been chosen as a solvent due to its low boiling point, which is 64.7 °C and requires a lower temperature to be evaporated; hence, the metabolites contained in the extract could be well-maintained (Sasidharan et al., 2011; Truong et al., 2019).

An efficient extraction procedure is essential to provide a high extraction rate and specificity for the interested compounds to analyse these metabolites (Sukrasno, 2014). A range of different solvent usage is involved in conventional extraction procedures, where the polarity of the compounds to be extracted is considered. For instance, the extraction of phenolic acids might need different types of solvents, such as alcohols, acetone, diethyl ether and ethyl acetate. The sole use of organic solvents is not recommended since the very polar phenolic acids would not be efficiently extracted (Zavala-Lopez & Garcia-Lara, 2017). Therefore, it has been suggested to use aqueous solutions of methanol or ethanol at a particular percentage, usually 70–80% alcohol, for the extraction (Sultana et al., 2009). For a safer and more environment-friendly solvent, ethanol is preferred over methanol, depending on the usage purpose of the extract. However, an improved method that employs a low volume of methanol can make it environmentally tolerable for extraction (Thouri et al., 2017). There are a few common solvents listed according to their polarity from least polar to most polar as follows: hexane < chloroform < ethyl acetate < acetone < methanol < water. The characterisation of a good solvent refers to how optimal it is in extracting and conserving the stability of the chemical structure of the desired compounds (Altemimi et al., 2017).

This phenomenon can be observed in the extraction of polyphenols from the plant materials. Polyphenols have a polarity range from polar to non-polar compounds; thus, their optimum extraction is usually achieved while using polar solvents, which can give a better solvation efficiency due to the interaction of hydrogen bonds between polar sites compared to non-polar solvents. Furthermore, non-polar solvents like acetone are less efficient in extracting polyphenols, major antioxidant compounds, due to being only proton acceptors compared to methanol and water, which are proton donors as well (Felhi et al., 2017).

A past study example reported by Perveen et al. (2019) showed the application of 70% methanol by mixing 70 ml of methanol with 30 ml of Millipore water according to the ratio methanol:aqueous (7:3) to extract metabolites (phenolic compounds and flavonoids) from bee pollen pallets using ultrasound-assisted extraction method. The results showed a high content of phenolic compounds and flavonoid glycosides extracted using aqueous methanol solvent at 70%, while eight polar secondary metabolites had been assigned via HPLC-ESI analysis and a new phenolic glycoside, namely 3,2,5-trihydroxycinnamic acid diglycoside, had been identified.

Maceration can be defined as a technique where powdered crude plant materials are mixed with a solvent (soaking) and allowed to stand at room temperature in a covered flask for a minimum duration of three days with regular agitation to allow the soluble matter to dissolve. The mixture is later subjected to filtration to separate the residue and the liquid containing soluble matter. It is a very simple conventional extraction method and has been broadly used in medicinal plant research to extract thermolabile components (Banu & Cathrine, 2015). The mechanism of this technique is by softening and breaking down the cell wall of the plant materials under low temperatures to release soluble phytochemical compounds. The type of extracted compounds from the plant samples will be determined by the choice of solvents used in the technique. Even though maceration is the simplest and easiest extraction method, this technique has its own advantages and disadvantages. The disadvantages of this technique include the increase of organic waste production when a large volume of solvents is used, which requires proper management of the waste, besides an increase in extraction time. The advantages come when the optimum extraction process can be achieved with the proper alteration in temperature and the choice of solvents is made appropriately. If the temperature is not objectionable, a reduction in the volume of the solvents might be required. Therefore, choosing good extraction solvents used in the soaking process in this technique is crucial. Some previous studies had been practising the maceration technique. The studies compared the yield of phytochemical compounds produced by different types of extraction methods using different types of extraction solvents (Azwanida, 2015; Zhang et al., 2018).

A review study reported by Zhang et al. (2018) showed high yields of phenols and anthocyanins extracted from chokeberry fruits under an optimised condition using 50% ethanol, which was achieved via the maceration method. Another study also reported that the maceration method with a lower temperature was the most effective and economically beneficial in extracting catechin from *Arbutus unedo* L. fruits owing to the high amount of its yield produced, which was similar to that of microwave-assisted extraction (MAE) method compared to ultrasound extraction. Another previous comparison study was conducted to compare the extraction rates of orientoside, luteolin, and flavonoids from *Cajanus cajan* leaves using microwave-assisted method, reflux extraction, ultrasound-assisted extraction and maceration method. The result showed that the maceration method gave the lowest extraction efficiency of the compounds.

According to a review study reported by Azwanida (2015) and Zhang et al. (2018), ethanol and hydroalcohol used in the extraction of phytochemical compounds from *Psidium guajava* L. leaves resulted in the highest extraction yield of alkaloids, saponins, carbohydrates, tannins, and flavonoids compared to the other solvents such as petroleum ether, chloroform, and water, thus indicating that polar solvents are more effective in the extraction of bioactive compounds from plant leaves. Next, the maceration technique using methanol and water showed different achievements in biological activities assessment of bioactive compounds extracted from *Garnicia atriviridis*, where the methanolic extract displayed higher antioxidant activity compared to the aqueous extract, while the performance of both extracts was vice versa in anti-hyperlipidemic activity. The efficiency of the maceration technique was favoured as a good choice of solvent for particular samples.

In addition, several extraction methods practise the same principle as maceration techniques, such as infusion, digestion, decoction, and percolation methods. Infusion is a dilute solution produced by soaking the solids of the readily soluble components of the crude samples in cold or boiling water in a shorter time compared to maceration. Digestion is an extraction method that applies gentle heat or slight warming during the maceration process of the mixture. It is used if a moderately raised temperature does not alter the active compounds of the samples, and the solvent efficacy of the menstruum is thus increased. The decoction is an extraction method where the crude samples are soaked and boiled in a particular volume of water for a defined time, usually within 15 min, and then the boiled mixture is cooled and strained. This method is suitable for extracting water-soluble and thermostable compounds, such as oil-soluble compounds generally obtained from hard parts of the plant materials such as roots, barks, stems, and rhizomes. Percolation also shares the same fundamental principle in maceration, but by using an equipment known as percolator. In percolation, the dried powdered materials are macerated with boiling water for 2 h in the percolator. The process involves extraction at a moderate rate, which is usually 6 drops per minute, followed by evaporation to obtain a concentrated extract form such as tinctures and fluid extract. Although it uses the same principle as maceration, this method is more effective owing to its continuous extraction process being continuous, where the saturated solvent is regularly replaced by fresh solvent (Banu & Cathrine, 2015).

Soxhlet extraction, also known as hot continuous extraction, employs the principle of reflux and syphoning to continuously extract plant samples with fresh solvent. The process is set up by adding finely pulverised plant material into a porous bag known as thimble, which is made up of strong filter paper or cellulose placed in the chamber of the Soxhlet apparatus. The extraction solvent in the bottom flask is heated, and its vapours will condense in the condenser. Then, the condensed vapours drip into the thimble containing the pulverised plant material and the extraction occurs by contact. When the liquid level in the chamber increases to the top of the syphon tube, the liquid contents are emptied into the bottom flask, and the process continues until no drop of solvent from the syphon tube leaves residue when evaporated (Azwanida, 2015).

The Soxhlet extraction method is only applied when the desired compound is limitedly soluble in a solvent, and its impurity is not soluble in that solvent. The advantages of using this method are that large amounts of extracts can be obtained with smaller quantities of solvent that are recycled, require small financial need, as well as being time and energy-saving processes since it is involved only in a single batch process compared to the maceration method. The disadvantages are that this technique is not suitable for extracting thermolabile compounds, and toxic emissions during extraction could occur due to exposure to hazardous and flammable liquid of the used organic solvents. In addition, the method is considered not eco-friendly since additional cost might be needed as the used solvents must possess high purity, and it is not truly environmentally safe since it can lead to pollution problems compared to other advanced techniques such as supercritical fluid extraction (SFE). Plus, the Soxhlet technique requires many factors to be pondered, including temperature, solvent-sample ratio and agitation speed (Azwanida, 2015).

These factors had also been reported in a previous study reviewed by Zhang et al. (2018), where the continuous extraction time under high temperature applied in the Soxhlet extraction increased the possibilities of thermal degradation. Degradation of catechins extracted from Cynomorium tea was observed in the Soxhlet extraction method due to the application of high temperature during the extraction process, while total polyphenols and total alkaloids concentrations were also decreased in this method at 70 °C compared to those extracted via maceration method at 40 °C.

Supercritical fluid extraction (SFE) is an extraction technique for samples using supercritical fluid (SF) as the extraction solvent. The solvent is known as dense gas, which shares the physical properties of both gas and liquid in terms of solubility at its critical point. SFE is the current advanced technology in extraction. Several factors need to be deliberated, such as temperature, pressure, sample volume, analyte collection, modifier (co-solvent) addition, flow and pressure control, as well as restrictors. SFE uses the principle of condensing gases, which is usually carbon dioxide (CO_2), into a dense liquid to minimise the use of organic solvents and increase yields. It is conducted by pushing the dense gas into its critical region by determination of temperature and pressure. The SF acts like a gas but has a solvating liquid feature. The CO_2 becomes SF at 31.1 °C and above under a pressure of 7380 kPa. Then, the liquid is pumped via a cylinder containing the sample to be extracted. After that, the extract-heavily loaded liquid is pumped into a separation chamber, where the extract is separated from the gas, and the gas is recycled. The advantages of SFE are that the CO_2 used is available at a low cost, possesses low toxicity, is ample and can extract thermolabile compounds (Azwanida, 2015; Doughari et al., 2009; Zhang et al., 2018).

Furthermore, the low polarity of the supercritical CO_2 (S-CO_2), due to it being the preferred liquid for SFE, allows it to extract non-polar compounds such as lipids and volatile oil. Although S-CO_2 is favoured by non-polar compounds, a modification can still be made by adding a small amount of ethanol or methanol to enable it to extract polar compounds. S-CO_2 also produces analytes in concentrated form when the CO_2 vaporises at ambient temperature. Other than carbon dioxide, argon (Ar) can be used in SFE as it is not expensive and more inert. The alteration of the solvents can be manipulated easily by altering the temperature and pressure or by adding a modifier to reduce extraction time. The end product (extract) may not contain solvent residues as CO_2 evaporates completely (Zhang et al., 2018).

The disadvantage of this technique is that the primary cost of the equipment is very expensive. The application of SFE can be seen in a study where the optimisation of S-CO_2 on *Wadelia calendulacea* reached its optimum yield at 25 MPa, 25 °C temperature, 10% modifier concentration and extraction time within 90 min. Next, the extraction of the essential oil from rosemary (*Rosmarinus ofcinalis*) using different extraction strategies, including SFE-CO_2, hydrodistillation and steam distillation, found that both yields of essential oil and antioxidant activity of SFE-CO_2 extract were higher than those from the other two approaches. Higher selective extraction of vinblastine, an antineoplastic compound from *Catharanthus roseus,* was achieved at 92% when modification of S-CO_2 was made by adding 2% ethanol at 300 bars at 40 °C compared to traditional extraction methods (Zhang et al., 2018).

6 Structural Elucidation of the Bioactive Compounds

There are several spectroscopic techniques that can be applied to determine the structure of compound molecules from plant extracts, such as UV–visible, infrared (IR), nuclear magnetic resonance (NMR), and mass spectroscopy. The basic principle of spectroscopy is passing electromagnetic radiation through an organic molecule that absorbs some of the radiation but not all. A spectrum can be produced by measuring the amount of the absorption of the electromagnetic radiation. Therefore, the spectra become specific to particular bonds in a molecule; hence, the structure of organic molecules can be identified based on these spectra (Popova et al., 2009).

UV–visible spectroscopy can be conducted for qualitative analysis and for identifying some classes of compounds in pure and crude mixtures. This application is preferable for qualitative analysis since the aromatic molecules in natural compounds are powerful chromophores in the UV range (Popova et al., 2009). On the other hand, phenolic compounds such as anthocyanins, tannins, polymer dyes, and phenols form complexes with iron that have been detected by ultraviolet/visible (UV–Vis) spectroscopy (Urbano et al., 2016). Furthermore, the UV–Vis spectroscopy methods were found to be less selective and give information on the composition of the total polyphenol content such as total phenolic extract (280 nm), flavones (320 nm), phenolic acids (360 nm) and anthocyanins (520 nm). The advantages of this method are that it is not time-consuming and presents reduced costs compared to other methods (Urbano et al., 2016).

Some of the frequencies will be absorbed when infrared light passes through a sample of an organic compound; however, some of the frequencies will be transmitted through the sample without any absorption. Infrared absorption is related to the vibrational changes that occur inside a molecule when it is visible to infrared radiation.

Hence, infrared spectroscopy can basically be defined as vibrational spectroscopy (Urbano et al., 2016). Another type of infrared spectroscopy is Fourier Transform Infrared Spectroscopy (FTIR), which is a high-resolution analytical instrument used to identify chemical compositions and elucidate structural components. It offers a rapid and non-destructive investigation for fingerprint plant extracts or powder (Urbano et al., 2016).

Next is nuclear magnetic resonance (NMR), which is primarily related to the magnetic properties of some atomic nuclei, especially the nucleus of hydrogen atom, proton, carbon and carbon isotope. NMR spectroscopy allowed many researchers to study molecules by recording the differences between a variety of magnetic nuclei, hence giving a clear picture of the positions of these nuclei in the molecule. Furthermore, it can display which atoms are present in neighbouring groups, which can eventually conclude how many atoms are present in each of these environments (Popova et al., 2009).

The organic molecules are blasted with either electrons or lasers in mass spectrometry and hence converted to charged ions, which are highly energetic. A mass spectrum is a plot of the relative abundance of a fragmented ion against the ratio of mass per charge of these ions. Therefore, the relative molecular mass (molecular weight) can be determined using mass spectrometry at high accuracy, and the exact molecular formula can be determined with a knowledge of places where the molecule has been fragmented (Christophoridou et al., 2005). For example, in a previous study reported by Cherkaoui et al. (2010), bioactive molecules from pith were isolated and purified by solvent extraction, column chromatography and HPLC. The techniques of UV–visible, IR, NMR, and mass spectroscopy were applied to characterise the structure of the bioactive molecule. Moreover, the molecules were hydrolysed, while their derivatives were characterised. Here, the mass spectrometry provided abundant information for the structural elucidation of the compounds when tandem mass spectrometry (MS) was applied. Therefore, the combination of HPLC and MS eases for fast and precise identification of the chemical compounds in botanical extracts, especially when the pure standard is unavailable (Altemimi et al., 2017). Currently, LCMS has been broadly used for the analysis of phenolic components, and Electrospray ionisation (ESI) has become an ideal source due to its high ionisation efficiency for the phenolic components (Altemimi et al., 2017).

7 Conclusions

In conclusion, it is important to pay attention to the production and post-harvest handling systems entirely to develop strategies for post-harvest disease control since many pre-harvest factors may also initiate infection and influence the development of post-harvest diseases. Fungicides are traditionally used to control post-harvest diseases. However, public demand to get eco-friendly and synthetic chemicals free of horticultural products is driving an urgency to develop new strategies to fulfil the demand. Plant extracts have been used as bio-fungicides and do not possess toxic effects on plants and soil to treat plant diseases, including fruit rot, due to the presence of anti-fungal and anti-microbial properties in the extracts. Besides treating fruit diseases, the plant extracts might also potentially increase the shelf life of the yields by reducing and controlling the post-harvest fungal decay of the fruits more safely and cheaply. The application of plant extracts to treat plant diseases need to be explored more as they can contribute to reducing environmental pollution due to the excessive use of synthetic chemical fungicide and their hazards to human health, as well as preventing the accumulation of toxic residual in the yields.

References

Abu Bakar, A. I., Nur Ain Izzati, M. Z., & Umi Kalsom, Y. (2013). Diversity of *Fusarium* species associated with post-harvest fruit rot disease of tomato. *Sains Malaysiana, 42*, 911–920.

Abubakar, M. G., Yerima, M. B., Zahriya, A. G., & Ukwuani, A. N. (2010). Acute toxicity and antifungal studies of ethanolic leaves, stem and pulp extract of *Tamarindus indica. Research Journal of Pharmaceutical, Biological and Chemical Sciences, 1*, 104–111.

Agrios, G. N. (2005). *Plant pathology* (5th ed.). Elsevier Acad Press.

Aini, Z., Sivapragasam, A., Vimala, P., & Mohamad Roff, M. N. (2005). Organic vegetable cultivation in Malaysia. In *Plant disease and its management* (pp. 1–22). Malaysian Agricultural Research and Development Institute (MARDI), Ministry of Agriculture and Agro-based Industries.

Aizat, W. M., Ahmad-Hashim, F. H., & Syed Jaafar, S. N. (2019). Valorization of mangosteen, "The Queen of Fruits", and new advances in postharvest and in food and engineering applications: A review. *Journal of Advanced Research, 20*, 61–70.

Alhassan, M. A., & Ahmed, Q. U. (2016). *Averrhoa bilimbi* Linn.: A review of its ethnomedicinal uses, phytochemistry, and pharmacology. *Journal of Pharmacy and Bioallied Science, 8*, 265–271.

Altemimi, A., Lakhssassi, N., Baharlouei, A., Watson, D. G., & Lightfoot, D. A. (2017). Review phytochemicals: Extraction, isolation, and identification of bioactive compounds from plant extracts. *Plants, 6*, 1–23.

Alvarez, A. M., & Nishijima, W. T. (1987). Postharvest diseases of papaya. *Plant Disease, 71*, 681–686.

Alvindia, D. G., & Hirooka, Y. (2011). Identification of *Clonostachys* and *Trichoderma* spp. from banana fruit surfaces by cultural, morphological and molecular methods. *Mycology, 2*, 109–115.

Amini, M., Safaie, N., Salmani, M. J., & Shams-Bakhsh, M. (2012). Antifungal activity of three medicinal plant essential oils against some phytopathogenic fungi. *Trakia Journal of Sciences, 10*, 1–8.

Anjum, M. A., Ahmed, N., Babita, B., Chauhan, H., & Gupta, P. (2016) Plant extracts in post-harvest disease management of fruits and vegetables—a review. *Journal of Food Processing and Technology, 7*, 592–596.

Anthony, S., Abeywickrama, K., Dayananda, R., Wigeratnam, S. W., & Arambewela, L. (2004). Fungal pathogens associated with banana fruit in Sri Lanka and their treatment with essential oils. *Mycopathologia, 157*, 91–97.

Aourach, M., Barbero, G. F., de Peredo, A. V. G., Diakite, A., Boukari, M. E., & Essalmani, H. (2021). Composition and antifungal effects of aqueous extracts of *Cymbopogon citratus, Laurus nobilis* and *Santolina chamaecyparissus* on the growth of *Fusarium oxysporum* f. sp. *lentis. Archives of Phytopathology and Plant Protection, 54*, 2141–2159.

Aziz, M. A., Rahman, S., Islam, M., & Begum, A. A. (2014). A comparative study on antibacterial activities and cytotoxic properties of various leaves extracts of Averrhoa bilimbi. *International Journal of Pharmaceutical Sciences and Research, 5*, 913–918.

Aziz, N. A. (2016). A review of the antimicrobial properties of three selected underutilized fruits of Malaysia. *International Journal of Pharmaceutical and Clinical Research, 8*, 1278–1283.

Azwanida, N. N. (2015). A review on the extraction methods use in medicinal plants, principle, strength and limitation. *Medicinal & Aromatic Plants, 4*, 1–6.

Babadoost, M. (2005). Phytophthora blight of cucurbits. *The Plant Health Instructor*. https://doi.org/10.1094/PHI-I-2005-0429-01

Bajpai, V. K., Rahman, A., & Kang, S. C. (2007). Chemical composition and anti-fungal properties of the essential oil and crude extracts of *Metasequoia glyptostroboides* Miki ex Hu. *Industrial Crops and Products, 26*, 28–35.

Bandara, B. M. R., Kumar, N. S., & Samaranayake, K. M. S. (1989). An antifungal constituent from the stembark of *Butea monosperma*. *Journal of Ethnopharmacology, 25*, 73–75.

Bansal, P., Paula, P., Nayaka, P. G., Pannakalc, S. T., Jian-hua, Z., Laatschd, H., Priyadarsinie, K. I., & Unnikrishnan, M. K. (2011). Phenolic compounds isolated from *Pilea microphylla* prevent radiation-induced cellular DNA damage. *Acta Pharmaceutica Sinica, 1*, 226–235.

Banu, K. S., & Cathrine, L. (2015). General techniques involved in phytochemical analysis. *International Journal of Advanced Research in Chemical Science, 2*, 25–32.

Bezerra, C. F., Rocha, J. E., Silva, M. K. N., Freitas, T. S., Sousa, A. K., Santos, A. T. L., Cruz, R. P., Ferreira, M. H., Silva, J. C. P., Machado, A. J. T., Carneiro, J. N. P., Sales, D. L., Coutinho, H. D. M., Ribeiro, P. R. V., Brito, E. S., & Morais-Braga, M. F. B. (2018). Analysis by UPLC-MS-QTOF and antifungal activity of guava (*Psidium guajava* L.). *Food and Chemical Toxicology, 119*, 122–132.

Bhattacherjee, A. K., Pandey, B. K., & Prakash, O. (2009). Persistence and dissipation of carbendazim residues in mango fruits after pre- and post-harvest applications. *Journal of Food Science and Technology Mysore, 46*, 347–349.

Bhattiprolu, S. L., & Bhattiprolu, G. R. (2006). Management of castor grey rot disease using botanical and biological agents. *Indian Journal of Plant Protection, 34*, 101–104.

Boukhatem, M. N., Ferhat, M. A., Kameli, A., Saidi, F., & Kebir, H. T. (2014). Lemongrass (*Cymbopogon citratus*) essential oil as a potent anti-inflammatory and antifungal drugs. *Libyan Journal of Medicine, 9*, 1–10.

Boyraz, N., & Özcan, M. (2005). Antifungal effect of some spice hydrosols. *Fitoterapia, 76*, 661–665.

Brauer, V. S., Rezende, C. P., Pessoni, A. M., De Paula, R. G., Rangappa, K. S., Nayaka, S. C., Gupta, V. K., & Almeida, F. (2019). Review antifungal agents in agriculture: Friends and foes of public health. *Biomolecules, 9*, 1–21.

Bussmann, R. W., Glenn, A., Meyer, K., Kuhlman, A., & Townesmith, A. (2010). Herbal mixtures in traditional medicine in Northern Peru. *Journal of Ethnobiology and Ethnomedicine, 6*, 1–10.

Chahardehi, A. M., Ibrahim, D., & Sulaiman, S. F. (2010). Antioxidant, antimicrobial activity and toxicity test of *Pilea microphylla*. *International Journal of Microbiology, 2010*, 1–7.

Cheng, S. S., Lin, H. Y., & Chang, S. T. (2005). Chemical composition and antifungal activity of essential oils from different tissue of Japanese Cedar (*Cryptomeria japonica*). *Journal of Agricultural and Food Chemistry, 53*, 614–619.

Cherkaoui, A., Hibbs, J., Emonet, S., Tangomo, M., Girard, M., Francois, P., & Schrenzel, J. (2010). Comparison of two matrix-assisted laser desorption ionization-time of flight mass spectrometry methods with conventional phenotypic identification for routine identification of bacteria to the species level. *Journal of Clinical Microbiology, 48*, 1169–1175.

Choi, G. J., Lee, S. W., Jang, K. S., Kim, J. S., Cho, K. Y., & Kim, J. C. (2004). Effects of chrysophanol, parietin, and nepodin of Rumex crispus on barley and cucumber powdery mildews. *Crop Protection, 23*, 1215–1221.

Choudhury, D., Dobhal, P., Srivastava, S., Saha, S., & Kundu, S. (2018). Role of botanical plant extracts to control plant pathogens: A review. *Indian Journal of Agriculture Research, 52*, 341–346.

Christophoridou, S., Dais, P., Tseng, L. H., & Spraul, M. (2005). Separation and identification of phenolic compounds in olive oil by coupling high-performance liquid chromatography with post-column solid-phase extraction to nuclear magnetic resonance spectroscopy (LC-SPE-NMR). *Journal of Agriculture and Food Chemistry, 53*, 4667–4679.

Corato, U., Maccioni, O., Trupo, M., & Di Sanzo, G. (2010). Use of essential oil of *Laurus nobilis* obtained by means of a supercritical carbon dioxide technique against post-harvest spoilage fungi. *Crop Protection, 29*, 142–147.

da Cruz Cabral, L., Pinto, V. F., & Patriarca, A. (2013). Application of plant derived compounds to control fungal spoilage and mycotoxin production in foods. *International Journal of Food Microbiology, 166*, 1–14.

Das, M., & Goswami, S. (2019). Antifungal and antibacterial property of guava (*Psidium guajava*) leaf extract: Role of phytochemicals. *International Journal of Health Sciences and Research, 9*(2), 39–45.

Dash, G. K., & Zakaria, Z. (2016). Pharmacognostic studies on *Persicaria odorata* (Lour.) Sojak. *Journal of Pharmacy Research, 10*, 377–380.

De Lima, V. L. A. G., Mélo, E. D. S., & Lima, L. D. S. (2001). Physicochemical characteristics of bilimbi (*Averrhoa bilimbi* L.). *Revista Brasileira De Fruiticultura, 23*, 421–423.

Der-Jiun, O., Iqbal, S., & Ismail, M. (2012). Proximate composition, nutritional attributes and mineral composition of *Peperomia pellucida* L. (Ketumpangan Air) grown in Malaysia. *Molecules, 17*, 11139–11145.

Doughari, J. H., Human, I. S., Bennade, S., & Ndakidemi, P. A. (2009). Phytochemicals as chemotherapeutic agents and antioxidants: Possible solution to the control of antibiotic resistant verocytotoxin producing bacteria. *Journal of Medicinal Plants Research, 3*, 839–848.

Eloff, J. N., & McGaw, L. J. (2014). Using African plant biodiversity to combat microbial infections. In A. Gurib-Fakim (Ed.), *Novel plant bioresources: Applications in food medicine and cosmetics* (pp. 163–173). John Wiley.

El-Sesy, M. E., & Mahran, B. N. A. (2020). The antibacterial and coagulant activity of *Psidium guajava* leaves extracts in purification of wastewater. *Biosciences Biotechnology Research Asia, 17*(1), 191–203.

Falah, S., Ayunda, R. D., & Faridah, D. N. (2015). Potential of lemongrass leaves extract (*Cymbopogon citratus*) as prevention for oil oxidation. *Journal of Chemical and Pharmaceutical Research, 7*(10), 55–60.

Farag, A., Ebrahim, H., El-Mazoudy, R., & Kadous, E. (2011). Developmental toxicity of fungicide carbendazim in female mice. *Birth Defects Research, 92*, 122–130.

Felhi, S., Daoud., A., Hajlaoui, H., Mnafgui, K., Gharsallah, N., & Kadri, A. (2017). Solvent extraction effects on phytochemical constituents' profiles, antioxidant and antimicrobial activities and functional group analysis of *Ecballium elaterium* seeds and peels fruits. *Food Science & Technology, Campinas, 37*, 483–492.

Feng, W., Zheng, X., Chen, J., & Yang, Y. (2008). Combination of cassia oil with magnesium sulphate for control of postharvest storage rots of cherry tomatoes. *Crop Protection, 27*, 112–117.

Garcia, L. (2011). A comparative study on the antifungal effects of tamarind (*Tamarindus indica*) and garlic (*Allium sativum*) extracts on banana anthracnose. *Journal of Nature Studies, 10*, 96–107.

Gatto, M. A., Ippolito, A., Linsalata, V., Cascarano, N. A., Nigro, F., Vanadia, S., & Di Venere, D. (2011). Activity of extracts from wild edible herbs against postharvest fungal diseases of fruit and vegetables. *Postharvest Biology and Technology, 61*, 72–82.

Gazwi, H. S. S. (2019). Preventive effect of lemongrass (*Cymbopogon citratus*) against oxidation in soybean oil. *Proceedings of the National Academy of Sciences, India Section B: Biological Sciences, 90*, 151–159.

Ghasemzadeh, A., Jaafar, H. Z. E., Baghdadi, A., & Tayebi-Meigooni, A. (2018). Alpha-mangostin-rich extracts from mangosteen pericarp: Optimization of green extraction protocol and evaluation of biological activity. *Molecules, 23*, 1–16.

Gyasi, E., Kotey, D. A., Adongo, B. A., Adams, F. K., Owusu, E. O., & Mohammed, A. (2022). Management of major seed-borne fungi of cowpea (*Vigna unguiculata* (L.) Walp) with four selected botanical extracts. *Advances in Agriculture, 2022*(8), 1–8.

Hartman, J. R. (2007). Peach fruit diseases. In *Plant Pathology fact sheet*. UK cooperative extension service, University of Kentucky, College of Agriculture. PPFS-FR-T-09

Hartman, J., & Kaiser, C. (2008). Strawberry fruit rots. In *Plant Pathology fact sheet*. UK cooperative extension service, University of Kentucky, College of Agriculture. PPFS-FR-S-08

Hashemi, P., Abolghasemi, M. H., Ghaisvand, A. R., Ahmadi, S., Hassan, V., & Yarahmadi, H. (2008). A comparative study of volatile component of *Echinophora cinerea*. *Journal of Chromatograph, 69*, 179–182.

Horst, R. K. (2001). Plant diseases and their pathogens. *Westcott's plant disease handbook* (pp. 65–530). Springer.

Ibrahim, M. A., & Emlee, A. M. (2020). Anti-fungal study on aqueous and ethanolic leaves extracts of *Piper sarmentosum*. *Matrix Science Pharma, 4*(1), 13–17.

Iordăchescu, G., Ploscuțanu, G., Pricop, E. M., Baston, O., & Barna, O. (2019). Postharvest losses in transportation and storage for fresh fruits and vegetables sector. *Journal of International Scientific Publications, 7*, 244–251.

Ismail, S. M., Hui, C. K., Aminuddin, A., & Agusman, A. (2018). *Piper sarmentosum* as an antioxidant: A systematic review. *Sains Malaysiana, 47*(10), 2359–2368.

Jasso de Rodríguez, D., Hernández-Castillo, D., Angulo-Sánchez, J. L., Rodríguez-García, R., Villarreal Quintanilla, J. A., & Lira-Saldivar, R. H. (2007). Antifungal activity *in vitro* of *Flourensia* spp. extracts on *Alternaria* sp., *Rhizoctonia solani* and *Fusarium oxysporum*. *Industrial Crops and Products, 25*, 111–116.

Ji, X., Avula, B., & Khan, I. A. (2007). Quantitative and qualitative determination of six xanthones in *Garcinia mangostana* L. by LC-PDA and LC-ESI-MS. *Journal of Pharmaceutical Biomedicine Analysis, 43*, 1270–1276.

Kantachot, C., Chantaranothai, P., & Simpson, D. A. (2010). A synopsis of the genus *Persicaria* (Polygonaceae) in Thailand. *Thai Forest Bulletin, 38*, 128–149.

Khallil, A. R. M. (2001). Phytofungitoxic properties in the aqueous extracts of some plants. *Pakistan Journal of Biological Science, 4*, 392–394.

Khan, A., Rahman, M., & Islam, M. S. (2008). Antipyretic activity of *Peperomia pellucida* leaves in rabbit. *Turki Journal of Biology, 32*, 37–41.

Khusnul, S. R., Virgianti, D. P., Fathurohman, M., & Pratita, A. T. K. (2019). Effect of Karuk leaves (*Piper Sarmentosum* Roxb.) and white galangal rhizome (*Alpinia galanga* L.) ethanol extract on the growth of *Microsporum gypseum* and *Candida albicans* in vitro. *Journal of Physics: Conference Series, 1179*, 012168.

Kumar, R., Mishra, A. K., Dubey, N. K., & Tripathi, Y. B. (2007). Evaluation of *Chenopodium ambrosioides* oil as a potential source of antifungal, antiaflatoxigenic and antioxidant activity. *International Journal of Food Microbiology, 115*, 159–164.

Latha, P., Anand, T., Ragupathi, N., Prakasam, V., & Samiyappan, R. (2009). Antimicrobial activity of plant extracts and induction of systemic resistance in tomato plants by mixtures of PGPR strains and Zimmu leaf extract against *Alternaria solani*. *Biological Control, 50*, 85–93.

Lawal, O. A., Ogundajo, A. L., Avoseh, N. O., & Ogunwande, I. A. (2017). Cymbopogon citratus. In V. Kuete (Ed.), *Medicinal spices and vegetables from Africa* (pp. 397–423). Academic Press.

Lee, S. E., Park, B. S., Kim, M. K., Choi, W. S., Kim, H. T., Cho, K. Y., Lee, S. G., & Lee, H. S. (2001). Fungicidal activity of pipernonaline, a piperidine alkaloid derived from long pepper, *Piper longum* L., against phytopathogenic fungi. *Crop Protection, 20*, 523–528.

Liu, X. S., Tong, Z. F., Zhen, L., Huang, D. X., Gao, X., & Xu, C. Y. (2009). Simultaneous analysis ofthiabendazole, carbendazim and 2-aminobenzimidazole in concentrated fruit juices

by liquid chromatography after a single mix-mode solid-phase extraction clean up. *Journal of Environmental Science and Health, 44*, 591–597.
Magro, A., Carolino, M., Bastos, M., & Mexia, A. (2006). Efficacy of plant extracts against stored products fungi. *Revista Iberoamericana De Micología, 23*, 176–178.
Mahmud, T. M. M. (2017). *Postharvest an unsung solution for food security* (pp. 1–91). Universiti Putra Malaysia.
Mailoa, M. N., Mahendradatta, M., Laga, A., & Djide, N. (2013). Antimicrobial activities of tannins extract from guava leaves (*Psidium guajava* L) on pathogens microbial. *International Journal of Scientific and Technology Research, 3*(1), 236–241.
Manikandan, R., Anand, A. V., Kumar, S., & Pushpa (2016). Phytochemical and in vitro antidiabetic activity of *Psidium Guajava* leaves. *Pharmacognosy Journal, 8*(4), 392–394.
Manoharachary, C., & Gourinath, A. (1988). Effects of plant extracts on four pathogenic fungi. In *5th International Congress of Plant Pathology*, Kyoto.
Mazid, M., Khan, T. A., & Mohammad, F. (2012). Medicinal plants of rural India: A review of use by Indian folks. *Indo Global Journal of Pharmaceutical Science, 2*, 286–304.
Mehta, S., & Sharma, K. (2016). Natural resources: An eco-friendly and safer alternate to control plant diseases. *International Journal of Pharmaceutical Sciences and Research, 7*, 4327–4340.
Melia, S., Novia, D., Juliyarsi, I., & Purwati, E. (2019). The characteristics of the pericarp of *Garcinia mangostana* (mangosteen) extract as natural antioxidants in rending. In *International Conference on Animal Production for Food Sustainability. IOP Conference Series: Earth and Environmental Science* (pp. 287–295).
Mohajer, S., Mat Taha, R., Yaacob, J. S., & Kumari, A. (2015). Post-harvest management of the pest and some important diseases of the fruits. *Recent trends in post-harvest technology and management* (pp. 1–12). Manglam Publications.
Mokhtar, S. I., & Abd Aziz, N. A. (2016). Antimicrobial properties of *Averrhoa bilimbi* extracts at different maturity stages. *Journal of Medical Microbiology and Diagnosis, 5*, 233–234.
Monsálvez, M., Zapata, N., Vargas, M., Berti, M., Bittner, M., & Hernández, V. (2010). Antifungal effects of n-hexane extract and essential oil of *Drimys winteri* bark against Take-All disease. *Industrial Crops and Products, 31*, 239–244.
Muchtaridi, M., Puteri, N. A., Milanda, T., & Musfiroh, I. (2017). Validation analysis methods of α-mangostin, β-mangostin and gartanin mixture in mangosteen (*Garcinia mangostana* L.) fruit rind extract from west java with HPLC. *Journal of Applied Pharmaceutical Science, 7*, 125–130.
Mutee, A. F., Salhimi, S. M., Yam, M. F., Lim, C. P., Abdullah, G. Z., Ameer, O. Z., Abdulkarim, M. F., & Asmawi, M. Z. (2010). In vivo anti-inflammatory and in vitro antioxidant activities of *Peperomia pellucida. International Journal of Pharmacology, 6*, 686–690.
Naeini, A., Ziglari, T., Shokri, H., & Khosravi, A. R. (2010). Assessment of growth-inhibiting effect of some plant essential oils on different Fusarium isolates. *Journal De Mycologie Médicale/ journal of Medical Mycology, 20*, 174–178.
Nakyinsige, K., Abdul Rahman, N. S., Salwani, M. S., Abd Hamid, A., Adeyemi, K. D., Sakimin, S. Z., & Sazili, A. Q. (2016). Effect of belimbing buluh (*Averrhoa bilimbi*) juice extract on oxidative stability and microbiological quality of spent chicken meat. *International Food Research Journal, 23*, 2675–2680.
Narasimhan, S., Maheshwaran, S., Abu-Yousef, I. A., Majdalawieh, A. F., Rethavathi, J., Das, P. E., & Poltronieri, P. (2017). Anti-bacterial and anti-fungal activity of xanthones obtained via semi-synthetic modification of α-Mangostin from *Garcinia mangostana. Molecules, 22*, 1–13.
Natarajan, M. R., & Lalithakumari, D. (1987). Antifungal activity of the leaf extract of *Lawsonia inermis* on *Dreschslera oryzae. Indian Phytopathology, 40*, 390–395.
Nduagu, C., Ekefan, E. J., & Nwankiti, A. O. (2008). Effect of some crude plant extracts on growth of *Colletotrichum capsici* (Synd) Butler & Bisby, causal agent of pepper anthracnose. *Journal of Applied Biosciences, 6*, 184–190.
Neela, F. A., Sonia, I. A., & Shamsi, S. (2014). Antifungal activity of selected medicinal plant extract on *Fusarium oxysporum* Schlecht, the causal agent of Fusarium wilt disease in tomato. *American Journal of Plant Science, 5*(18), 2665–2671.

Nelson, S. (2005). Rhizophus rot of jackfruit. *Plant Disease, 29*, 1–2.

Noriel, L. M., & Robles, R. P. (1990). Fungicidal activity of *Portulaca oleracea* extract against *Helminthosporium maydis* Wisik and Miyake in corn (Zea mays L.). *Philippine Journal of Weed Science, 17*, 26–32.

Okigbo, R. N., Putheti, R., & Achusi, C. T. (2009). Post-harvest deterioration of cassava and its control using extracts of *Azadirachta indica* and *Aframomum melegucta. E-Journal Chemistry, 6*, 1274–1280.

Okigbo, R. N., Ezebo, R. O., & Ugwu, S. C. (2018). Antifungal attributes of extracts of *Ocimum gratissimum, Zingiber officinale*, and *Cymbopogon citratus* on rot fungi of soursop fruit. *Clinical Journal of Nutrition and Dietetics, 1*(1), 1–7.

Ovalle-Magallanes, B., Eugenio-Perez, D., & Pedraza-Chaverri, J. (2017). Review: Medicinal properties of mangosteen (*Garcinia mangostana* L.): A comprehensive update. *Food Chemistry and Toxicology, 109*, 102–122.

Parajuli, R. R., Tiwari, R. D., Chaudhary, R. P., & Gupta, V. N. (2005). Fungitoxicity of the essential oils of some aromatic plants of Manang against *Alternaria brassicicola. Scientific World, 3*, 39–43.

Perveen, S., Orfali, R., Al-Taweel, A. M., Khan, A., Alghanem, B., & Shaibah, H. (2019). Simultaneous identification of phenolic and flavonoid contents in bee pollen by HPLC-ESI-MS data. *Biomedical Research, 30*, 770–774.

Pinto, E., Ribeiro, S. L., Cavaleiro, C., Palmeira, A., & Gonçalves, M. J. (2007). In vitro susceptibility of some species of yeasts and filamentous fungi to essential oils of *Salvia officinalis. Industrial Crops and Products, 26*, 135–141.

Popova, I. E., Hall, C., & Kubátová, A. (2009). Determination of lignans in flaxseed using liquid chromatography with time-of-flight mass spectrometry. *Journal of Chromatography A, 1216*, 217–229.

Qin, W., Huang, S., Li, C., Chen, S., & Peng, Z. (2010). Biological activity of the essential oil from the leaves of *Piper sarmentosum* Roxb. (Piperaceae) and its chemical constituents on *Brontispa longissimi* (Gestro) (Coleoptera: Hispidae). *Pesticide Biochemistry and Physiology, 96*, 132–139.

Rahman, S. F. S. A., Sijam, K., & Omar, D. (2016). *Piper sarmentosum* Roxb: A mini review of ethnobotany, phytochemistry and pharmacology. *Journal of Analytical and Pharmaceutical Research, 2*(5), 00031.

Rai, N. K., Leepikatuli, B., Sharma, K., & Singh, U. P. (2000). Effect of plant extracts on spore germination of some fungi. *Indian Journal of Plant Pathology, 18*, 44–47.

Rana, S. M. M., Billah, M. M., Hossain, M. S., Saifuddin, A. K. M., Islam, S. K. M. A., Banik, S., Naim, Z., & Raju, G. S. (2014). Susceptibility of microorganism to selected medicinal plants in Bangladesh. *Asian Pacific Journal of Tropical Biomedicine, 4*, 911–917.

Ravikumar Patil, H. S., Makari, H. K., & Gurumurthy, H. (2007). In vitro antimicrobial activity of ethanol extract of *Thevetia peruviana. Electronic Journal of Environmental, Agricultural and Food Chemistry, 6*, 2318–2322.

Rhimi, W., Mohammed, M. A., Zarea, A. A. K., Greco, G., Tempesta, M., Otranto, D., & Cafarchia, C. (2022). Antifungal, antioxidant and antibiofilm activities of essential oils of *Cymbopogon* spp. *Antibiotics, 11*(6), 829–841.

Ridzuan, P. M., Aini, H. H., Norazian, M. H., Shah, A., Roesnita, & Aminah, K. S. (2013). Antibacterial and antifungal properties of *Persicaria odorata* leaf against pathogenic bacteria and fungi. *The Open Conference Proceedings Journal, 4*, 71–74.

Rimal Isaac, R. S., Sakthivel, G., & Murthy, C. (2013). Green synthesis of gold and silver nanoparticles using *Averrhoa bilimbi* fruit extract. *Journal of Nanotechnology, 2013*, 1–6.

Rohman, A., Rafi, M., Alam, G., Muchtaridi, M., & Windarsih, A. (2019). Chemical composition and antioxidant studies of underutilized part of mangosteen (*Garcinia mangostana* L.) fruit. *Journal of Applied Pharmaceutical Science, 9*, 47–52.

Sakilan, J. M., Demayo, C. G., & Opanasopit, P. (2019). Phytochemical analysis and determination of anti-microbial, anti-oxidant, and anti-cancer activity of the leaf ethanolic extracts of

Piper sarmentosum Roxb. *Lapuyan Zamboanga Del Sur, Philippines. International Journal of Pharmaceutical Science and Research, 10*(12), 5715–5722.
Salomone, A., Scaritto, G., Sacco, A., Cabras, G., & Angioni, A. (2008). Inhibitory effects of the main compounds of oregano essential oil against some pathogenic fungi. Modern fungicide and antifungal compounds V. *Deutsche Phytomedizinische Gesellschaft, 44*, 345–360.
Sandra. (2016). Modified atmosphere packaging of fresh produce: Current status and future needs. *LWT-Food Science and Technology, 43*, 381–392.
Sasidharan, S., Chen, Y., Saravanan, D., Sundram, K. M., & Latha, Y. (2011). Extraction, isolation and characterization of bioactive compounds from plants' extracts. *African Journal of Traditional Complement Alternative Medicine, 8*, 1–10.
Sawadogo, I., Paré, A., Kaboré, D., Montet, D., Durand, N., Bouajila, J., Zida, E. P., Sawadogo-Lingani, H., Nikiéma, P. A., Nebié, R. H. C., & Bassole, I. H. (2022). Antifungal and antiaflatoxinogenic effects of *Cymbopogon citratus*, *Cymbopogon nardus*, and *Cymbopogon schoenanthus* essential oils alone and in combination. *Journal of Fungi, 8*(2), 117–129.
Seebold, K. (2010). Fruit rots of cucurbit. In *Plant Pathology Fact Sheet* (pp. 1–5).
Shi, Y. N., Liu, F. F., Jacob, M. R., Li, X. C., Zhu, H. T., Wang, D., Cheng, R. R., Yang, C. R., Xu, M., & Zhang, Y. J. (2017). Antifungal amide alkaloids from the aerial parts of *Piper flaviflorum* and *Piper sarmentosum*. *Planta Medica, 83*, 143–150.
Shuping, D. S. S., & Elof, J. N. (2017). The use of plants to protect plants and food against fungal pathogens: A review. *African Journal of Traditional Complement Alternative Medicine, 14*, 120–127.
Siddiqui, B. S., Afshan, F., Gulzar, T., & Hanif, M. (2004). Tetracyclic triterpenoids from the leaves of *Azadirachta indica*. *Phytochemistry, 65*, 2363–2367.
Singh, B. K., Yadav, K. S., & Verma, A. (2017). Impact of post-harvest diseases and their management in fruit crops: An overview. *Journal of Bio Innovation, 5*, 749–760.
Sitara, U., Naseem, J., & Sultana, N. (2008). Antifungal effect of essential oils on in vitro growth of pathogenic fungi. *Pakistan Journal of Botany, 40*, 409–414.
Soliman, F. M., Fathy, M. M., Salama, M. M., & Saber, F. R. (2016). Comparative study of the volatile oil content and antimicrobial activity of *Psidium guajava* L. and *Psidium cattleianum* Sabine leaves. *Bulletin of Faculty of Pharmacy, Cairo University, 54*(2), 219–225.
Soradech, T. A., Johansson, A., & Gref, R. (2016). Toxic effects of some conifer resin acids and tea tree oil on human epithelial and fibroblast cells. *Toxicology, 107*, 99–109.
Souza, E. L. D., Lima, E. D. O., Freire, K. R. D. L., & Sousa, C. P. D. (2005). Inhibitory action of some essential oils and phytochemicals on the growth of various moulds isolated from foods. *Brazilian Archives of Biology and Technology, 48*, 245–250.
Stuardo, M., & San Martín, R. (2008). Antifungal properties of quinoa *Chenopodium quinoa* Willd alkali treated saponins against *Botrytis cinerea*. *Industrial Crops and Products, 27*, 296–302.
Sukrasno. (2014). Changes in secondary metabolite contents following crude drug preparation. *Procedia Chemistry, 13*, 57–62.
Sultana, B., Anwar, F., & Ashraf, M. (2009). Effect of extraction solvent or technique on the antioxidant activity of selected medicinal plant extracts. *Molecules, 14*, 2167–2180.
Sun, X., Chen, W., Dai, W., Xin, H., Rahmand, K., Wang, Y., Zhang, J., Zhang, S., Xu, L., & Han, T. (2020). *Piper sarmentosum* Roxb.: A review on its botany, traditional uses, phytochemistry, and pharmacological activities. *Journal of Ethnopharmacology, 263*, 112897.
Tegegne, G., Pretorius, J. C., & Swart, W. J. (2008). Antifungal properties of *Agapanthus africanus* L. extracts against plant pathogens. *Crop Protection, 27*, 1052–1060.
Thouri, A., Chahdoura, H., El Arem, A., Hichri, A. O., Hassin, R. B., & Achour, L. (2017). Effect of solvents extraction on phytochemical components and biological activities of Tunisian date seeds (var. Korkobbi and Arechti). *BMC Complementary and Alternative Medicine, 17*, 1–10.
Toivonen, P. M. A., Mitcham, E. J., & Terry, L. A. (2014). Postharvest care and the treatment of fruits and vegetables. In G. R. Dixon & D. E. Aldous (Eds.), *Horticulture: Plants for people and places* (Vol. 1, pp. 465–483). Springer Science + Business Media Dordrecht.

Tripathi, P., & Dubey, N. K. (2004). Exploitation of natural products as alternative strategy to control post-harvest fungal rotting of fruits and vegetables. *Postharvest Biology and Technology, 32*, 235–245.

Truong, D. H., Nguyen, D. H., Anh Ta, N. T., Bui, A. V., Do, T. H., & Nguyen, H. C. (2019). Evaluation of the use of different solvents for phytochemical constituents, antioxidants, and in vitro anti-inflammatory activities of *Severinia buxifolia*. *Journal of Food Quality, 2019*, 1–9.

Twumasi, P., Ohene-Mensah, G., & Moses, E. (2014). The rot fungus *Botryodiplodia theobromae* strains cross infect cocoa, mango, banana and yam with significant tissue damage and economic losses. *African Journal of Agricultural Research, 9*, 613–619.

Urbano, G., Lázaro, I., Rodríguez, I., Reyes, J. L., Larios, R., & Cruz, R. (2016). Electrochemical and spectroscopic study of interfacial interactions between chalcopyrite and typical flotation process reagents. *International Journal of Minerals, Metallurgy, and Materials., 23*, 127–136.

Vilela, G. R., de Almeida, G. S., D'Arce, M. A. B. R., Moraes, M. H. D., Brito, J. O., da Silva, M. F. d. G. F., Silva, S. C., de Stefano Piedade, S. M., Calori-Domingues, M. A., & da Gloria, E. M. (2009). Activity of essential oil and its major compound, 1,8-cineole, from *Eucalyptus globulus* Labill., against the storage fungi *Aspergillus flavus* Link and *Aspergillus parasiticus* Speare. *Journal of Stored Products Research, 45*, 108–111.

Vince Carlo, A. C., Mark Renill, V. H., & Clint Erick, A. (2017). In vitro screening of anti-fungal effect of *Averrhoa bilimbi* leaves methanolic extract against *Microsporum canis*. Lyceum of the Philippines-St. Cabrini. *College of Allied Medicine Research, 2*, 58–66.

Wei, L. S., Wee, W., Siong, J. Y. F., & Syamsumir, D. F. (2011). Characterization of anticancer, antimicrobial, antioxidant properties and chemical compositions of *Peperomia pellucida* leaf extract. *Acta Medica Iranica, 49*, 671–674.

Yadav, A. R., & Mohite, S. K. (2020). Screening of in-vitro anti-inflammatory and antifungal assay of *Psidium guajava* leaf extracts. *Research Journal of Topical and Cosmetic Science, 11*(2), 62–64.

Yanpirat, P., & Vajrodaya, S. (2015). Antifungal activity of Persicaria odorata extract against anthracnose caused by *Colletotrichum capsici* and *Colletotrichum gloeosporioides* malays. *Malaysian Applied Biology, 44*, 69–73.

Yegen, O., Begger, B., & Heitefuss, R. (1992). Studies on the fungitoxic effect for extracts of six selected plants from Turkey on phytopathogenic fungi. *Zeitschrift Fur Pflanzenkrankheiten Und Pflanzenschütz, 99*, 349–359.

Zabka, M., Pavela, R., & Slezakova, L. (2009). Antifungal effect of *Pimenta dioica* essential oil against dangerous pathogenic and toxinogenic fungi. *Industrial Crops and Products, 30*, 250–253.

Zakaria, L., Chik, M. W., Heng, K. W., & Salleh, B. (2012). *Fusarium* species associated with fruit rot of banana (Musa spp.), papaya (Carica papaya) and guava (*Psidium guajava*). *Malaysian Journal of Microbiology, 8*(2), 127–130.

Zaker, M. (2016). Natural plant products as eco-friendly fungicides for plant diseases control—a review. *The Agriculturists, 14*, 134–141.

Zavala-López, M., & García-Lara, S. (2017). An improved microscale method for extraction of phenolic acids from maize. *Plant Methods, 13*, 1–11.

Zhang, Q. W., Lin, L. G., & Ye, W. C. (2018). Techniques for extraction and isolation of natural products: A comprehensive review. *Chinese Medicine, 13*, 1–26.

Zhang, Z. Y., Dai, G. H., Zhuge, Y. Y., & Li, Y. B. (2008). Protective effect of *Robinia pseudoacacia* Linn. extracts against cucumber powdery mildew fungus, Sphaerotheca Fuliginea. *Crop Protection, 27*, 920–925.

Zhou, J., Xiong, K., Yang, Y., Ye, X., Liu, J., & Li, F. (2015). Deleterious effects of benomyl and carbendazim on human placental trophoblast cells. *Reproductive Toxicology, 51*, 64–71.

Ziv, O., & Zitter, T. A. (1992). Effects of bicarbonates and film-forming polymers on cucurbit foliar diseases. *Plant Disease, 76*, 513–517.

Zulfa, Z., Chia, C. T., & Rukayadi, Y. (2016). *In vitro* antimicrobial activity of *Cymbopogon citratus* (lemongrass) extracts against selected foodborne pathogens. *International Food Research Journal, 23*(3), 1262–1267.

Zulhaimi, H. I., Rosli, I. R., Kasim, K. F., Muhammad Akmal, H., Nuradibah, M. A., & Sam, S. T. (2017). A comparative study of *Averrhoa bilimbi* extraction method. In *Paper presented in Electronic and Green Materials International Conference 2017 (EGM 2017)* (pp. 1–5), published by AIP Publishing, 1885.

Current Thinking on Infection and Control Strategies of Basal Stem Rot Disease of Oil Palm in Malaysia

Mohd Aswad Abdul Wahab and Khairulmazmi Ahmad

Abstract *Ganoderma* causes basal stem rot and upper stem rot infection of oil palm, and the former is the most frequently found and devastating disease in oil palm cultivation in Indonesia and Malaysia. The primary route of infection through root contact has been generally regarded as the sole mode of infection by the pathogen. Other root-infecting basidiomycete diseases of tree and perennial crops are spread by vegetative growth. After basidiospores of *Heterobasidion* spp. colonize stump surfaces, the subsequent spread of the fungus from infected stumps to neighbouring trees occurs via root contacts. This vegetative infection of trees creates a clonal spread over a wide area. *Armillaria* species cause serious root diseases of a wide range of trees and other woody species. For example, in Kenya and Indonesia they affect tea plantations. This fungus can spread by root contact and by cord-like rhizomorphs, which can directly penetrate healthy roots to cause disease. In contrast, various studies using vegetative compatibility and molecular analysis of *Ganoderma* isolates indicate high genetic variability within oil palm plantations and raise an argument on the issue of disease epidemiology. Currently there are two opinions on the mode of infection: (i) Healthy palm roots encounter a source of inoculum, such as infected roots from adjacent palms; this become infected and mycelial growth progresses along or through the root towards the bole leading to stem rot of the base of the palm. (ii) Basidiospores are dispersed throughout the plantations onto wounded palm tissues and cause direct infection, for example via wounds created on harvesting or pruning of palm petioles (fronds). (iii) Basidiospores land on and colonize palm debris to create an inoculum to be contacted by roots. There is evidence to support the first two routes, which complicates disease control. Therefore, understanding *G. boninense* infection and disease spread are critical to develop a successful disease management against BSR.

Keywords Basal stem rot · Oil palm · *Ganoderma boninense*

M. A. Abdul Wahab (✉) · K. Ahmad
Department of Plant Protection, Faculty of Agriculture, Universiti Putra Malaysia, 43400 Serdang, Selangor, Malaysia
e-mail: mohdaswad@upm.edu.my

K. Ahmad
e-mail: khairulmazmi@upm.edu.my

M. Wong (ed.), *Advances in Tropical Crop Protection*,
https://doi.org/10.1007/978-3-031-59268-3_10

1 Introduction

1.1 Oil Palm Industry and Economic Importance

The oil palm *(Elaeis guineensis Jaq.)* has remained as a major crop in Southeast Asia, especially in Malaysia and Indonesia in the last 25 years. Oil palm is a valuable economic crop and one of the most important oil tree species in tropical regions because of the high yield of raw materials it produces: palm oil and palm kernel oil. It is by far the most widely-produced tropical oil and constitutes more than thirty percent of total edible oil production worldwide (FAO, 2023). From oil palm two grades of oil can be produced: crude palm oil (CPO) comes from grinding fruit pulp to release standard quality oil used in biscuits, margarine and cooking oils, and palm kernel oil (PKO) which is obtained from grinding the nut to release high quality oil used in pharmaceuticals (Murphy et al., 2021). Soh et al. (2009) reported palm oil as the largest internationally traded vegetable oil with its main markets in China, European Union, Pakistan, India, Japan and Bangladesh. It is used primarily in food (80%), e.g. as cooking oil, margarine, vanaspati or vegetable ghee and shortenings, and the remaining 20% is used as oleochemicals replacing mineral oil to feed the detergents, cosmetics, pharmaceutical/nutraceutical, plastics and lubricants industries. Currently, the oil palm industry is a major sector in the economy of Malaysia. It has grown rapidly from a minor crop to overtake rubber as the most widely grown crop in the late 1980s. The Malaysian oil palm industry recorded an impressive performance from 2012 to 2022 (Parveez et al., 2023). Export earnings of oil palm products rose to a record £16 billion in that year. The industry also saw exciting developments shaping up in the local biofuel industry with the launching of the Envo Diesel (palm olein blend with diesel) and the first integrated palm oil biodiesel plant using MPOB technology.

In 2006, oil palm became the world's leading oil crop, with a production of 37.1 million tonnes of palm oil and more than 4.3 million tonnes of palm kernel oil as opposed to 35.3 million tonnes of soybean oil (Glycine max (L.) Merr.), which dropped to second place (Maluin et al., 2020). The production of crude palm oil (CPO) increased by 16.3% to 19 million tonnes in 2023 from 15.9 million tonnes in 2006. In Malaysia alone, oil palm planted areas grew from 54,000 ha in 1960 to 5.7 million hectares in 2016 (Nambiappan et al., 2018; Parveez et al., 2023). Nevertheless, Indonesia has taken over as the world's largest producer, supplying approximately 50% of world palm oil volume. Cultivation in Indonesia alone has grown enormously from just under 2000 ha of oil palm production in the 1970s increasing to almost 14 million ha in 2005 (Budidarsono et al., 2013a, b). However, pest and fungal diseases have remained a major threat to oil palm cultivation worldwide.

Fungal diseases of the oil palm can cause very serious losses in production of the CPO and crude palm kernel oil (CPKO). Stem and spear-rotting pathogens caused more than 50% mortality in old stands established after coconuts. Because there are 25 years of productive life for oil palms, losses, especially if early on, come to several hundred thousand dollars per hectare (Ariffin, 2000). The most important diseases are basal stem rot (*Ganoderma boninense*), vascular wilt caused by *Fusarium oxysporum f.sp. elaeidis* (Foe), sudden wilt (*Phytomonas staheli*) and emerging diseases such as Phytoplasma, Cercospora leaf spot and spear rot (unknown pathogen) (Jazuli et al., 2022).

Recently, the threat of bud rot disease caused by *Phytopthora palmivora* also has been identified in South America whereby this disease affecting the quality of the fresh fruit bunch (Torres et al., 2016). While, vascular wilt disease is the main challenge in west and central Africa oil palm cultivation. In Southeast Asia, Malaysia and Indonesia, the most serious and a major threat to the oil palm industry is BSR disease. Some of these diseases can be very devastating and are directly responsible for hampering development of oil palm cultivation in countries where they occur.

2 Basal Stem Rot (BSR) Disease of Oil Palm

As far as the disease problem to oil palm in Southeast Asia is concerned, BSR is the only disease requiring an urgent solution in Malaysia. It has been known to attack oil palm since the early years when the crop was introduced into this country. Losses due to BSR can be seen through the direct reduction in oil palm numbers in the stand and also through the reduction in the number and weight of fruit bunches from both standing diseased palms and those with sub-clinical infections (Turner, 1981).

Basal stem rot (BSR) caused by *Ganoderma boninense* were first described in 1915 in the Republic of Congo, West Africa as a disease of senescing palms (Wakefield, 1920a, 1920b). The disease was first reported in peninsular Malaysia in 1931 infecting oil palms of over 25 years (Thompson, 1931). As such BSR was considered not to be economically important. Originally, Thompson (1931) identified the causal agent of BSR as *Ganoderma lucidum*. Later, Steyaert (1980) identified six *Ganoderma* species isolated from oil palm fields, *G. boninense*, *G. miniatocinctum*, *G. chalceum*, *G. tornatum*, *G. zonatum* and *G. xylonoides*. As with the genus *Ganoderma* in general, considerable uncertainty surrounding identity of the causal agent of BSR has led to many authors simply referring to the disease as '*Ganoderma* rot of palms' (Corley & Tinker, 2015). However, recent studies have indicated that *Ganoderma boninense* is the primary cause of the disease in Southeast Asia and other identified species are synonyms or due to secondary invasion of existing infections (Ariffin et al., 2000; Pilotti & Bridge, 2002a, 2002b; Pilotti, 2005a).

Since the 1960s, oil palm has remained important as a plantation crop in South East Asia. During this time BSR incidence has increased, frequently infecting younger palms (10–15 years), leading to suggestions that monoculture of oil palm has had a role in rising disease incidence (Utomo & Niepold, 2000). Most severe losses from BSR occur in Indonesia and Malaysia, while less severe losses being recorded in the Democratic Republic of Congo, Ghana, Nigeria, Cameroon, Angola, Zimbabwe, Tanzania, Papua New Guinea and the Solomon Islands and are not economically important. The disease has also been reported in Thailand where cultivation of the crop has occurred more recently with 2 out of 2000, 20-year-old oil palms showed typical symptoms (Likhitekaraj & Tummakate, 2000). Increased incidence has been reported in Malaysia, where Singh observed infected palms as young as 12–24 months after planting with increased rates in 4–5 year old palms in coastal alluvial soils (Singh et al., 1991). Furthermore, Lim et al. (1992) reported an average of 50% yield losses from 80% of 13 year old plantings in Malaysian coastal areas. A recent survey by Rao et al. (2003) further demonstrates the increasing problem of *Ganoderma* stem rot of oil palm. They reported typical disease levels of 30% by the time palms were 13 years of age in inland and peat soils of Malaysia. Whereas in the coastal areas incidence of the disease has been known for many years, these inland areas have had typically low or negligible incidence of BSR or USR until recently, and losses reaching levels comparable to those on the coast increases concern for sustainability.

The threat of *G. boninense* to the oil palm industry in this country warrants new and more aggressive approaches in finding solution to the disease. The situation is made even more critical with the active replanting of second-generation oil palms that is currently being carried out in some plantations, including areas with a bad history of *G. boninense*.

2.1 Symptoms

Symptoms of *G. boninense* infection can vary depending on age of the palm and environmental conditions. Foliar symptoms usually only appear after more than 50% of the area of the basal region of the palm has been decayed. The symptoms first manifest as opened fronds (leaves) being smaller and paler than normal. Follow by unopened leaves and lead to multiple unopened fronds, termed 'spears' based on the morphology. As infection advances the palm looks pale in appearance and affected leaves eventually die with leaves drooping and fracturing at the base (see Fig. 1).

Infected young palms will normally die within 6–24 months after first appearance of symptoms, whereas mature palms can take 2–3 years or more to die. However, these symptoms are not diagnostic of BSR as similar symptoms can also be seen during water stress (Turner, 1981). In fact, a palm may be heavily infected with 60% of the stem degraded with no foliar symptoms or basidiocarps. Therefore, there is not always a correlation between foliar symptoms, formation of basidiocarps and degree of infection, making it difficult to diagnose early infections. In Malaysia it was reported that 13–22% of palms showing no foliar symptoms or yield loss in 22-year-old palms were actually infected with BSR on inspection of bole transverse sections (Corley & Tinker, 2015). Hasan and Turner (1998) also noted the prevalence of sub-clinical infections during experiments using bait seedlings where comparisons were made between infection rates of seedlings planted adjacent to BSR and apparently healthy trees. In one instance, 40% of seedlings planted adjacent to an apparently healthy mature palm (two palms distant from the nearest BSR infected neighbour) became infected within 2 years, with later manifestation of symptoms on the standing mature palm indicating that the palm may have had a sub-clinical infection. Thus,

Fig. 1 Infected and healthy mature oil palms. **a** BSR-infected tree with drooping fronds and one-sided chlorosis, typical of advanced infection. **b** healthy canopy of mature uninfected oil palm. Fronds are uniformly green and arrayed to maximize light capture (photo by Author Mohd Aswad)

sub-clinical infections are difficult to identify and may go unnoticed for long periods; the only truly diagnostic feature of the disease remains the presence of basidiocarps at the palm base.

Basidiocarps appearance on the stem surface indicates extensive internal rotting; analysis of cross-sections through rotting areas shows soft discoloured areas of rot where extensive colonization of the tissue and degradation of cell wall polymers has occurred. There is a distinct dark brown margins at the advancing edge of invasion and is termed the ‘reaction zone’ (Turner, 1981). Pearce (1996) describes reaction zones as ‘a static barrier that serves to restrict or slow fungal ingress through a combination of factors including deposition of phytoalexin-like compounds such as phenolics and polyphenolics as well as microenvironmental factors’. The dark colouration of reaction zones may in part be due to fungal stress responses inducing formation of pigmented, thick cell walled hyphae. In saprophytic decay wood, margins of colonization are often delimited by ‘pseudo-sclerotial plates’ and have been shown to be produced by other *Ganoderma* species including *G. australe*, responsible for white-rot of wood tissue in Chile (Agosin et al., 1990), and similar formations have been formed by *G. boninense* on oil palm blocks (Rees et al, 2007). Pseudo-sclerotial plates have been suggested to play a role in compartmentalizing decay tissue, protecting the nutrient source and the fungus itself, from desiccation and competitors in saprophytic wood degradation. Immediately in front of this reaction zone, stem tissue is a pale yellow colour and may be caused by some form of host defence responses. BSR decay patterns are disorganized with a greater occurrence of highly defined pseudo-sclerotial plates. This suggests there may be multiple points of infection and more inter-specific and intra-specific competition in the basal lesion. It is this extensive decay of woody palm tissues that result in foliar symptoms, presumably due to a disrupted vascular system, and ultimately palm death and toppling due to compromised structural integrity.

2.2 Epidemiology of BSR Disease

Little is known of the epidemiology of the pathogen in BSR. The primary route of infection of BSR disease appears to be through root contact with inoculum sources in the soil. Various infection trials using oil palm seedlings and large *Ganoderma*-colonized rubber-wood blocks have provided data supporting this view (Breton et al., 2006; Hasan & Turner, 1998; Lim & Fong, 2005; Lim et al., 1992; Navaratnam , 1964; Sariah et al., 1994). Additionally, Rees et al. (2007) showed that by attaching infested wood blocks to roots much smaller inoculum can be used, allowing infection to occur through unwounded roots and progression and rate of invasion to be followed. Disagreement against root infection as the primary means of BSR comes from genetic studies which show a high degree of diversity of *G. boninense* in oil palm plantations (Miller et al., 2000; Pilotti, 2005; Pilotti et al., 2004, 2021; Wong et al., 2021). However, there is no direct evidence of infection from basidiospores inoculum occur in the field, although monokaryotic mycelium derived from a single basidiospore can colonize sterilized oil palm wood blocks (Hasan & Flood, 2003).

2.3 Root to Root Contact

BSR of oil palm was claimed to occur as a result of contact between healthy roots and diseased tissues left buried in the soil (Turner, 1981). In addition, it was also believed that wounded tissues or dead roots promote infection by *Ganoderma.* The fungus then grows along the infected root and reaches the bole of the palm causing the base of the stem to rot. The role of mycelial contact in disease establishment was primarily proved with several artificial inoculation methods with large inoculum which induced infection of an oil palm seedling roots in a control environment; the fungus was then re-isolated from disease lesions. Navaratnam and Chee (1965) artificially inoculated 22 oil palm seedlings with very large 750 cm3 blocks of naturally infected oil palm tissue and 15 of them were infected with BSR after 9 months. They claimed that a large inoculum source was essential for *G. boninense* infection (Navaratnam & Chee, 1965). Later Lim et al. (1992) used dikaryotic *G. boninense* colonized oat grain inoculated onto a single root and obtained 100% infection of palm roots after 9–10 weeks. He reported that the quality and purity of *G. boninense* inoculum determine the infection potential of the inoculum and not the inoculum size. However, there are no further reports of success using this method. Root Inoculation Technique (RIT) was developed by Ariffin et al., (2000), which involved inoculating primary roots of oil palm seedlings in a polybag which were exposed and inserted into a test tube containing *Ganoderma* inoculum grown in POPW medium (mixture of oil palm wood sawdust, paddy, ammonium sulphate, calcium sulphate, sucrose and peptone). They reported that *G. boninense* progression of root was <2 cm/month, and mean death of various progenies was 26% after 12 months inoculation with the technique. Currently, inoculation method involves growing *G. boninense* on sterilized rubber wood blocks (RWB) (*Hevea brasiliensis*) which are then placed beneath seedlings growing in soil polybags and roots are wrapped over the RWB inoculum (As'wad et al., 2011; Najihah et al., 2015; Khairuddin, 1990; ; Nur Sabrina et al., 2012; Rees et al., 2007; Sariah et al., 1994). It was first introduced by Khairuddin (1990) using a 750 cm3 colonized RWB with dikaryotic *G. boninense* mycelium, which gave successful infection of oil palm in Malaysia (Khairuddin, 1990). Similarly, Sariah et al. (1994) obtained 100% infection of oil palm seedlings with 5 × 5 × 15 cm RWB inoculum. In Indonesia Rees et al. (2007) showed that all oil palm seedlings inoculated with 6 × 6 × 6 cm RWB inoculum were severely infected 12 month after inoculation. However, this artificial inoculation of seedlings requires a large RWB. This has been evident from many studies which have used sizes ranging from 6 × 6 × 6, 6 × 3 × 3, 5 × 5 × 15 and 6 × 6 × 12 cm^3 (Idris et al., 2006; Rees et al., 2007; Sariah et al., 1994). Rees et al. (2007) reported that only 30% of seedlings were infected with a small 6 × 3 × 3 RWB inoculum after 15 months. However, by attaching *G. boninense* RWB inoculum with parafilm closely onto roots a smaller inoculum 3 × 3 × 3 cm^3 managed to induce infection (Cooper et al., 1988; Rees et al., 2009).

Currently RWB technique is used widely for infection by *G. boninense* (Muniroh et al., 2014; Nur Sabrina et al., 2012; Rees et al., 2007; Sariah et al., 2004; Tan et al., 2013) to study various control measures in the nursery (Alizadeh et al., 2011; Najihah et al., 2015). It is unknown whether these methods provide the optimum methods for the key behind finding disease resistance i.e. screening new crosses for BSR resistance. The RWB methods are all slow; inoculum preparation requires much time and space. Furthermore, the enormous size of RWB used might overcome any potential resistance of oil palm. A simple and more rapid infection technique, with a small inoculum size would provide a better method for resistance screening of BSR disease to facilitate oil palm breeding.

BSR in the field has sometimes been linked with multiple, natural infections of different roots by *G. boninense* within a single palm (Ariffin et al., 2000; Flood et al., 2005; Rees et al., 2007). The source of disease inoculum is important in root to root contact for development of BSR disease in the field (Flood et al., 2005; Hassan & Turner 1998). Bait seedlings planted close to BSR infected palm stumps became infected within 6–28 months and eventually died. Stumps left in the ground were more infectious than felled trunks left on the surface and buried shredded palm tissue was more infectious than shredded palm tissue left on the surface (Hassan & Turner, 1998). Untreated stumps left in the ground resulted in 38% infection of seedlings adjacent to stumps from BSR palms compared with 6% infection in seedlings planted adjacent to healthy palm stumps. Higher infection rates of seedlings adjacent to diseased stumps and trunk tissue suggest root contact as a likely route for infection. Also, it is probable that palm roots come into contact with buried colonized oil palm material more readily than with surface material. In addition, Flood et al. (2005) showed young palms would become infected quickly by BSR when exposed to large amount of infected debris close to their planting points in the field. Seedlings showed 60% infection after 52 months when planted 0.5 m from buried trunks, 25% infection at 1 m distance from buried trunks and 0% infection at 1.5 m distance. All seedlings planted adjacent to non-buried trunks remained uninfected within the observation period (Flood et al., 2005).

It is important to note that the survival of *G. boninense* inoculum in soils needs to be addressed. How long inoculum can survive in the soil and in what form the pathogen survives have never been determined before. However, Rees et al. (2009) observed that *G. boninense* forms an extensive, tough, melanized mycelium or pseudo-sclerotium around colonized roots and suggested that this melanized structure could play an important role in survival in soils (Cooper et al., 2011; Rees et al., 2007). Therefore, determination of the role of melanized structures in the survival of *G. boninense* may provide insights on how *G. boninense* inoculum in the field (i.e. colonized debris or roots) from previous planting contributes to new infections.

Shading and consequent soil temperature in the plantation have a dramatic effect on BSR infection. Severe disease occurred after 8 months on inoculated seedlings under shade, but not on seedlings exposed to sun. Soil temperatures in sunlight frequently rose above 40 °C and reached 45 °C, whereas in shade they never exceeded 32 °C (Rees et al., 2007). *G. boninense* mycelia with a growth optimum of 25–28 °C cannot tolerate high temperature for growth or survival and therefore fails to establish infection under exposed conditions. The need for intimate contact between *G. boninense* mycelium and oil palm roots probably relates to the pathogen's inability to grow from a source of inoculum through the soil; it appears to be a very weak competitor against soil microflora (Rees et al., 2007).

Penetration and colonization of oil palm roots by *G. boninense* are essential for the pathogen to establish infection and develop BSR disease. Artificial inoculation methods such as RWB showed *Ganoderma* mycelium could penetrate the roots of oil palm seedlings on contact, colonized and reaching the bole, eventually causing typical symptoms of BSR (Rees et al., 2007; Sariah et al., 1994). Rees et al. (2009) showed that colonization of *G. boninense* can occur through unwounded roots following intimate contact. Histopathological investigations of *Ganoderma*-infected roots showed tissues like xylem, phloem, pith and parenchymal cells were all infected at the advanced stage of the fungal pathogenesis (Ariffin et al., 2000). In contrast, Rees et al. (2009) showed colonization by *G. boninense* progresses mainly through inner, thin-walled cortex and no colonization was observed in the vascular tissue of the roots.

2.4 Direct Infection and Dissemination by Basidiospores

Currently, the role of *Ganoderma* basidiospores in disease initiation and spread of infection remains unclear (Ariffin et al., 2000; Rees et al., 2012; Sanderson et al., 2000). The role of basidiospores was first suggested by Thompson (1931) following BSR incidence in first generation oil palms on cleared virgin jungle areas. Later Sanderson et al. (2000) also reported occurrence of BSR incidence from two oil palm planting areas in Solomon Islands, which were a forest previously. A strong evidence of the involvement of basidiospores was reported by Sanderson et al. (2000) based on high incidence of BSR in first generation oil palms, on land previously cultivated with paddy for 10 years. It was a commercial paddy cultivation area where heavy machinery was used to prepare the land and regular application of chemicals from the air was applied over the area. In this instance, it is difficult to implicate roots coming into contact with *G. boninense* inoculum buried in the soil.

In other important basidiomycete tree pathogen such as *Heterobasidion annosum*, Basidiospores initiate primary infection of *H. annosum* on freshly exposed wood surfaces such as stump tops or stem and root wounds and once established the disease spreads through root to root contact (Baumgartner et al. 2011). *H. annosum* basidiospores play an important role in in disease initiation and spread of the butt rot disease in a coniferous forests of Northern Hemisphere (Garbelotto & Gonthier, 2013). *Ganoderma boninense* basidiospores may play a significant roles in spreading BSR in plantations (Pilotti et al., 2003; Rees et al., 2012; Sanderson, 2005). It was reported that basidiospores were involved in the epidemiology of BSR in the field, based on genetic analysis involving somatic compatibility testing and molecular markers (Miller et al., 2000; Pilotti et al., 2000, 2002a, b, 2003; Rees et al., 2012; Zakaria et al., 2009). These studies showed high genetic diversity of *G. boninense* within and between population of BSR in Malaysia, Indonesia and Papua New Guinea. High genotypic variations may have occurred through sexual recombination following dispersal of basidiospores, which strongly implies involvement of basidiospores in BSR establishment (Miller et al., 2000; Pilotti et al., 2018; Rees et al., 2012; Zakaria et al., 2009).

Abundant *G. boninense* basidiocarps form around the stem of BSR-infected palms (Cooper et al., 2011; Hushiarian et al., 2013; Idris et al., 2006; Rees et al., 2009;), which subsequently will release myriad basidiospores to the atmosphere within affected plantations. Sanderson (2005) recorded a stream of basidiospores released from a basidiocarp using a video camera with a telephoto zoom in an oil palm plantation in PNG. He reported that approximately 2 million basidiospores were released from a 100×50 mm basidiocarp per minute throughout the day from 7 am to 4 pm. He suggested that the astronomical number of basidiospores available in the plantation environment must contribute to BSR disease. Similarly, vast number of *G. boninense* spores 2–11 000/m3 were sampled day and night in Indonesian plantations, which could potentially establish infection to wounds and colonize palm debris in the plantation (Rees et al., 2012). However, Ho and Nawawi (1986) observed that although abundant basidiospores were produced in the field, the majority of oil palms remain healthy and they suggested that basidiospores may require very specific conditions to establish infection (Hushiarian et al., 2013). However, there is currently no literature on the properties of *G. boninense* basidiospores that would influence their spread and survival as a viable propagules to establish BSR or USR in the field.

Wound sites created by plantation management may provide an opportunity for infection by *G. boninense* basidiospores along with physical damage by invertebrate pests such as *Oryctes rhinoceros* and also by rats. In order to harvest oil palm fruit it is necessary to sever the peduncle (fruit stalk) and petiole (frond) to free the fruit bunch. Several harvests are made from one tree in a year, causing multiple large wound sites. A tough abscission layer forms at the base of a cut frond and likely poses a considerable challenge for infection progress. However, as the tree grows in height fruit becomes more difficult to harvest and the probability of wounding the trunk becomes greater. This point of entry by basidiospores was further supported following detection of *G. boninense* DNA in 10–15% (200 total) of palm tissue beneath the cut frond bases of oil palm (Bridge et al., 2000; Panchal & Bridge, 2005). This may facilitate infection by allowing direct penetration of the pathogen into the palm trunk (Rees et al., 2007).

However, direct infection of basidiospores on or into oil palm has so far failed to initiate infection of BSR and USR disease. Inoculation of cut frond bases and oil palm seedlings with *G. boninense* spores failed to establish BSR disease (Idris et al., 2006; Ramasamy, 1972). Similarly, Hasan et al. (2005) reported that basidiospores inoculation onto different oil palm tissues of cut young male inflorescence stalks, cut ripe bunch stalks, freshly cut fronds and fronds damaged by wedges in the field failed to initiate USR infection based on a 3 years observation. They also reported that although fresh basidiospores were used and considerable care using basidiospores was taken, infection did not occur (Hasan et al., 2005). Turner (1981) suggested that the spores may provide inadequate inoculum to establish infection, therefore requiring a suitable substrate to build up potential inoculum, such as cut stumps of trees or palms left to rot in the field, which may become infection foci (Flood et al., 2005). Rees et al. (2012) showed by cryo-SEM that *G. boninense* basidiospores germinated within the xylem vessels of freshly cut fronds. They also showed that spores could travelled down the xylem vessels and stop at vessel end walls with a maximum of c. 10 cm (Rees et al., 2012). Spores contaminating cut frond surfaces are drawn into xylem as a result of negative tension within functional vessels (Cooper, 2011). These vessels would provide protection to the basidiospores from dehydration, microbial competition and solar radiation to germinate and potentially establish BSR following petiole colonization (Rees et al., 2012).

3 Control Strategies of Basal Stem Rot Disease

Despite continuous research in combating the problem, the resolution remains stagnant. The current control strategies for BSR include good cultural practices, chemical methods and biological procedures. These methods are intended to (a) curtail the incidence of BSR after replanting and (b) increase the productive life of the infected oil palm. Surprisingly, the extent to which they are effective remains unknown. Earlier the most traditional control method is to eliminate the pathogen from the field was the clearing method that had been practiced by farmers in oil palm plantations. The clearing method was done by removing and burning the infected or dead plant materials (Noor, 2003). The method has now been suspended due to environmental concerns such as air pollution from burning plant materials. The new environmentally friendly technique has been introduced and implemented in an oil palm plantation, the technique called zero burning technique without employing fire (ASEAN Secretariat, 2003).

3.1 Good Agricultural Practices

This involves elimination measures such as clean clearing and windrowing intending to minimize the spread of the inoculum. Infected palms can be chipped, pulverized and stacked in windrows between planting to promote natural decomposition, however, the procedures are expensive. Windrows may not prevent fungal pathogens survival which may infect healthy palm. Minimizing wounds to oil palm is an obvious method to avoid spread. The open burning technique was used but is banned in Malaysia under the Environmental Quality Act 1974. The Indonesian government has issued numerous laws on land fires, especially after the fires in 1997–98 (Nurhidayah, 2013; Tan, 1999). However, regulatory frameworks suffer from weak enforcement. Also, exemptions allow open burning, although an amendment placed a complete ban on burning on any peat soils. The recent fires and haze in 2019 indicated that the issue remains problematic. In addition, this method also increases the disease incidence by providing the decomposed material as a medium for the reproduction of pests and diseases particularly *rhinoceros* beetles and *Ganoderma* causing Basal stem rot disease (Noor, 2003). BSR disease can cause severe losses to the new oil palm stand unless adequate preventive measures are taken before and during the application of the zero-burning technique (ASEAN Secretariat, 2003).

The disease may be reduced by clearing oil palm before they reach the greatest susceptibility (Rees et al., 2009). As reported, removal of *Ganoderma* inoculum sources during replanting contributes to lower BSR incidence on oil palm in the first 9 years after planting whereas the incidence of BSR was maintained under 5% (Priwiratama et al., 2020). Digging trenches around infected palms (Wakefield, 1920a, 1920b) may have some value but requires scientific testing although disease spread by basidiospore dispersal would render the method useless (Pilotti et al.,

2018). Basidia prevention and removal may be useful as basidiospores are now recognized as the primary source of infection (Pilotti et al., 2018; Sanderson et al., 2000). Infected palms, trunk bases and root rings are required to be removed from the plantations. Removing basidiomata from diseased palms and painting them with carbolineum (fungicidal paste) (Turner, 1981) may be effective, although modern studies are required. Controlling BSR employing surgery of the diseased part of the outer stem has been proposed (Turner, 1981). A protectant chemical (coal tar or Thiram) was used to treat the open surfaces preventing further decay. Paints or fungicide as dressing can be a potential control. The efficacy of this method requires study urgently (Ariffin et al., 2000) and was more successful on younger palms than those older than 12 years. Soil mounding around the base of mature diseased palms can delay the development of BSR (Turner, 1981). Surgery-mounding of Ganoderma-infected palms can prolong the lifetime of the infected palms for 2 to 3 years (Priwiratama et al., 2020). However, these procedures require assessment in a rigorous scientific manner in field conditions to validate the method as a potential BSR management technique.

3.2 Chemical Control

The use of fungicides requires careful consideration and results need reporting from field evaluation. This method, in combination with soil mounding, appears to be effective treatments, and the cost-benefits require reassessment. Pressurized trunk injection using hexaconazole may be useful (Mohammed et al., 2014). Melanized mycelium, basidiospores and pseudosclerotia are resistant to fungicides and may explain why they are not generally effective (Susanto et al., 2005) additionally, the defense mechanisms in plants may be inhibited by the fungicides (Oostendrop et al., 2001). Mineral fertilizers play an important role in overall plant health. Sariah and Zakaria (2000a, 2000b) demonstrated calcium nitrate in combination with the biological control agent (BCA), *Trichoderma* sp., suppressed disease. Rahamah Bivi et al. (2014) demonstrated impressive results in vitro: Calcium chloride + copper-EDTA + salicylic acid (SA) caused a significant decrease in disease symptoms. Continuous supplementation of calcium/copper/SA may be key to enhancing disease resistance in oil palm (Bivi et al., 2016) and higher lignin content was also apparent, indicating how resistance to infection could occur. EDTA is a potent inhibitor of ligninolytic enzymes of *G. boninense* (Siddiqui et al., 2019). This work requires urgent development. Silicon has beneficial effects on growth, yield and disease resistance in various plants (Wang et al., 2017). Najihah and co-workers (2015) elucidated that silicon oxide, potassium silicate, calcium silicate, sodium silicate, and sodium metasilicate reduced the BSR severity in oil palm seedling. The accumulation of silica in host cell walls may have altered the ultrastructure of the roots and shoots hence deterring the pathogen from infection in an encouraging approach.

There is no single proven method for BSR disease management, and even a combination of methods is quite ineffective. Hence, the inhibition of oxidative and hydrolytic enzymes affecting the pattern of wood degradation and induction of resistance can be a potent weapon against *G. boninense* infection. The other possible method is inhibition of lignolytic enzymes by *G. boninense* by using phenolic compounds. The ability of several phenolic compounds has been tested to inhibit the production of lignolytic enzymes (Surendran et al., 2018a, b, c). The study shows that the presence of the phenolic compound in the plant could inhibit the production of lignolytic enzymes by *G. boninense* (Surendran et al., 2018a, b, c). The nursery studies that were done by Surendran et al (2018b) on using benzoic acid and salicylic acid as a selected phenolic compound in oil palm seedling inoculated with *G. boninense* inoculum (Surendran et al., 2018a). This study was done to test and observe the ability of those compounds in controlling *G. boninense*. The result shows that both compounds give an effect on the growth of the pathogen where benzoic acid shows the greatest inhibitory effect (Surendran et al., 2018a). The phenolic compounds increase the productivity of the defense enzyme of polyphenol oxidative which increases the lignin content in oil palm seedlings that leads to suppression of the infection (Surendran et al., 2018a). Moreover, Surendran et al., (2018b, c) emphasized inhibition of wood-degrading enzymes from the fungus to prevent BSR disease. The selected phenolic compounds inhibit the growth as well as the production of wood-degrading enzymes indicating their potential use as chemical control against *G. boninense* for disease management of BSR (Surendran et al., 2017). Therefore, the use of benzoic acid as one of the phenolic compounds to control *G. boninense* causing the degradation of wood in oil palm can be used as one of the potential control strategies. But these are interesting initial investigations and fieldwork is required. Moreover, environmental impact of using phenolic compounds in the field needs consideration as they are often toxic.

3.3 Biological Control

A great deal of research has been undertaken on biocontrol application (BCA) of *G. boninense* but with little proven practical application. BCA often suffer from a lack of effectiveness in the field due to susceptibility to hostile environmental conditions. Microbial populations from the oil palm rhizospheres and on the sporophores may have the potential to control *G. boninense* (Susanto et al., 2005). BCA-induced tolerance to stress through enhance root or plant development, solubilization of inorganic nutrients making them available to palms and inducing resistance in host plants may assist control (Susanto et al., 2005). *Trichoderma koningii* was tested against *G. boninense* (Soepena & Pawirosukarto, 2000), but no data are available on its effectiveness in the field. *T. harzianum* alone or in combination with mycorrhizal preparation, dried palm oil mill effluent and calcium nitrate were tested on oil palm seedlings in a nursery trial and had a significant effect (Sariah & Zakaria, 2000a, 2000b). A field trial indicated that treatment with *T. harzianum* and *T. viride* was superior to *Bacillus* sp in that the disease incidence was lower in a field treated with the agents than untreated fields (Susanto et al., 2005). Similarly, *Trichoderma* induced the production of fungal cell wall-degrading enzymes, such as glucanases and chitinases (Naher et al., 2011), as a possible defense mechanism in the oil palm, and these enzymes may degrade the cell wall of the invading fungus hence controlling the disease as discussed for other plant pathogens in Latgé (2007). Arbuscular Mycorrhizal fungi (AMF) are associated with the roots of oil palm and it has been speculated that they may inhibit *G. boninense* (Sundram et al., 2015). In general, AMF competes with the plant pathogens for the nutrients and space and may induce the plant defense system by inducing siderophores as discussed in other plant-pathogen systems (Brundrett, 2002). Endophytic microorganisms have received attention as methods to control plant diseases as they may assist in inhibiting disease (Kobayashi & Palumbo, 2000). The inhibitory effect of endophytic Pseudomonas aeruginosa from healthy oil palm roots against *G. boninense* was studied in vitro (Bivi et al., 2010). Esyanti et al. (2017) utilized four endophytic fungi; *T. harzianum*, *T. longibrachiatum*, *Lasiodiplodia venezuelensis* and a *Dothidiomycetes* sp. which were able to induce pathogen-related protein in the oil palm which partially explains control observed. Actinomycetes isolated from empty fruit bunches of oil palms showed an antagonistic effect against *G. boninense*. The isolates were identified as *Nocardiopsis* sp., *Streptomyces violaceorubidus* and *Streptomyces* sp., with percentages of inhibition of 91.4%, 86.4% and 69.1%, respectively (Ting et al., 2014). These organisms are useful as they are unlikely to cause disease of oil palm. However, field trials are required in all cases. Biological control studies using basidiomycetes to prevent stump infections in forest trees have been conducted (Roy et al., 2003). No such study has been conducted on oil palm trunks infected with BSR. Twenty-five white-rot hymenomycetes were isolated from healthy oil palm (Naidu et al., 2015) and 8 showed antagonistic effects against *G. boninense*. Pathogenicity tests towards oil palm were conducted which indicated they were not pathogenic (Naidu et al., 2018). Yet, novel conditions under variation in climate may allow them to become pathogenic so great care would be required if used in plantations. Theoretically, biocontrol is an advance compared to other approaches and numerous bio-fungicides are available in the market. The efficacy of the preparations in the field is a major concern and they have not cured BSR and the disease remains highly serious. This seems to be related to the abundance of Ganoderma inoculum in the field, particularly in Ganoderma-endemic areas. Failures of BCA may result from interaction with non-target organisms, variation in the rhizosphere reducing efficacy, lack of ability to colonization in different types of soil, sensitivity to climate, problems competing with oversized populations of other microbes and the genetic diversity of the target pathogen (Meyer & Roberts, 2002). Their utilization as biofertilizers is challenging from (a) poor storage, (b) sensitivity to transport and handling and (c) complex application requirements (Vidhyasekaran et al., 1997). A potential further problem may be that some BCA fungi produce mycotoxins which could enter the environment or final product (i.e. palm oil). Only few BCA have been commercialized due to their unstable behaviour in the field. Available commercial products include Draz-M, a formulation of an arbuscular mycorrhiza that prolongs productivity of 25-year-old infected oil palms and increased their oil yield by 42 and 68% (Sariah & Zakaria, 2000a, 2000b). Similarly, *T. koningii* has been developed for use commercially in Sumatra as a preventative or curative treatment in the field (Soepena & Pawirosukarto, 2000).

4 Conclusions

The sustainability of oil palm cultivation is of immediate concern to farmers, processors and the countries involved in palm oil production due to threatening effects of BSR. *Ganoderma boninense* degrades the lignin component of the palm employing an arsenal of extracellular enzymes to enable the energy-rich cellulose to be utilized. When the cellulose is degraded the palm is vulnerable to collapse. Deeper knowledge is required to understand how *G. boninense* degrades oil palm lignocellulose. However, to commence the infection it is the more amenable carbohydrates that will allow initial infection and inhibition of the enzymes involved requires investigation. The crucial role of basidiospores in infection is accepted and should not be overlooked. The current control methods proved to be inefficient if not ineffective. There are too many reports limited to in vitro or nursery trials which may have dubious scientific merit. There are only a few investigations where control methods have been tested on a large scale in plantations and proven to control the disease beyond any doubt: These are likely to be the most suitable methods for general application. One might expect a preparation would be available from the promising findings, but this is not the case. We have entered a new paradigm where all methods require to be re-evaluated. Finally, there requires much greater focus in developing the leads (lab-scale research) into actual products that are proven to work consistently.

Acknowledgements The authors would like to express gratitude to the Fundamental research grant scheme (FRGS): FRGS/1/2019/WAB01/UPM/02/28 by the Ministry of Higher Education (MoHE), Malaysia, for funding for this project.

References

Agosin, E., Blanchette, R. A., Silva, H., Lapierre, C., Cease, K. R., Ibach, R. E., Abad, A. R., & Muga, P. (1990). Characterization of palo podrido, a natural process of delignification in wood. *Applied and Environmental Microbiology, 56*(1), 65–74.

Alizadeh, F., Abdullah, S. N. A., Khodavandi, A., Abdullah, F., Yusuf, U. K., & Chong, P. P. (2011). Differential expression of oil palm pathology genes during interactions with Ganoderma boninense and Trichoderma harzianum. *Journal of Plant Physiology, 168*(10), 1106–1113.

Ariffin. (2000). Major diseases of oil palm. In B. Yusof, B. S. Jalani, & K. W. Chan (Eds.), *Advances in Oil Palm Research* (Vol. 1, pp. 596–622). Malaysian Palm Oil Board.

Ariffin, D., Idris, A., & Singh, G. (2000). Status of Ganoderma in oil palm. In J. Flood, P. Bridge, & M. Holderness (Eds.), *Ganoderma diseases of perennial crops* (pp. 49–68).

ASEAN Secretariat. (2003). *Implementation of the ASEAN policy on zero burning guidelines for the implementation of the ASEAN policy on zero burning.*

As'wad, A. M., Sariah, M., Paterson, R. R. M., Abidin, M. Z., & Lima, N. (2011). Ergosterol analyses of oil palm seedlings and plants infected with Ganoderma. *Crop Protection, 30*(11), 1438–1442.

Baumgartner, K., Coetzee, M. P., & Hoffmeister, D. (2011). Secrets of the subterranean pathosystem of Armillaria. *Molecular Plant Pathology, 12*(6), 515–534.

Breton, F., Hasan, Y., Lubis, Z., & De Franqueville, H. (2006). Characterization of parameters for the development of an early screening test for basal stem rot tolerance in oil palm progenies.

Bridge, P. D., O'grady, E. B., Pilotti, C. A., & Sanderson, F. R. (2000). Development of molecular diagnostics for the detection of Ganoderma isolates pathogenic to oil palm. *Ganoderma diseases of perennial crops*, 225–234.

Bivi, M. R., Farhana, M. S. N., Khairulmazmi, A., & Idris, A. (2010). Control of Ganoderma boninense: A causal agent of basal stem rot disease in oil palm with endophyte bacteria in vitro. *International Journal of Agriculture and Biology, 12*(6), 833–839.

Bivi, M. S. H. R., Paiko, A. S., Khairulmazmi, A., Akhtar, M. S., & Idris, A. S. (2016). Control of basal stem rot disease in oil palm by supplementation of calcium, copper, and salicylic acid. *The Plant Pathology Journal, 32*(5), 396.

Brundrett, M. C. (2002). Coevolution of roots and mycorrhizas of land plants. *New Phytologist, 154*(2), 275–304.

Budidarsono, S., Susanti, A., & Zoomers, A. (2013a). Oil palm plantations in Indonesia: The implications for migration, settlement/resettlement and local economic development. *InTech. Doi, 10*(5772/53586), 622.

Budidarsono, S., Susanti, A., & Zoomers, A. (2013b). Oil palm plantations in Indonesia: The implications for migration, settlement/resettlement and local economic development. *Biofuels-Economy, Environment and Sustainability, 1*, 173–193.

Cooper, R. M. (2011). Fusarium wilt of oil palm: a continuing threat to South East Asian plantations. *Planter, 87*(1023), 409–418.

Cooper, R. M., Longman, D., Campbell, A., Henry, M., & Lees, P. (1988). Enzymic adaptation of cereal pathogens to the monocotyledonous primary wall. *Physiological and Molecular Plant Pathology, 32*(1), 33–47.

Cooper, R. M., Flood, J., & Rees, R. W. (2011). *Ganoderma boninense* in oil palm plantations: Current thinking on epidemiology, resistance and pathology. *The Planter, 87*(February), 515–526.

Corley, R. V. H., & Tinker, P. B. (2015). *The oil palm* (5th ed.). Blackwell Science.

Esyanti, R. R., Susanto, A., & Aryantha, I. N. P. (2017). The chitinase activity of oil palm ('Elaeis guineensis' Jacq.) roots against fungal endophytes and pathogenic 'Ganoderma boninense'. *Plant Omics, 10*(5), 247–251.

FAOSTAT. (2023). *Palm oil: Production, trade, and producer's price data.* http://www.fao.org/fao stat/ en/#data

Flood, J., Keenan, L., Wayne, S., & Hasan, Y. (2005). Studies on oil palm trunks as sources of infection in the field. *Mycopathologia, 159*(1), 101–107.

Garbelotto, M., & Gonthier, P. (2013). Biology, epidemiology, and control of Heterobasidion species worldwide. *Annual Review of Phytopathology, 51*, 39–59.

Hasan, Y., & Flood, J. (2003). Colonisation of rubber wood and oil palm blocks by monokaryons and dikaryons of *Ganoderma boninense* - implications to infection in the field. *Planter, 79*(922), 31–38.

Hasan, Y., & Turner, P. (1998). The comparative importance of different oil palm tissues as infection sources for basal stem rot in replantings. *Planter, 74*(864), 119–135.

Hasan, Y., Foster, H. L., & Flood, J. (2005). Investigations on the causes of upper stem rot (USR) on standing mature oil palms. *Mycopathologia, 159*(1), 109–112.

Ho, Y. W., & Nawawi, A. (1986). Isolation, growth and sporophore development of *Ganoderma boninense* from oil palm in malaysia. *Pertanika, 9*(1), 69–73.

Hushiarian, R., Yusof, N. A., & Dutse, S. W. (2013). Detection and control of *Ganoderma boninense*: Strategies and perspectives. *Springerplus, 2*, 1–12.

Idris, A. S., Kushairi, D., Ariffin, D., & Basri, M. (2006). Technique for inoculation of oil palm germinated seeds with Ganoderma. *MPOB Info Series, 314*, 1–4.

Jazuli, N. A., Kamu, A., Chong, K. P., Gabda, D., Hassan, A., Abu Seman, I., & Ho, C. M. (2022). A review of factors affecting Ganoderma basal stem rot disease progress in oil palm. *Plants, 11*(19), 2462.

Khairudin, H. (1990). Basal Stem Rot of Oil Palm: Incidence, Etiology and Control. Master of Agriculture Science Thesis, Universiti Pertanian Malaysia, Selangor, Malaysia, Selangor, Malaysia.

Kobayashi, D., & Palumbo, J. (2000). *Bacterial endophytes and their effects on plants and uses in agriculture Microbial endophytes* (pp. 199–233).

Latgé, J. P. (2007). The cell wall: A carbohydrate armour for the fungal cell. *Molecular Microbiology, 66*(2), 279–290.

Likhitekaraj, S., & Tummakate, A. (2000). Basal stem rot of oil palm in Thailand caused by *Ganoderma*. In J. Flood, P. D. Bridge, & M. Holderness (Eds.), *Ganoderma Diseases of perennial crops* (pp. 69–71). CABI.

Lim, H. P., & Fong, Y. K. (2005). Research on basal stem rot (BSR) of ornamental palms caused by basidiospores from *Ganoderma boninense. Mycopathologia, 159*(1), 171–179.

Lim, T. K., Chung, G. F., & Ko, W. H. (1992). Basal stem rot of oil palm caused by *Ganoderma boninense. Plant Pathology Bulletin, 1*(3), 147–152.

Maluin, F. N., Hussein, M. Z., & Idris, A. S. (2020). An overview of the oil palm industry: Challenges and some emerging opportunities for nanotechnology development. *Agronomy, 10*(3), 356.

Meyer, S. L., & Roberts, D. P. (2002). Combinations of biocontrol agents for management of plant-parasitic nematodes and soilborne plant-pathogenic fungi. *Journal of Nematology, 34*(1), 1.

Miller, R. N. G., Holderness, M., & Bridge, P. D. (2000). Molecular and morphological characterization of Ganoderma in oil-palm plantings. *Ganoderma diseases of perennial crops* (pp. 159–182). CABI.

Mohammed, C. L., Rimbawanto, A., & Page, D. E. (2014). Management of basidiomycete root- and stem-rot diseases in oil palm, rubber and tropical hardwood plantation crops. *Forest Pathology*. https://doi.org/10.1111/efp.12140

Muniroh, M. S., Sariah, M., Abidin, M. Z., Lima, N., & Paterson, R. R. M. (2014). Rapid detection of Ganoderma-infected oil palms by microwave ergosterol extraction with HPLC and TLC. *Journal of Microbiological Methods, 100*, 143–147.

Murphy, D. J., Goggin, K., & Paterson, R. R. M. (2021). Oil palm in the 2020s and beyond: Challenges and solutions. *CABI Agric Biosci, 2*, 39. https://doi.org/10.1186/s43170-021-00058-3

Naher, L., Ho, C. L., Tan, S. G., Yusuf, U. K., & Abdullah, F. (2011). Cloning of transcripts encoding chitinases from Elaeis guineensis Jacq. and their expression profiles in response to fungal infections. *Physiological and Molecular Plant Pathology, 76*(2), 96–103.

Naidu, Y., Idris, A. S., Nusaibah, S. A., Norman, K., & Siddiqui, Y. (2015). In vitro screening of biocontrol and biodegradation potential of selected hymenomycetes against Ganoderma boninense and infected oil palm waste. *Forest Pathology, 45*(6), 474–483. https://doi.org/10.1111/efp.12197

Naidu, Y., Siddiqui, Y., Rafii, M. Y., Saud, H. M., & Idris, A. S. (2018). Inoculation of oil palm seedlings in Malaysia with white-rot hymenomycetes: Assessment of pathogenicity and vegetative growth. *Crop Protection*. https://doi.org/10.1016/j.cropro.2018.02.018

Najihah, N. I., Hanafi, M. M., Idris, A. S., & Hakim, M. A. (2015). Silicon treatment in oil palms confers resistance to basal stem rot disease caused by Ganoderma boninense. *Crop Protection, 67*, 151–159.

Nambiappan, B., Ismail, A., Hashim, N., Ismail, N., Shahari, D. N., Idris, N. A. N., Omar, N., Salleh, K.M., Hassan, N. A. M., & Kushairi, A., (2018). Malaysia: 100 years of resilient palm oil economic performance. *Journal of Oil Palm Research, 30*(1), 13-25.

Navaratnam, S., & Chee, K. (1965). Root inoculation of oil palm seedlings with *Ganoderma sp. Plant Disease, 49*, 1011–1012.
Noor, M. M. (2003). Zero burning techniques in oil palm cultivation: An economic perspective. *Oil Palm Industry Economic Journal, 3*(1), 16–24.
Nurhidayah, L. (2013). Legislation, regulations, and policies in Indonesia relevant to addressing land/forest fires and transboundary haze pollution: A critical evaluation. *Asia Pacific Journal of Environmental Law, 16*, 215.
Nur Sabrina, A. A., Sariah, M., & Zaharah, A. R. (2012). Suppression of basal stem rot disease progress in oil palm (*Elaeis quineensis*) after copper and calcium supplementation. *Pertanika Journal of Tropical Agriculture Science, 35*, 13–24.
Oostendorp, M., Kunz, W., Dietrich, B., & Staub, T. (2001). Induced disease resistance in plants by chemicals. *European Journal of Plant Pathology, 107*(1), 19–28.
Panchal, G., & Bridge, P. D. (2005). Following basal stem rot in young oil palm plantings. *Mycopathologia, 159*, 123–127.
Parveez, G. K. A., Omar, A. R., Ahmad, M. N., Mat Taib, H., Mohd-Bakri, M. A., Sitti-Rahma, A. H., Tuan-Ismail, T. N. M., Loh, S. K., Ong-Abdullah, M., Zakaria, K., & Zainab, I. (2023). Oil palm economic performance in Malaysia and R&D progress in 2022. *Journal of Oil Palm Research, 35*, 193–216.
Pearce, R. B. (1996). Antimicrobial defences in the wood of living trees. *New Phytologist, 132*(2), 203–233.
Pilotti, C. A., Sanderson, F. R., Aitken, E. A. B., & Bridge, P. D. (2000). Genetic variation in Ganoderma spp. from Papua New Guinea as revealed by molecular (PCR) methods (p. 195). In J. Flood, P. D. Bridge, & M. Holderness (Eds.), *Ganoderma diseases of perennial crops*. CABI Publishing.
Pilotti, C. A., Sanderson, F. R., Aitken, E. A. B., & Armstrong, W. (2004). Morphological variation and host range of two Ganoderma speices from Papua New Guinea. *Mycopathologia, 158*, 251–265.
Pilotti, C. A. (2005). Stem rots of oil palm caused by *Ganoderma boninense*: Pathogen biology and epidemiology. *Mycopathologia, 159*(1), 129–137.
Pilotti, C., & Bridge, P. D. (2002a). Sexuality and interactions of monokaryotic and dikaryotic mycelia of *Gånoderma boninense. Mycological Research, 106*(11), 1315–1322.
Pilotti, C. A., & Bridge, P. D. (2002b). Basal stem rot: Probing the facts. *The Planter, 78*, 365–370.
Pilotti, C. A., Gorea, E. A., & Bonneau, L. (2018). Basidiospores as sources of inoculum in the spread of Ganoderma boninense in oil palm plantations in Papua New Guinea. *Plant Pathology, 67*(9), 1841–1849.
Pilotti, C. A., Killah, G., Rama, D., Gorea, E. A., & Mudge, A. M. (2021). A preliminary study to identify and distinguish southern tropical populations of *Ganoderma boninense* from oil palm via mating assays, sequence data, and microsatellite markers. *Mycologia, 113*(3), 574–585.
Pilotti, C. A., Sanderson, F. R., & Aitken, E. A. B. (2002). Sexuality and interactions of monokaryotic and dikaryotic mycelia of *Ganoderma boninense. Myological Research, 11*, 1315–1322.
Pilotti, C. A., Sanderson, F. R., & Aitken, E. A. B. (2003). Genetic structure of population of Ganoderma boninense on oil palm. *Plant Pathology, 52*, 455–463.
Priwiratama, H., Prasetyo, A. E., & Susanto, A. (2020, March). Incidence of basal stem rot disease of oil palm in converted planting areas and control treatments. In *IOP Conference Series: Earth and Environmental Science* (Vol. 468, No. 1, p. 012036). IOP Publishing.
Rahamah Bivi, M., Siti Noor Farhana, M., Khairulmazmi, A., Idris, A., Susilawati, K., & Sariah, M. (2014). Assessment of plant secondary metabolites in oil palm seedlings after being treated with calcium, copper ions and salicylic acid. *Archives of Phytopathology and Plant Protection, 47*, 1120–1135.
Ramasamy, S. (1972) Cross-infectivity and decay ability of *Ganoderma* species parasitic to rubber, oil palm and tea. In *Bachelor of Agriculture Science Project Report*. Universiti Malaya, Malaysia.
Rao, V., Lim, C. C., Chia, C. C., & Teo, K. W. (2003). Studies on *Ganoderma* spread and control. *Planter, 79*(927), 367–383.

Rees, R. W., Flood, J., Hasan, Y., & Cooper, R. M. (2007). Effects of inoculum potential, shading and soil temperature on root infection of oil palm seedlings by the basal stem rot pathogen *Ganoderma boninense. Plant Pathology, 56*(5), 862–870.

Rees, R. W., Flood, J., Hasan, Y., Potter, U., & Cooper, R. M. (2009). Basal stem rot of oil palm (Elaeis guineensis); Mode of root infection and lower stem invasion by Ganoderma boninense. *Plant Pathology, 58*(5), 982–989.

Rees, R. W., Flood, J., Hasan, Y., Wills, M. A., & Cooper, R. M. (2012). Ganoderma boninense basidiospores in oil palm plantations: Evaluation of their possible role in stem rots of *Elaeis guineensis. Plant Pathology, 61*(3), 567–578.

Roy, G., Laflamme, G., Bussières, G., & Dessureault, M. (2003). Field tests on biological control of Heterobasidion annosum by Phaeotheca dimorphospora in comparison with Phlebiopsis gigantea. *Forest Pathology, 33*(2), 127–140.

Sanderson, F., Pilotti, C., & Bridge, P. (2000). Basidiospores: their influence on our thinking regarding a control strategy for basal stem rot of oil palm. In J. Flood, P. D. Bridge, M. Holderness (Eds.), *Ganoderma diseases of perennial crops* (pp. 113–119).

Sanderson, F. R. (2005). An insight into spore dispersal of Ganoderma boninense on oil palm. *Mycopathologia, 159*(1), 139–141.

Sariah, M., & Zakaria, H. (2000a). The use of soil amendments for the control of basal stem rot of oil-palm seedlings (pp. 89–99). In J. Flood, P. D. Bridge, M. Holderness (Eds.), Ganoderma diseases of perennial crops (pp. 101–112). CAB International Wallingford. https://doi.org/10.1079/9780851993881.0101

Sariah, M., & Zakaria, H. (2000b). The use of soil amendments for the control of basal stem rot of oil-palm seedlings. *Ganoderma diseases of perennial crops* (pp. 89–99). CABI.

Sariah, M., Hussin, M. Z., Miller, R. N. G., & Holderness, M. (1994). Pathogenicity of *Ganoderma boninense* tested by inoculation of oil palm seedlings. *Plant Pathology, 43*(3), 507–510.

Siddiqui, Y., Surendran, A., Masya, E., Agricultural, F., & Sdn, S. (2019). *Inhibition of lignin degrading enzymes of Ganoderma spp .: An alternative control of basal stem rot disease of oil palm* (pp. 523–530).

Singh, G., Darus, A., & Sukaimi, J. (1991). *Ganoderma*-the scourge of oil palm in the coastal area. In *Proceedings of Ganoderma workshop*, Bangi, Selangor, Malaysia, 11 Sept 1990 (pp. 7–35).

Soepena, H., & Pawirosukarto, S. (2000). Control strategy for basal stem rot (Ganoderma) on oil palm. In J. Flood, P.D. Bridge, & M. Holderness (Eds.), *Ganoderma diseases of perennial crops* (pp. 83–88). CABI Publishing.

Soh, C. S., Wong, C. K., Ho, Y. W., & Choong, C. W. (2009). Oil palm. In J. Vollmann, & I. Rajcan (Eds.), *Oil crop, handbook of plant breeding* (Vol. 4, pp. 333–367).

Steyaert, R. L. (1980). Study of some Ganoderma species. In *Bulletin du Jardin Botanique National De Belgique/Bulletin Van De Nationale Plantentuin Van België* (pp. 135–186).

Sundram, S., Meon, S., Seman, I. A., & Othman, R. (2015). Application of arbuscular mycorrhizal fungi with Pseudomonas aeruginosa UPMP3 reduces the development of Ganoderma basal stem rot disease in oil palm seedlings. *Mycorrhiza, 25*, 387–397.

Surendran, A., Siddiqui, Y., Saud, H. M., Ali, N. S., & Manickam, S. (2017). The antagonistic effect of phenolic compounds on ligninolytic and cellulolytic enzymes of Ganoderma boninense, causing basal stem rot in oil palm. *International Journal of Agriculture and Biology, 19*(6), 1437–1446. https://doi.org/10.17957/IJAB/15.0439

Surendran, A., Siddiqui, Y., Ali, N. S., & Manickam, S. (2018a). Inhibition and kinetic studies of cellulose- and hemicellulose-degrading enzymes of Ganoderma boninense by naturally occurring phenolic compounds. *Journal of Applied Microbiology, 124*(6), 1544–1555. https://doi.org/10.1111/jam.13717

Surendran, A., Siddiqui, Y., Saud, H. M., Ali, N. S., & Manickam, S. (2018b). Inhibition and kinetic studies of lignin degrading enzymes of Ganoderma boninense by naturally occurring phenolic compounds. *Journal of Applied Microbiology, 125*(3), 876–887. https://doi.org/10.1111/jam.13922

Surendran, A., Siddiqui, Y., Manickam, S., & Ali, A. (2018c). Role of benzoic and salicylic acids in the immunization of oil palm seedlings-challenged by Ganoderma boninense. *Industrial Crops and Products, 122*(March), 358–365.

Susanto, A., Sudharto, P. S., & Purba, R. Y. (2005). Enhancing biological control of basal stem rot disease (Ganoderma boninense) in oil palm plantations. In *Mycopathologia.*

Tan, A.K.-J. (1999). Forest fires of Indonesia: State responsibility and international liability. *The International and Comparative Law Quarterly, 48*(4), 826–855.

Tan, Y. C., Yeoh, K. A., Wong, M. Y., & Ho, C. L. (2013). Expression profiles of putative defence-related proteins in oil palm (*Elaeis guineensis*) colonized by *Ganoderma boninense. Journal of Plant Physiology, 170*(16), 1455–1460.

Thompson, A. (1931). *Stem-rot of the oil palm in Malaya.* Bulletin Department of Agriculture, Science Series, 6.

Ting, A. S. Y., Hermanto, A., & Peh, K. L. (2014). Indigenous actinomycetes from empty fruit bunch compost of oil palm: Evaluation on enzymatic and antagonistic properties. *Biocatalysis and Agricultural Biotechnology, 3*(4), 310–315.

Torres, G., Sarria, G., Martínez, G., Varón, F., Drenth, A., & Guest, D. (2016). Bud rot caused by Phytophthora palmivora: A destructive emerging disease of oil palm. *Phytopathology, 106*(4), 320–329.

Turner, P. (1981). *Oil palm diseases and disorders.* Oxford University Press.

Utomo, C., & Niepold, F. (2000). Development of diagnostic methods for detecting *Ganoderma*-infected oil palms. *Journal of Phytopathology, 148*(9–10), 507–514.

Vidhyasekaran, P., Ponmalar, T. R., Samiyappan, R., Velazhahan, R., Vimala, R., Ramanathan, A., Paranidharan, V., & Muthukrishnan, S. (1997). Host-specific toxin production by Rhizoctonia solani, the rice sheath blight pathogen. *Phytopathology, 87*(12), 1258–1263.

Wakefield, E. M. (1920a). Diseases of the oil palm in West Africa. In: J. Flood, P. D. Bridge, & M. Holderness (Eds.), *Ganoderma diseases of perennial crops.* CABI Publisher.

Wakefield, E. M. (1920b). Diseases of the oil palm in West Africa. *Bulletin of Miscellaneous Information (royal Botanic Gardens, Kew), 1920*(9), 306–308.

Wang, M., Gao, L., Dong, S., Sun, Y., Shen, Q., & Guo, S. (2017). Role of silicon on plant–pathogen interactions. *Frontiers in Plant Science, 8*, 701.

Wong, W. C., Tung, H. J., Fadhilah, M. N., Midot, F., Lau, S. Y. L., Melling, L., Goh, Y. K. (2021). Genetic diversity and gene flow amongst admixed populations of *Ganoderma boninense*, causal agent of basal stem rot in African oil palm (Elaeis guineensis Jacq.) in Sarawak (Malaysia), Peninsular Malaysia, and Sumatra (Indonesia). Mycologia, 113(5), 902–917.

Zakaria, L., Ali, N. S., Salleh, B., & Zakaria, M. (2009). Molecular analysis of Ganoderma species from different hosts in Peninsula Malaysia. *Journal of Biological Sciences, 9*(1), 12–20.

Revolution of the Next-Generation Sequencing and Its Application in Phytobacterial Diseases: Unraveling the Culprits

Adam Zafdri Md. Zali, Dzarifah Zulperi, Siti Izera Ismail, Norsazilawati Saad, Muhammad Asyraf Md. Hatta, Mansor Hakiman, Mohd Termizi Yusof, Syari Jamian, and Sumaiyah Abdullah

Abstract The genome-sequencing revolution advances exponentially with a rich and diverse history and application. Next-generation sequencing (NGS) has vastly improved sequencing output while being significantly faster and less expensive than Sanger sequencing. Although next-generation sequencing has been around for

A. Z. Md. Zali · D. Zulperi (✉) · S. I. Ismail · N. Saad · S. Jamian · S. Abdullah
Department of Plant Protection, Faculty of Agriculture, Universiti Putra Malaysia, 43400 Serdang, Selangor, Malaysia
e-mail: dzarifah@upm.edu.my

S. I. Ismail
e-mail: izera@upm.edu.my

N. Saad
e-mail: norsazilawati@upm.edu.my

S. Jamian
e-mail: syari@upm.edu.my

S. Abdullah
e-mail: sumaiyah@upm.edu.my

D. Zulperi
Laboratory of Sustainable Resources Management, Institute of Tropical Forestry and Forest Products, Universiti Putra Malaysia, 43400 Serdang, Selangor, Malaysia

M. A. Md. Hatta
Department of Agricultural Technology, Faculty of Agriculture, Universiti Putra Malaysia, 43400 Serdang, Selangor, Malaysia
e-mail: m.asyraf@upm.edu.my

M. Hakiman
Department of Crop Science, Faculty of Agriculture, Universiti Putra Malaysia, 43400 Serdang, Selangor, Malaysia
e-mail: mhakiman@upm.edu.my

M. T. Yusof
Department of Microbiology, Faculty of Biotechnology and Biomolecular Sciences, Universiti Putra Malaysia, 43400 Serdang, Selangor, Malaysia

M. Wong (ed.), *Advances in Tropical Crop Protection*,
https://doi.org/10.1007/978-3-031-59268-3_11

nearly two decades, it is still relevant to describe highly parallel or high-throughput sequencing methods that produce data at the genome-scale. NGS has enabled researchers to expedite their studies in life sciences in general and comprehend phytobacterial disease from a more extensive and deeper perspective. This chapter explores the evolution and developments of NGS from historical and current perspectives, contemporary concepts, technologies, and their application in phytobacterial diseases.

Keywords Next-generation sequencing technologies · Phytobacterial diseases · Sanger sequencing · Second-generation sequencing · Third-generation sequencing

1 Introduction

Deoxyribonucleic acid, known as DNA, comprises two polynucleotide chains that wrap around each other to create a double helix bond. Within the DNA structure lies the code required to assemble protein and noncoding RNA (Thompson & Milos, 2011). Understanding the sequence of DNA has allowed researchers to explore the structure and function of proteins and ribonucleic acid (RNA) and gain insight into the underlying causes of illness (Kumar et al., 2019). The first molecule to be sequenced was tRNA by Robert Holley in 1965 of *Saccharomyces cerevisiae*, followed by Walter Fiers' first complete protein-coding gene sequence of coat protein from bacteriophage MS2 1972 (McGinnis et al., 2016). The development of Sanger sequencing allowed Human Genome Project to begin in 1990, and the first complete draft was reported in 2003 (Levy & Myers, 2016). Numerous next-generation sequencing technologies have emerged, and the associated subject of bioinformatics has grown to become a significant scientific discipline. As high-throughput next-generation sequencing platforms now support genome-wide sequencing applications, genomics research has shifted toward DNA sequencing as the critical molecular tool (Mardis, 2011).

Rapid technological progress and the development of bioinformatics tools allowed the first commercially viable next-generation sequencing technology. This technology was invented in 2005 by Nick McCooke at Solexa. Next-generation sequencing (NGS) is a sophisticated technology that allows for deep, high-throughput, in-parallel DNA sequencing of hundreds to millions of DNA molecules in contrast to Sanger sequencing, which only sequences a single DNA fragment at one time (Behjati & Tarpey, 2013). Regarding time and cost, NGS is superior to Sanger sequencing and considerably defies Moore's law (Kumar et al., 2019). The miniaturization of sequencing processes and the development of microfluidics and enhanced detection systems enabled NGS to parallelize many sequencing processes (Ari & Arikan, 2016). Building amplified sequencing libraries before sequencing characterizes second-generation sequencing technologies. In contrast, third-generation sequencing applies single DNA molecules rather than amplified

e-mail: mohdtermizi@upm.edu.my

DNA, which removes flaws in the DNA sequence produced during the amplification process (Attia & Saeed, 2015; Mardis, 2011).

Phytobacterial diseases are those that affect plants and are caused by bacterial pathogens. These diseases can significantly impact agriculture, leading to yield losses and reduced crop quality (Mostafa et al., 2015). Some of the common bacterial pathogens causing phytobacterial diseases include *Xanthomonas*, *Pseudomonas*, *Erwinia*, and *Burkholderia*. One example of a bacterial pathogen and host involved in phytobacterial disease is *Burkholderia glumae*, which causes bacterial panicle blight in rice. This disease can lead to significant yield losses, particularly in areas with high rainfall and humidity. Next-generation sequencing (NGS) has become an essential tool in the study of phytobacterial diseases. NGS allows insights into the genetic makeup of bacterial pathogens and their interactions with plant hosts (Mostafa et al., 2015). By analyzing the genetic information of bacterial pathogens, scientists can identify virulence factors, antibiotic resistance genes, and other factors that contribute to pathogenicity (Buermans & den Dunnen, 2014). NGS has also been used to study the diversity of bacterial populations in phytobacterial diseases. For example, researchers have used NGS to analyze the microbiome of plants, identifying the presence of both pathogenic and beneficial bacteria (Franca et al., 2002). This information can be used to develop strategies to control phytobacterial diseases by promoting the growth of beneficial bacteria. The application of NGS in studying phytobacterial diseases has dramatically expanded our understanding of these diseases and provided valuable insights into their biology and control (Buermans & den Dunnen, 2014; Franca et al., 2002).

1.1 First-Generation DNA Sequencing

Between 1967 and 1977, the genomic field saw the inception of DNA sequencing through the pioneering efforts of Sanger and Maxam–Gilbert techniques, marking the emergence of first-generation sequencing (Kumar et al., 2019). Sanger and Coulson introduced the chain-termination method in 1975, while Maxam and Gilbert developed the chemical chain degradation approach from 1976 to 1977 (Kanzi et al., 2020).

1.2 Maxam–Gilbert Sequencing

Maxam–Gilbert sequencing relies on the stepwise cleavage of nucleotides and is most effective with short nucleotide sequences (Dorado et al., 2017). A significant advantage of Maxam–Gilbert sequencing over Sanger sequencing is its ability to directly purify DNA, while Sanger sequencing requires DNA cloning before each run (El-Bondkly, 2014).

The Maxam–Gilbert technique needs radioactive labeling at one end and the purification of the DNA segment to be sequenced. This is done by denaturing the DNA chain into single-stranded protein and labeled on the 5' end using phosphorus-32 (radioactive isotope) (Dorado et al., 2017). The chemical treatment produces breakdowns in small quantities of one or two of the four nucleotide-based reactions (G, A + G, C, C + T) (Franca et al., 2002). As a result, a sequence of tagged fragments is produced, beginning with the radio-labeled end and arranged side by side according to their size in gel electrophoresis. Finally, the fragment was visualized by exposure to X-ray film for autoradiography. This resulted in a series of dark bands, each corresponding to a radio-labeled DNA fragment from which the sequence may be deduced. However, there are some downsides to Maxam–Gilbert sequencing. For starters, it is time-consuming as this approach can only sequence 200–300 bases of DNA every few days (El-Bondkly, 2014). This method is also considered dangerous because of the use of large amounts of radioactive material such as hydrazine, a known neurotoxin, and radioactive chemical.

1.3 Sanger Sequencing

Sanger sequencing is based on the chain-termination method, sequencing by synthesis method, or dideoxynucleotide method. It entails utilizing one strand of double-stranded DNA as a template for sequencing. The principle of Sanger sequencing is similar to the polymerase chain reaction (PCR) process in which DNA is denatured and annealed to an oligonucleotide primer. The difference in Sanger sequencing with conventional PCR is the addition of modified nucleotides called dideoxynucleotides triphosphates (ddNTPs) in DNA polymerase (Heather & Chain, 2016). These ddNTPs have ddATP, ddTTP, ddGTP, and ddCTP for each DNA base with a unique fluorescent label. DNA polymerase added dNTPs to the expanding DNA strand during the extension step of standard PCR by catalyzing the creation of a phosphodiester bond between the free 3'-OH group of the last nucleotide and the 5'-phosphate of the next nucleotide. However, ddNTPs lack the 3'-OH group necessary for phosphodiester bond synthesis. The extension is terminated when DNA polymerase incorporates ddNTPs at random. Thus, chain-termination PCR produces millions to billions of oligonucleotide copies of the target DNA sequence, each terminated at an arbitrary length of 5'-ddNTPs. All chain-terminated oligonucleotides within the sequencing machine are sorted by size using single capillary gel electrophoresis and autoradiography. A computer later scanned each position of the electrophoretic band of the capillary gel, in turn utilizing ddNTPs' unique fluorescence to determine the identity of each terminal ddNTP (see Fig. 1).

In summary, a laser stimulates the fluorescent tags in each band, and a computer detects the light produced as a consequence. Since each ddNTP is labeled with a unique fluorescent label, the light emitted is directly linked to identifying the terminal ddNTPs. The result is called a chromatogram, which displays each nucleotide's fluorescence peak and the template DNA's length (Mardis, 2011). The first genomes

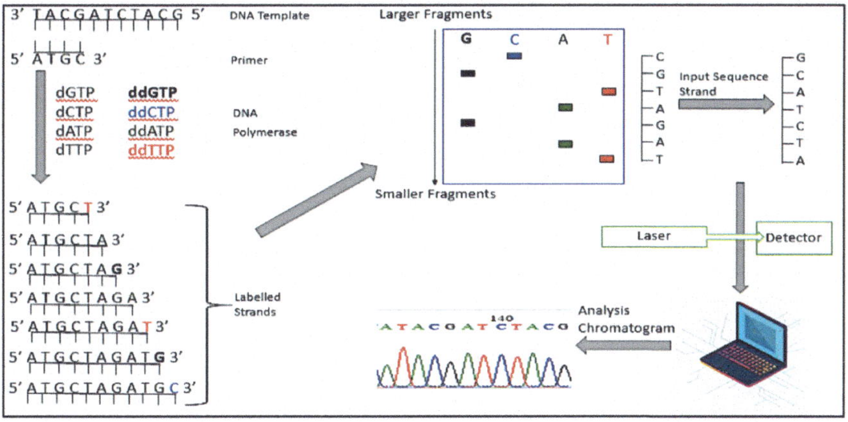

Fig. 1 Overview workflow of the Sanger termination method

sequenced by Sanger sequencing were the phiX174 genome, which is 5374 bp, and the bacteriophage λ genome, with a 48,501 bp in 1975 and 1980, respectively (Kchouk et al., 2017).

After years of improvement, ABI Prism 370 was built in 1987 for automated Sanger sequencing by Applied Biosystems company. The process is based on fluorescent dyes, and capillary electrophoresis led to more accurate and fast sequencing. In 1995 AB370xl could reach an output of 2.88 million bases per day, and the read length reached 900 bases (Levy & Myers, 2016). In 2001, automated sequencing equipment, software employing capillary sequencing machines, and Sanger sequencing technology provided the primary tools for completing the Human Genome Project, costing almost $2.7 billion (Slatko et al., 2018). This effort substantially aided in establishing national novel sequencing equipment that increased speed and accuracy while decreasing cost and labor. Aside from that, thermal cycle sequencing reduces the amount of input DNA required, and thermostable polymerases are used to efficiently and correctly integrate terminator dyes into the developing DNA strands (Ambardar et al., 2016).

2 The Next-Generation Sequencing

After more than 40 years of first-generation sequencing, the NGS technology allows millions or billions of DNA molecules to be sequenced in parallel, enhancing throughput sequencing. In addition, the NGS technology encompasses second-generation and third-generation sequencing methods, referred to as short-read and real-time long-read technologies (see Fig. 2). Regardless of the enormous diversity in NGS techniques, the procedure follows three significant steps: sample preparation, nucleic acid sequencing, and bioinformatic analyses.

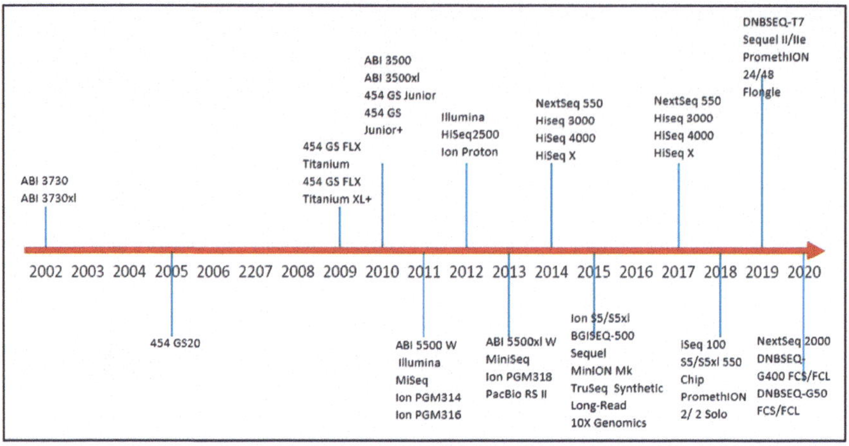

Fig. 2 Timeline of the next-generation sequencing technology revolution

2.1 *Second-Generation Sequencing*

The second-generation platform attempts to amplify a single strand of a DNA library from the DNA source of interest and sequence the amplified strands. DNA libraries are created by annealing platform-specific linkers to blunt-ended fragments produced directly from DNA samples with a sequencing adapter (Mostafa et al., 2015).

The second-generation sequencing platform comprises two approaches; (1) sequencing by ligation (SBL) and (2) sequencing by synthesis (SBS). From the SBL perspective, the DNA ligase used to identify the nucleotide present at a particular location in a DNA sequence instead creates a second strand of the DNA. A pool of probe oligonucleotides labeled with fluorescent dyes is introduced and hybridized to the target DNA sequence (Del Vecchio et al., 2017). When the probe bases match the target DNA sequence, it will ligate to an adjacent oligonucleotide for imaging (Goodwin et al., 2016; Slatko et al., 2018). From the SBS perspective, this approach extends Sanger sequencing without the dideoxy terminators in conjunction with multiple synthesis cycles, imaging, and techniques for incorporating more nucleotides into the expanding chain. A signal, such as fluorescence or a change in ionic concentration, detects a nucleotide's incorporation into DNA and is clonally amplified on a solid surface. The presence of thousands of identical copies of a DNA fragment in a specific region ensures that the signal can be distinguished from background noise (Ambardar et al., 2016).

2.2 *Roche/454 Sequencing*

In 2005, Roche introduced the 454 Genome-Sequencing Technology, marking the inaugural commercially feasible second-generation sequencing system (Liu et al., 2012). This sequencer employed a high-throughput sequencing approach based on pyrosequencing. Unlike the approach of using dideoxynucleotides to halt chain elongation, pyrosequencing relies on the detection of chemiluminescence resulting from pyrophosphate release during nucleotide incorporation. This emitted light serves as an indicator, revealing whether a particular DNA base has been integrated into the newly synthesized DNA strand (Behjati & Tarpey, 2013).

The underlying principle of this sequencing technology involves the initial denaturation of DNA samples, which are then fragmented randomly into pieces spanning 400 to 1000 base pairs in length. Each of these fragments is linked to an amplification bead. These beads possess a synthetic oligonucleotide primer on their surfaces, designed to be complementary to the DNA fragment, effectively tethering a single fragment to each bead. The ensuing in vitro DNA amplification process employs polymerase chain reaction (PCR) within individual emulsion droplets, resulting in a remarkable amplification of a million copies for each DNA fragment situated on the bead's surface (Liu et al., 2012).

In the next step, each bead is combined with specific components including DNA polymerase, ATP sulfurylase, luciferase, luciferin, adenosine-5'-phosphosulfate (APS), and deoxynucleotide triphosphates (dNTPs). These beads are placed onto a plate referred to as a Picotiter plate (PTP) containing multiple wells. The dNTPs (representing the DNA bases A, C, T, and G) match the bases of the DNA fragment and as a result, pyrophosphate molecules are released in a quantity equivalent to the integrated nucleotides. This pyrophosphate is then converted into ATP, which further triggers the conversion of luciferin into oxyluciferin. This transformation generates visible light through a chemiluminescent reaction (Gupta & Gupta, 2014; Mostafa et al., 2015).

The Roche/454 platform can sequence DNA fragments of up to 1000 base pairs in length. The Genome Sequencer FLX Plus instrument (GS FLX plus) can produce 1 million reads in a single run, completed within a day. An advantage of the Roche 454 system is its ability to generate lengthy sequence reads, contributing to the creation of longer contigs and reduced gaps during the genome assembly process (Ambardar et al., 2016). Nevertheless, this approach comes with certain limitations. For instance, difficulties in accurately sequencing insertions and deletions stemming from repetitive sequences of the same nucleotide (homopolymer repeats). Furthermore, the method is associated with the generation of relatively shorter base lengths per run, averaging around 700 million base pairs per second (Mbps). Additionally, the reagents required for this technique are relatively expensive (Liu et al., 2012; Walker et al., 2014).

2.3 ABI/SOLiD Sequencing

SOLiD (Sequencing by Oligonucleotide Ligation Detection) was developed by Applied Biosystems (ABI) in 2006. It was announced as a second-generation sequencing system that uses the sequencing by ligation (SBL) method (Qiang-long et al., 2014). Sequencing by ligation is a DNA sequencing method that uses the enzyme DNA ligase to determine the nucleotide at a certain point in a DNA sequence. This sequencer is known as polony sequencing, where the principle of ligation sequencing is used for mate-paired library construction by tagging adapters to DNA fragments, bead deposition to construct "polonies", and amplifying emulsion PCR. These beads are immobilized onto a glass surface with 8-mer fluorescent-labeled at the end (Buermans & den Dunnen, 2014). The fluorescent octamer is subsequently ligated to DNA fragments on the glass surface, and the color produced by the fluorescent 8-mer is recorded. The 8-mer ends are cleaved chemically to allow the next ligation cycle to begin. Finally, the letter of the nucleotide's base is examined by the color produced by two consecutive ligation processes (Kchouk et al., 2017).

The first ABI Solid sequencer launched was the SOLiD 5500 W system with one flow chip capable of generating sequence data up to 80–160 GB total throughput with a read at initially 35 bp, an output of 3 GB per run, and sequence reads accuracy up to 99.85% (Kulski, 2016). Upgraded SOLiD 5500xl achieved read length, accuracy, instrument throughput, and reads per run improvements of 75 bp, 99.99%, 160–320 GB, and 30 GB per run, respectively. Sequencing costs around $\$40 \times 10^{-9}$ per base based on reagent consumption exclusively to customers (Alekseyez et al., 2020). The advantage of this technique is that the sequencing data are accurately read compared to other sequencing methods since each base is read twice, making it the ideal way for SNP analysis (Ambardar et al., 2016). In contrast, the drawbacks of this technology are fewer data output than Illumina and a shorter read length which is 75 bases only. Also, the whole run takes 6 to 7 days, compared to pyrosequencing, which only needs one day to complete the entire run (Slatko et al., 2018).

2.4 Illumina/Solexa Sequencing

Illumina purchased the Solexa Genome Analyzer in 2006. In 2007, it was marketed as a second-generation sequencing instrument that employs the short-read sequencing by synthesis (SBS) approach with optical base calling (Kulski, 2016). During the first phase of the Illumina sequencing process, DNA samples are randomly broken into 450–500 bp fragments, and adapters are ligated to both ends of each fragment. The adapters are then employed as substrates for repeated amplification synthesis processes on a glass plate containing oligonucleotide sequences complementary to a ligated adapter (Slatko et al., 2018). Once attached to the solid plate, the fragments are amplified by "PCR bridge amplification" from one, or both ends into miniaturized colonies called polonies, yielding several identical copies of each sequence. Each cluster contains millions to billions of clonal template DNA fragments of the same original sequence that can be sequenced concurrently. The last step is to determine the nucleotide on the sequences, where Illumina employs reversible terminator (RT) nucleotide where the RT nucleotide is fluorescently labeled and protected at 3′-OH groups (2-cyanoethyl) (Hari & Parthasarathy, 2018).

Illumina presently has five benchtop platforms and two production scale platforms, each with various uses, including in vitro diagnostics. With a growing capacity, benchtop sequencers (iSeq, MiniSeq, MiSeq, NextSeq 550 series, and NextSeq 1000 and 2000) provide smaller studies spanning from bacterial genomes to transcriptome and whole-exome sequencing. Production-scale platforms (HiSeq and NovaSeq) are primarily utilized for large-scale whole-exome sequencing and whole-genome-sequencing studies (Del Vecchio et al., 2017). The first sequencer produced by Illumina is the Illumina Genome Analyzer IIx system, capable of producing 35 bp length short reads in the form of paired-end (PE) reads, with the newest GAIIx series reaching 85 GB/run (Liu et al., 2012). Novaseq6000, the latest release to the Illumina sequencer collection, associated the patterned flow cells with a two-color nucleotide detection system, allowing for increased speed, flexibility, and throughput for ultra-deep whole-exome and whole-genome sequencing as well as tumor-normal profiling. The advantage of Illumina sequencing over other sequencing is low incorporation bias during nucleotide incorporation. All four reversible terminator-bound dNTPs are present simultaneously and separately in molecules compared to pyrosequencing. As a result, only one type of nucleotide is available at once (Ambardar et al., 2016). Finally, all nucleotide bases are called directly from signal intensity measurement during every cycle, reducing raw error rates (Hui, 2014). Illumina sequencing system has the highest output at ~1.2 to 6000 GB, with low cost per base, increased diversity of applications, and library preparation configurations (Kumar et al., 2019).

2.5 Ion Torrent Sequencing

Life Technologies introduced the Ion Torrent Personal Genome Machine (PGM) based on semiconductor sequencing (Rothberg et al., 2011). Unlike other second-generation technologies, the ion torrent sequencing system did not employ fluorescently labeled nucleotides and was the first next-generation sequencing platform without an electro-optical sensor (Chen et al., 2020). It can produce longer read lengths of up to 600 bp with a faster operation, high accuracy of up to 99.99%, and complete sequence within 8 h (Fujimoto et al., 2014).

However, the drawback of using Ion Torrent is that the error rate can increase as the homopolymer sequences' length (more than six bases), affecting indel performance and high cost of sequencing compared to other platforms (Singh & Bhatia, 2020). This technology also appears to have poor overall throughput and scalability compared to the Illumina sequencer (Chen et al., 2020).

2.6 Nanoball Sequencing

In 2015, Complete Genomic and BGI developed a next-generation sequencer, the BGISEQ-500 with DNA nanoball sequencing (DNBS) method, which used hybrid sequencing by hybridization and ligation (Xu et al., 2019). BGI sequencing technology uses DNA nanoball creation, combinatorial probe-anchor synthesis (cPAS), and combinatorial probe-anchor ligation (cPAL) method during sequencing.

Nanoball sequencing initiates with DNA fragments that are fragmented and linked to the first of four adapter sequences. This fragment is then amplified, circularized, and cleaved by a type II endonuclease. The remaining three adapters are attached subsequently, creating a circular template with distinct adapter regions interspersed among template sequences (Liu et al., 2019). This circular template forms the library template. This library template is further subjected to rolling circle amplification, yielding a considerable number of DNA nanoball concatemers (approximately 100 to 1000 copies). These DNA nanoballs are immobilized onto a sequencing flow cell, organized by binding to positively charged aminosilane in a precise manner. This organized setup enables the high-density sequencing of DNA nanoballs (Fehlmann et al., 2016). In the cPAS technique, fluorescently labeled terminated substrates (ddNTPs) are utilized. Laser excitation of these probes on the DNA nanoballs captures images, which are then transformed into digital signals (Liu et al., 2019).

BGI nanoball sequencing has low amplification error rates and a high-density designed array, resulting in high sequencing accuracy and reduced duplication rates (Fehlmann et al., 2016). Furthermore, DNA nanoball sequencing also has high single nucleotide polymorphism 96.5–97% and insertion high deletion detection accuracy compared to other second-generation sequencing due to the PCR-free library construction method (Poplin et al., 2018). However, on the other hand, BGI nanoball sequencing is the same as other second-generation platforms, which cannot resolve satellite DNA and interspersed repeat sequences and some structural variation analysis (Kumar et al., 2019). In addition, the read length of DNA nanoball sequencing is 50–100 bp, which is relatively short and can lead to a massive gap during assembly (Levy & Myers, 2016).

2.7 Third-Generation Sequencing

Second-generation sequencing methods often need a PCR amplification step, which leads to time-consuming and costly operations. Thus, third-generation long-read sequencing has provided new insight into sequencing by allowing for sequencing without PCR amplification of the DNA, allowing for shorter preparation times while reducing the possibility of errors and eliminating the need for costly computationally and time-consuming assembly steps (Kanzi et al., 2020). Furthermore, third-generation sequencing is capable of producing lengthy reads surpassing several kilobases (>1 kb to 2 Mb) for the solution of assembly problems, copy number changes, structural variants, mapping certainty, improved de novo assembly, and repetitive sections of complex genomes (Xiao & Zhou, 2020).

2.8 Single Molecule Real-Time Sequencing (Pacific Biosciences)

The Single-molecule real-time sequencing (SMRT) technique employs a synthesis-by-synthesis approach and real-time fluorescence detection of tagged nucleotides during their incorporation. Pacific Biosciences (Menlo Park, CA, USA) introduced this sequencing method in 2009, enabling the sequencing of lengthy fragments ranging from 20 to 50 Kb using a combination of synthesis by synthesis and optical detection (Zhou et al., 2016). Pacific Biosciences has developed three commercial sequencing systems: the original RSII model, the Sequel system, and more recently, in 2019, the Sequel II and Sequel IIe systems (https://www.pacb.com/products-and-services/sequel-system/).

The SMRT principle commences with the creation of a template called the SMRT-bell library. This involves ligating a circular double-stranded DNA template to a hairpin adapter, establishing a common primer-binding site at both ends of the DNA fragment. These SMRTbell libraries are then individually placed into microfabricated nanostructures referred to as zero-mode waveguides (ZMWs) (Rhoads & Au, 2015). Within SMRT cells, numerous ZMWs are embedded, each containing a DNA polymerase anchored at the base. During sequencing, fluorescently labeled nucleotides (distinguished by four distinct colors) are incorporated by the DNA polymerase. After incorporation, the fluorescent dye is cleaved from the nucleotide, allowing it to diffuse away from the sensor area before the next labeled nucleotide is incorporated. As nucleotides are integrated, they emit light signals that are captured in real-time by a camera within the machine. Each fluorescent-labeled nucleotide produces a unique signal associated with a specific ZMW, enabling the determination of the base sequence (Ardui et al., 2018; Tamaki et al., 2011).

In terms of throughput, the PacBio RS II employs chips containing 150,000 zero-mode waveguide (ZMW) wells, yielding up to 1 Gb of sequence per single-molecule real-time (SMRT) cell, with an average read length of 20 kb (Slatko et al., 2018). The more recent PacBio Sequel II/Sequel IIe sequencers feature 8 million ZMWs and can generate approximately 5,000,000 reads per SMRT cell, averaging around 15 kb per read (160 Gb/SMRT cell) (Tamaki et al., 2011). PacBio's SMRT imaging and detection technique measures the rate of nucleotide addition during synthesis, termed the inter-pulse duration (IPD) (Ardui et al., 2018). The substantial read length addresses the challenges posed by short-read technologies in handling repetitive sequences and incomplete assembly. This capability significantly enhances read mapping, assembly, and resolution of repetitive regions and structural variations (Ozsolak, 2012). However, a limitation of this technology is its relatively lower data yield, generating around 70 to 160 GB of data per cell, which is comparatively less than second-generation technologies (Kumar et al., 2019). Nevertheless, the introduction of the Sequel and Sequel II platforms, featuring increased ZMWs per SMRT cell and upgraded sequencing chemistry, has led to improvements in reducing error rates, achieving higher read depths, and lowering reagent costs per base.

2.9 Nanopore Sequencing

In 2014 Oxford Nanopore Technologies (ONT) introduced the MinION, a portable sequencer that utilizes a grid of membrane-embedded biological nanopores, revolutionizing the field by enabling rapid, real-time long-read sequencing of nucleic acids. This advancement provides enhanced resolution for structural genomic variations and repeat content analysis (Bowden et al., 2019). The nanopore device's membrane creates a separation between two ionic solutions, allowing an electric current to pass through and facilitating ion exchange (Makalowski & Shabardina, 2020). The MinION, measuring four inches in size, connects to a personal laptop via a USB 3.0 interface, offering a portable and tethered third-generation sequencing solution. ONT's nanopore DNA sequencing portfolio includes the MinION, introduced as the first commercially available platform in 2015, followed by GridION in 2017, PromethION in 2018, and the notable PrimethION48 in 2019 (https://nanoporetech.com/products/comparison).

The standout advantage of this sequencer is its ability to generate extensive read lengths (>4 Mb at 1 bp/ns) without requiring amplification or cloning, setting it apart from other next-generation sequencing methods (Makalowski & Shabardina, 2020). Additionally, this technology offers significant cost savings due to its distinctive analytical capabilities (Bowden et al., 2019). The MinION sequencer is compact, lightweight, and exceptionally portable, easily plugging into a laptop for sequencing. It enables real-time data display on-screen without waiting for full sequence analysis. Moreover, Nanopore sequencing stands out for its absence of fluorescent tags and reduced enzyme presence, contributing to less temperature sensitivity during the sequencing process (Delahaye & Nicolas, 2021). However, it is worth noting that MinION initially exhibited a higher error rate of around 12%, encompassing approximately 3% mismatches, 4% insertions, and 5% deletions. Through advancements in flow cell technology, ONT has enhanced its technology, achieving a raw read accuracy exceeding >99.3%.

2.10 Synthetic Long Reads

In contrast to third-generation technology, synthetic long reads (SLRs) rely on associating existing short-read fragments based on barcoding information assigned to individual transcripts during library preparation. They are reconstructed from clusters of existing short-read sequences (Liu et al., 2021). This technology was developed to improve the quality and lower the costs associated with third-generation sequencing (Adewale, 2020). Currently, three approaches are available for generating synthetic long-read sequences; Illumina synthetic long-read sequencing technology, 10X genomics, and LoopSeq by Loop Genomics (https://www.illumina.com/science/technology/next-generation-sequencing/long-read-sequencing.html).

Illumina and 10X Genomics techniques involve breaking down long DNA fragments into smaller sections and assigning unique DNA barcodes. These fragments are then partitioned using microtiter wells or gel beads (Mantere et al., 2019). This partitioning allows for the sequencing of smaller DNA fragments within each well or bead. By identifying shared barcodes, the data from split barcodes are combined, and short DNA fragments are linked back to their original large DNA fragment. This approach aids in assembling the genome, including the segmentation of repetitive regions. However, some challenging regions may result in gaps and shorter assembled sequences due to the need for local assembly and handling of complex regions (Stapleton et al., 2016).

When comparing Illumina and 10X Genomics approaches, it's noteworthy that Illumina doesn't require specialized instruments, unlike 10X Genomics, which utilizes instruments like Chromium and GemCode for partitioning long DNA fragments (Amarasinghe et al., 2020). Through a microfluidic system, 10X Genomics employs its Chromium technology to divide large DNA fragments (60–100 kb) into small droplets named Gel Bead in Emulsions (GEMs). These GEMs contain oligonucleotides and distinct barcodes for amplifying and sequencing short DNA fragments. This process ensures multiple fragments from the same genomic region are amplified together to minimize gaps during alignment and form a reads cloud (Amarasinghe et al., 2020). Subsequently, the unique barcode sequences are used to align and assemble the short reads, associating them with their original long DNA fragment to generate synthetic linked reads.

LoopSq uses unique molecular identifiers (UMIs) to tag DNA/RNA molecules rather than well/micro fluids identifiers. UMIs are distributed randomly and intramolecularly along the length of barcoded molecules (Liu et al., 2021). Following the barcoding step, the library is prepared by combining a biotinylated DNA probe with a cDNA template. After hybridization and PCR amplification, UMIs are randomly transposed to different internal positions of the template (Callahan et al., 2021). Sequences adjacent to the UMIs insertion are immediately converted into a standard short-read library and sequenced on an Illumina PE150 sequencer. Following sequencing, each heap of short-reads sequences with identical barcodes is grouped and assembled back into the long-reads tagged by those UMIs (Callahan et al., 2021), allowing for the sequencing of single, long adjacent DNA molecules from short sequencing reads, even in the template that contains mixtures of highly homologous long molecules like gene isoforms or RNA variants (Liu et al., 2021). Figure 3 provides a summarized overview of instruments and principles of first-, second-, and third-generation DNA sequencing technologies.

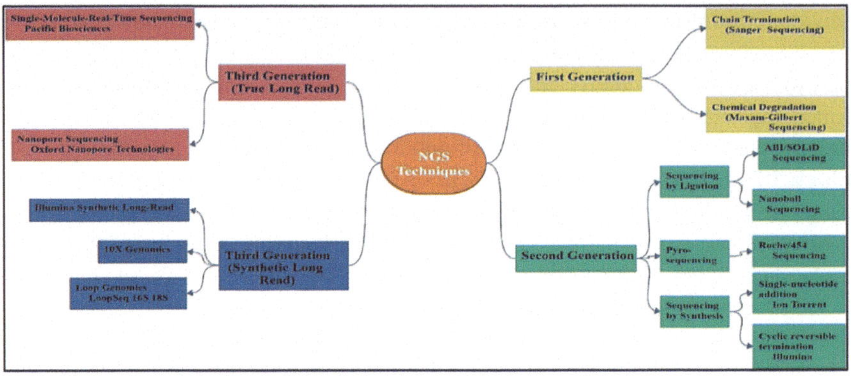

Fig. 3 Summary of next-generation sequencing technologies over the past decades

3 Application of Next-Generation Sequencing in Phytobacterial Diseases

3.1 Second-Generation Sequencing

Midha and Patil (2014) applied the 454 pyrosequencing for studies genomic of *Xanthomonas axonopodis* pv. *citri* that causes citrus bacterial canker and its ecological relatives to gain insight into the evolutionary history of these pathogens by using whole-genome-sequencing methods. The data reported might help decipher the unique variation and virulence of *X. axonopodis* pv. *citri*. In another study, Tamaki et al. (2011) utilized 454 pyrosequencing for microbial ecological studies using the 16S rRNA gene. Using 454 pyrosequencing, McCann et al. (2013) conducted a genomic analysis of *Pseudomonas syringae* pv. *actinidae* causes canker disease to Kiwifruit (*Actinidia* spp.). The study aimed to provide insight into the origin of the emergent plant disease and identify the outbreak-specific gene set associated with the virulent infection system via SNP identification, phylogenetic analyses, and identification of the core and accessory genomes (McCann et al., 2013). Remenant et al. (2012) published the whole genome sequence of *Rastolnia solanacearum* that caused a tomato bacterial wilt outbreak in the southeastern United States and annotated 360 unique genes. Among the unique genes annotated, 82% encode for uncharacterized proteins, and 9% encode proteins of extrachromosomal origin adaptable to Caribbean environmental circumstances (Remenant et al., 2012).

Erwinia amylovora is associated with the fire blight disease of apple and pear trees. Sebaihia et al. (2010) reported a complete genome sequence of *E. amylovora* (ATCC 49946) utilizing dye terminator chemistry using the ABI3730 automated sequencer. In addition, the study compared the results obtained with genomes of closely related Enterobacteria to understand the pathoadaptation of the bacterium to host plants. These studies also revealed that *E. amylovora* WGS is 3.5 Mb long and contains T3SS-related protein and sorbitol-metabolizing clusters involving virulence and host plants' survival advantage (Sebaihia et al., 2010). Furthermore, *Vibrio vulnificus*, a human foodborne pathogen, has been shown to induce typical disease symptoms in *Arabidopsis thaliana* under favorable circumstances via SOLiD DNA and RNA-sequencing. The results demonstrated the identification of candidate clade-specific virulence genes. They differentially expressed genes (DEGs) of salicylic acid, jasmonic acid, and ethylene-dependent pathways responsive to chemical and hormone stimuli of plant resistance (Gulig et al., 2010; Park et al., 2019). Lee et al. (2005) investigated the genome sequence of *X. oryzae*. This pathogen causes bacterial leaf blight in rice to gain insight into the relationship between the pathogen and a susceptible host. Whole-genome shotgun sequencing of *X. oryzae* was performed using a BigDye terminator and an ABI3700 automated sequencer, where the RTX toxin genes (*rtxA* and *rtxC*) are responsible for producing phenylacetic acid, trans-3-methylthio-acrylic acid, and 3-methylthio-propionic acid, which cause plant wilting and discoloration (Lee et al., 2005).

Bart et al. (2012) identified conserved effector and virulence components from 65 strains of *Xanthomonas axonopodis* pv. *manihotis* that caused cassava bacterial blight using Illumina-based draft genome sequencing. This study revealed the *FliC* and *Ax21* genes are essential in regulating virulence factors toward host plants (Bart et al., 2012). Another study by Weng et al. (2021) identified the Type IV pili (T4P) gene of *Xylella taiwanensis* as a vital virulence gene in causing pear leaf scorch of pear plants. The complete genome sequence of *X. taiwanensis* was sequenced using Illumina MiSeq sequencer v2, where annotations contain 49 tRNA, and 2176 coding sequences (CDSs) (Weng et al., 2021). In addition, RNA-sequencing of *Plasmopara viticola* causing grapevine downy mildew performed using HiSeq 2000 and HiSeq 3000 disclosed rapidly evolving and lineage-specific putative host adaptation of this phytopathogen gene (Dussert et al., 2019). Ibrahim et al. (2020) used the Illumina HiSeq 4000 to conduct a draft genome sequencing of *Pantoea stewartii* strain SQT1, Malaysia's pathogen responsible for jackfruit bronzing disease. The study discovered 86 coding sequences correlated to secretion systems, pathogenicity, and resistance factors of the *P. stewartii* bacterium (Ibrahim et al., 2020).

Sarris et al. (2013) investigate the genomics of various strains of *Pseudomonas cannabina* pv. *alisalensis* to better understand host–pathogen interaction mechanisms, host adaptability, pathogen diversity, and the genetic basis of virulence of the destructive pathogen. Four strains of *P. cannabina* pv. *alisalensis*, causes leaf and stem spots in hemp (Cannabis sativa) were sequenced on an ION Torrent PGM 314 chip. One carbamoyl transferase of the *NodU* family, two acyltransferases, a beta-lactamase, and a multidrug efflux system transmembrane protein were identified, indicating the presence of *P. cannabina* pv. *alisalensis* metabolism and virulence factors inside the host (Sarris et al., 2013). Another study by Darrasse et al. (2013) reported a draft genome of *Xanthomonas axonopodis* pv. *glycines* associated with bacterial pustule disease of soybean. The study exhibited that *XagR* and *rpfF* genes are involved in the bacterium's virulence and exoenzyme production (Darasse et al., 2013). Figure 4 shows the summarized application of second-generation sequencing in phytobacterial disease.

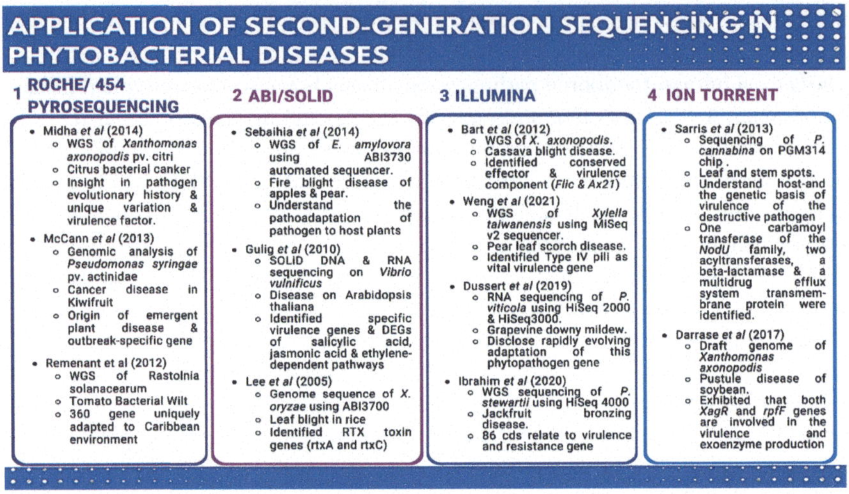

Fig. 4 Summary of application of second-generation sequencing in phytobacterial diseases

3.2 Third-Generation Sequencing

Chen et al. (2021) conducted a study using PacBio RS II to sequence the genome of *Ralstonia solanacearum* to understand the pathogenesis interaction mechanism between the pathogen and peanut (*Arachis hypogaea* L.) plant, which led to bacterial wilt disease. Results disclosed that Type III effector (T3Es) and *RipP1* protein are vital and play a significant role in gene transfer and pathogenicity of *R. solanacearum* (Chen et al., 2021). Yang et al. (2019) performed whole-genome sequencing of *Acidovorax citrulli*, a fruit blotch bacterium of cucurbit crops. PacBio sequencing detected the presence of a ~53 Kb long plasmid, with a crucial role in the production of this pathogen's Type IV secretion system.

Fujikawa et al. (2021) employed the GridION X5 system (Nanopore sequencing) and DNBSEQ-G400 (Nanoball sequencing) to perform whole-genome sequencing on *Pseudomonas cannabin*a pv. *alisalensis* and *Pseudomonas syringae* pv. *maculicola* causes bacterial leaf blight and bacterial leaf spot disease in crucifer plants. The study was conducted to identify unique genes and pathogenicity factors of these pathogens toward crucifers (Fujikawa et al., 2021). In other studies, Xu et al (2021) characterized *Clavibacter nebrakensis* in corn using an amplicon-based Nanopore detection system (MinION flowcells) targeting purine permease gene and bacteremic profiling of necrosis lesion on leaf-infected samples. The study showed that the average number of Nanopore readings increased exponentially in the real-time PCR results, revealing the occurrence of *C. nebrakensis* in bacterial communities of damaged leaf samples (Xu et al., 2021). Hu et al (2021) reported the complete genomic sequencing of *Xanthomonas perforans* isolated from spinach water using Nanopore PromethION to predict gene structure and function and annotate the secretion systems of this bacterium (Hu et al., 2021). *Pseudomonas coronafaciens* pv.

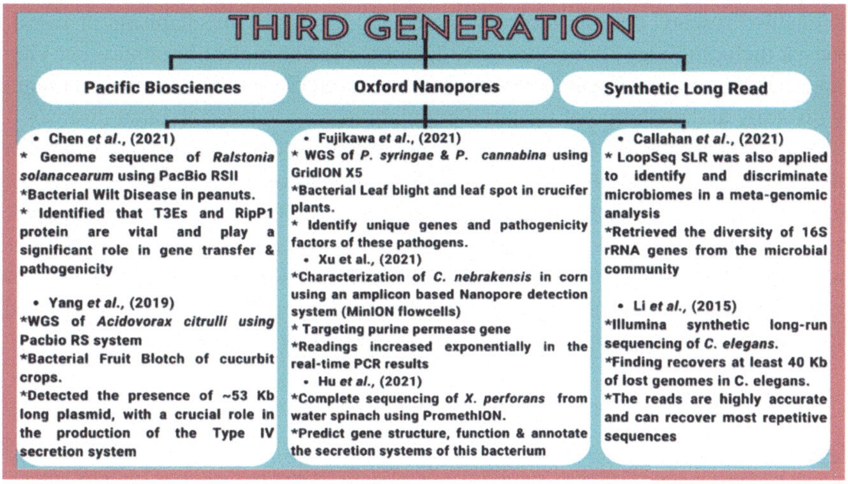

Fig. 5 Summary of application of third-generation sequencing in phytobacterial diseases

oryzae have been reported to cause bacterial halo blight in rice. Baltrus and Clark (2020) published a complete genome sequence of *P. coronafaciens* from rice infected with bacterial halo blight disease using long reads sequenced Nanopore MinION in Flongle flow cells. Results revealed that this pathogen contains a single circular chromosome and a single circular plasmid with 5220 genes comprising 4976 coding proteins, 63 tRNAs, four noncoding proteins, and three complete sets of rRNAs (Baltrus & Clark, 2020).

On the other hand, LoopSeq SLR was also applied to identify and discriminate microbiomes in a metagenomic analysis of the human microbiome. This technology perfectly retrieved the diversity of 16S rRNA genes from the microbial community in the research. The full-length sequences of 16S rRNA could reliably categorize organisms down to the species level. In another study, Li et al. (2015) used Illumina synthetic long-run sequencing to recover missing sequences from the *Caenorhabditis elegans* genome. The findings discovered that Illumina SLR recovers at least 40 Kb of lost genomes in *C. elegans*. The reads are highly accurate and can recover most repetitive sequences (Li et al., 2015). Figure 5 shows the application and summarized of third-generation sequencing application in phytobacterial disease.

4 Conclusions

Next-generation sequencing (NGS) technology has significantly advanced our understanding of plant phytobacterial diseases. With the advent of second and third-generation sequencing platforms, it is now possible to identify and characterize a wide range of bacterial pathogens with unprecedented accuracy and speed. NGS has enabled researchers to understand better the molecular mechanisms of pathogenesis, including identifying virulence factors, effector molecules, and other vital components of bacterial pathogenicity. Moreover, NGS provided new insights into the diversity and evolution of bacterial populations associated with plant diseases.

The ability to rapidly sequence the genomes of bacterial pathogens associated with plant diseases has also enabled the development of new diagnostic tools for the early detection of these pathogens. Additionally, NGS-based approaches have opened up new avenues for developing sustainable disease management strategies and designing more effective treatments. Overall, the NGS technology represents a powerful tool for understanding phytobacterial diseases in plants, and it is expected to play an increasingly important role in plant pathology in future.

References

Adewale B. A. (2020). Will long-read sequencing technologies replace short-read sequencing technologies in the next 10 years? *African Journal of Laboratory Medicine, 9*(1), 1340. https://doi.org/10.4102/ajlm.v9i1.1340

Alekseyev, Y. O., Fazeli, R., Yang, S., Basran, R., Maher, T., Miller, N. S., & Remick, D. A. (2020). Next-generation sequencing primer—how does it work and what can it do? *Academic Pathology, 5*. https://doi.org/10.1177/2374289518766521

Amarasinghe, S. L., Su, S., Dong, X., Zappia, L., Ritchie, M. E., & Gouil, Q. (2020). Opportunities and challenges in long-read sequencing data analysis. *Genome Biology, 21*(1), 30. https://doi.org/10.1186/s13059-020-1935-5

Ambardar, S., Gupta, R., Trakroo, D., Lal, R., & Vakhlu, J. (2016). High Throughput sequencing: An overview of sequencing chemistry. *Indian Journal of Microbiology, 56*(4), 394–404. https://doi.org/10.1007/s12088-016-0606-4

Ardui, S., Ameur, A., Vermeesch, J. R., & Hestand, M. S. (2018). Single molecule real-time (SMRT) sequencing comes of age: Applications and utilities for medical diagnostics. *Nucleic Acids Research, 46*(5), 2159–2168. https://doi.org/10.1093/nar/gky066

Ari, S., & Arikan, M. (2016). Next-generation sequencing: Advantages, disadvantages, and future. In *Plant omics: Trends and applications* (pp. 109–135). https://doi.org/10.1007/978-3-319-31703-8_5

Attia, T. H., & Saeed, M. A. (2015). Next generation sequencing technologies: A short review. *Journal of Next Generation Sequencing & Applications, 1*(2). https://doi.org/10.4172/2469-9853.S1-006

Baltrus, D. A., & Clark, M. (2020). Complete genome sequence of *Pseudomonas coronafaciens* pv. *oryzae* 1_6. *Microbiology Resource Announcements, 9*(3), e01564–19. https://doi.org/10.1128/MRA.01564-19

Bart, R., Cohn, M., Kassen, A., McCallum, E. J., Shybut, M., Petriello, A., & Staskawicz, B. J. (2012). High-throughput genomic sequencing of cassava bacterial blight strains identifies conserved effectors to target for durable resistance. *Proceedings of the National Academy of Sciences, 109*(28), E1972–E1979. https://doi.org/10.1073/pnas.1208003109

Behjati, S., & Tarpey, P. S. (2013). What is next generation sequencing? *Archives of Disease in Childhood. Education and Practice Edition, 98*(6), 236–238. https://doi.org/10.1136/archdischild-2013-304340

Bowden, R., Davies, R. W., Heger, A., Pagnamenta, A. T., de Cesare, M., Oikkonen, L. E., Parkes, D., Freeman, C., Dhalla, F., Patel, S. Y., Popitsch, N., Ip, C. L. C., Roberts, H. E., Salatino, S., Lockstone, H., Lunter, G., Taylor, J. C., Buck, D., Simpson, M. A., & Donnelly, P. (2019).

Sequencing of human genomes with nanopore technology. *Nature Communications, 10*(1), 1869. https://doi.org/10.1186/s13073-021-00883-1
Buermans, H. P. J., & den Dunnen, J. T. (2014). Next generation sequencing technology: Advances and applications. *Biochimica et Biophysica Acta (BBA) - Molecular Basis of Disease, 1842*(10), 1932–1941. https://doi.org/10.1016/j.bbadis.2014.06.015
Callahan, B. J., Grinevich, D., Thakur, S., Balamotis, M. A., & Yehezkel, T. B. (2021). Ultra-accurate microbial amplicon sequencing with synthetic long reads. *Microbiome, 9*(1). https://doi.org/10.1186/s40168-021-01072-3
Chen, K., Wang, L., Chen, H., Zhang, C., Wang, S., Chu, P., Li, S., Fu, H., Sun, T., Liu, M., Yang, Q., Zou, H., & Zhuang, W. (2021). Complete genome sequence analysis of the peanut pathogen *Ralstonia solanacearum* strain Rs-P.362200. *BMC Microbiology, 21*(1), 118. https://doi.org/10.3389/fgene.2022.966092
Chen, P., Yu, X., Huang, H., Zeng, W., He, X., Liu, M., & Huang, B. (2020). Evaluation of Ion Torrent next-generation sequencing for thalassemia diagnosis. *Journal of International Medical Research, 48*(12), 300060520967778. https://doi.org/10.1177/0300060520967778
Darrasse, A., Bolot, S., Serres-Giardi, L., Charbit, E., Boureau, T., Fisher-Le Saux, M., Briand, M., Arlat, M., Gagnevin, L., Koebnik, R., Noël, L. D., Carrère, S., & Jacques, M. A. (2013). High-quality draft genome sequences of *Xanthomonas axonopodis* pv. *glycines* Strains CFBP 2526 and CFBP 7119. *Genome Announcements, 1*(6). https://doi.org/10.1128/genomeA.01036-13
Del Vecchio, F., Mastroiaco, V., & Di Marco, A. (2017). Next-generation sequencing: Recent applications to the analysis of colorectal cancer. *Journal of Translational Medicine, 15*, 246. https://doi.org/10.1186/s12967-017-1353-y
Delahaye, C., & Nicolas, J. (2021). Sequencing DNA with nanopores: Troubles and biases. *PLoS ONE, 16*(10), e0257521. https://doi.org/10.1371/journal.pone.0257521
Dorado, G., Galvez, S., Budak, H., Unver, T., & Hernandez, P. (2017). Nucleic-acid sequencing. In *Reference module in biomedical sciences* (pp. 443–460). https://doi.org/10.1016/B978-0-12-801238-3.08998-4
Dussert, Y., Mazet, I. D., Couture, C., Gouzy, J., Piron, M.-C., Kuchly, C., & Delmotte, F. A. (2019). High-quality grapevine downy mildew genome assembly reveals rapidly evolving and lineage-specific putative host adaptation genes. *Genome Biology and Evolution, 11*(3), 954–969. https://doi.org/10.1093/gbe/evz048
El-Bondkly, A. M. A. (2014). Sequence analysis of industrially important genes from trichoderma. In V. K. Gupta, M. Schmoll, A. Herrera-Estrella, R. S. Upadhyay, I. Druzhinina, & M. G. Tuohy (Eds.), *Biotechnology and biology of trichoderma* (pp. 377–392). Elsevier.
Fehlmann, T., Reinheimer, S., Geng, C., Su, X., Drmanac, S., & Alexeev, A. (2016). CPAS-based sequencing on the BGISEQ-500 to explore small noncoding RNAs. *Clinical Epigenetics, BioMed Central, 8*, 123. https://doi.org/10.1186/s13148-016-0287-1
Franca, L. T. C., Carrilho, E., & Kist, T. B. L. (2002). A review of DNA sequencing techniques. *Quarterly Reviews of Biophysics, 35*(2), 169–200. https://doi.org/10.1017/s0033583502003797
Fujikawa, T., Takikawa, Y., & Inoue, Y. (2021). Complete and draft genome sequences of the cruciferous pathogens *Pseudomonas cannabina* pv. *alisalensis* and *Pseudomonas syringae* pv. *maculicola*. *Microbiology Resource Announcements, 10*(17), e00149–21. https://doi.org/10.1128/MRA.00149-21
Fujimoto, M., Moyerbrailean, G. A., Noman, S., Gizicki, J. P., Ram, M. L., Green, P. A., Ram, J. L., & Gilbert, J. A. (2014). Application of ion torrent sequencing to the assessment of the effect of alkali ballast water treatment on microbial community diversity. *PLoS ONE, 9*(9), e107534. https://doi.org/10.1371/journal.pone.0107534
Goodwin, S., McPherson, J. D., & McCombie, W. R. (2016). Coming of age: Ten years of next-generation sequencing technologies. *Nature Reviews Genetics, 17*(6), 333–351. https://doi.org/10.1038/nrg.2016.49
Gulig, P. A., Crécy-Lagard, V., de Wright, A. C., Walts, B., Telonis-Scott, M., & McIntyre, L. M. (2010). SOLiD sequencing of four *Vibrio vulnificus* genomes enables comparative genomic

analysis and identification of candidate clade-specific virulence genes. *BMC Genomics, 11*(1), 512. https://doi.org/10.1186/1471-2164-11-512

Gupta, A. K., & Gupta, U. D. (2014). Next generation sequencing and its applications. In *Animal biotechnology* (pp. 345–367). https://doi.org/10.1016/B978-0-12-416002-6.00019-5

Hari, R., & Parthasarathy, S. (2018). Next generation sequencing data analysis. In *Reference module in life sciences*. https://doi.org/10.1016/B978-0-12-809633-8.20093-9

Heather, J. M., & Chain, B. (2016). The sequence of sequencers: The history of sequencing DNA. *Genomics, 107*(1), 1–8. https://doi.org/10.1016/j.ygeno.2015.11.003

Hu, M., Li, C., Zhou, X., Xue, Y., Wang, S., Hu, A., Chen, S., Mo, X., & Zhou, J. (2021). Microbial diversity analysis and genome sequencing identify *Xanthomonas perforans* as the pathogen of bacterial leaf canker of water spinach (*Ipomoea aquatic*). *Frontiers in Microbiology, 12*, 752760. https://doi.org/10.3389/fmicb.2021.752760

Hui, P. (2014). Next-generation sequencing: Chemistry, technology and applications. *Topic in Current Chemistry, 336*, 1–18. https://doi.org/10.1007/128_2012_329

Ibrahim, R., Ismail-Suhaimy, N. W., Shu-Qing, T., Ismail, S. I., Ina-Salwany, M. Y., Yusof, M. T., Hakiman, M., Karam, D. S., & Zulperi, D. (2020). Draft genome sequencing data of a pathogenic *Pantoea stewartii* subspecies *stewartii* strain SQT1 causing bronzing disease of jackfruit in Malaysia. *Data in Brief, 30*, 105634. https://doi.org/10.1016/j.dib.2020.105634

Illumina, Inc. (n.d.). *Sequencing platforms*. Illumina, Inc. https://www.illumina.com/systems/sequencing-platforms.html

Kanzi, A. M., San, J. E., Chimukangara, B., Wilkinson, E., Fish, M., Ramsuran, V., & de Oliveira, T. (2020). Next generation sequencing and bioinformatics analysis of family genetic inheritance. *Frontiers in Genetics, 11*, 544162. https://doi.org/10.3389/fgene.2020.544162

Kchouk, M., Gibrat, J. F., & Elloumi, M. (2017). Generations of sequencing technologies: From first to next generation. *Biology and Medicine, 9*(3). https://doi.org/10.4172/0974-8369.1000395

Kulski, J. K. (2016). Next-generation sequencing—an overview of the history, tools, and "omic" applications. In: Kulski, J. K. (Ed.), *Next generation sequencing—advances, applications and challenges* (pp. 3–60). IntechOpen. https://doi.org/10.5772/61964

Kumar, K., Cowley, M., & Davis, R. (2019). Next-generation sequencing and emerging technologies. *Seminars in Thrombosis and Hemostasis, 45*(7), 661–673. https://doi.org/10.1055/s-0039-1688446

Lee, B. M., Park, Y. J., Park, D. S., Kang, H. W., Kim, J. G., Song, E. S., Park, I. C., Yoon, U. H., Hahn, J. H., Koo, B. S., Lee, G. B., Kim, H., Park, H. S., Yoon, K. O., Kim, J. H., Jung, C. H., Koh, N. H., Seo, J. S., & Go, S. J. (2005). The genome sequence of *Xanthomonas oryzae* pathovar *oryzae* KACC10331, the bacterial blight pathogen of rice. *Nucleic Acids Research, 33*(2), 577–586. https://doi.org/10.1093/nar/gki206

Levy, S. E., & Myers, R. M. (2016). Advancements in next-generation sequencing. *Annual Review of Genomics and Human Genetics, 17*(1), 95–115. https://doi.org/10.1146/annurev-genom-083115-022413

Li, R., Hsieh, C. L., Young, A., Zhang, Z., Ren, X., & Zhao, Z. (2015). Illumina synthetic long read sequencing allows recovery of missing sequences even in the "Finished" *C. elegans* Genome. *Scientific reports, 5*, 10814. https://doi.org/10.1038/srep10814

Liu, L., Li, Y., Li, S., Hu, N., He, Y., Pong, R., Lin, D., Lu, L., & Law, M. (2012). Comparison of next-generation sequencing systems. *Journal of Biomedicine and Biotechnology*, 1–11.

Liu, S., Wu, I., & Yu, Y. P. (2021). Targeted transcriptome analysis using synthetic long read sequencing uncovers isoform reprograming in the progression of colon cancer. *Communications Biology, 4*, 506. https://doi.org/10.1155/2012/251364

Liu, W., Xiong, J., Shen, Z., Ni, M., Liu, J., Ullah, K., Wang, W., Wei, D., Wu, J., & Liu, X. (2019). Spatial resolving method on DNA nanoballs with polarization parametric indirect microscopy. *Journal of Nanophotonics, 13*, 1. https://doi.org/10.1117/1.JNP.13.016005

Makalowski, W., & Shabardina, V. (2020). Bioinformatics of nanopore sequencing. *Journal of Human Genetics, 65*, 61–67. https://doi.org/10.1038/s10038-019-0659-4

Mantere, T., Kersten, S., & Hoischen, A. (2019). Long-read sequencing emerging in medical genetics. *Frontier in Genetics, 10*, 426. https://doi.org/10.3389/fgene.2019.00426

Mardis, E. R. (2011). Decade's perspective on DNA sequencing technology. *Nature, 470*, 198–203. https://doi.org/10.1038/nature09796

McCann, H. C., Rikkerink, E. H. A., Bertels, F., Fiers, M., Lu, A., Rees-George, J., & Templeton, M. D. (2013). Genomic analysis of the kiwifruit pathogen *Pseudomonas syringae* pv. *actinidiae* provides insight into the origins of an emergent plant disease. *PLoS Pathogens, 9*(7), e1003503. https://doi.org/10.1371/journal.ppat.1003503

McGinnis, J., Laplante, J., Shudt, M., & George, K. S. (2016). Next generation sequencing for whole genome analysis and surveillance of influenza A viruses. *Journal of Clinical Virology, 79*, 44–50. https://doi.org/10.1016/j.jcv.2016.03.005

MGI Tech. (n.d.). *Products*. MGI Tech. from https://en.mgi-tech.com/products/

Midha, S., & Patil, P. B. (2014). Genomic insights into the evolutionary origin of *Xanthomonas axonopodis* pv. *citri* and its ecological relatives. *Applied and Environmental Microbiology, 80*(20), 6266–6279. https://doi.org/10.1128/AEM.01654-14

Mostafa, E. M., Aly, S., & Bolbol, D. (2015). Overviews of "next-generation sequencing". *Research and Reports in Forensic Medical Science*, 1–5. https://doi.org/10.2147/RRFMS.S57998

Oxford Nanopore Technologies. (n.d.). *Products*. Oxford Nanopore Technologies. https://nanoporetech.com/products

Ozsolak, F. (2012). Third-generation sequencing techniques and applications to drug discovery. *Expert Opinion on Drug Discovery, 7*(3), 231–243. https://doi.org/10.1517/17460441.2012.660145

Pacific biosciences of California, Inc. (n.d.). *Sequencing systems*. Pacific Biosciences of California, Inc. https://www.pacb.com/sequencing-systems/

Park, Y.-S., Kim, S.-K., Kim, S.-Y., Kim, K. M., & Ryu, C.-M. (2019). The transcriptome analysis of the *Arabidopsis thaliana* in response to the *Vibrio vulnificus* by RNA-sequencing. *PLoS ONE, 14*(12), e0225976. https://doi.org/10.1371/journal.pone.0225976

Poplin, R., Chang, P. C., Alexander, D., Schwartz, S., Colthurst, T., Ku, A., Newburger, D., Dijamco, J., Nguyen, N., Afshar, P. T., Gross, S. S., Dorfman, L., McLean, C. Y., & DePristo, M. A. (2018). A universal SNP and small-indel variant caller using deep neural networks. *Nature Biotechnology, 36*(10), 983–987. https://doi.org/10.1038/nbt.4235

Qiang-long, Z., Shi, L., Peng, G., & Fei-shi, L. (2014). High-throughput sequencing technology and its application. *Journal of Northeast Agricultural University, 21*, 84–96. https://doi.org/10.1016/S1006-8104(14)60073-8

Remenant, B., Babujee, L., Lajus, A., Médigue, C., Prior, P., & Allen, C. (2012). Sequencing of K60, type strain of the major plant pathogen *Ralstonia solanacearum*. *Journal of Bacteriology, 194*(10), 2742–2743. https://doi.org/10.1128/JB.00249-12

Rhoads, A., & Au, K. F. (2015). PacBio sequencing and its applications. *Genomics, Proteomics & Bioinformatics, 13*, 178–289. https://doi.org/10.1016/j.gpb.2015.08.002

Rothberg, J. M., Hinz, W., Rearick, T. M., Schultz, J., Mileski, W., Davey, M., Leamon, J. H., Johnson, K., Milgrew, M. J., Edwards, M., Hoon, J., Simons, J. F., Marran, D., Myers, J. W., Davidson, J. F., Branting, A., Nobile, J. R., Puc, B. P., Light, D., ... Bustillo, J. (2011). An integrated semiconductor device enabling non-optical genome sequencing. *Nature, 475*(7356), 348–352. https://doi.org/10.1038/nature10242

Sarris, P. F., Trantas, E. A., Baltrus, D. A., Bull, C. T., Wechter, W. P., Yan, S., & Goumas, D. E. (2013). Comparative genomics of multiple strains of *Pseudomonas cannabina* pv. a*lisalensis*, a potential model pathogen of both monocots and dicots. *PLoS ONE, 8*(3), e59366. https://doi.org/10.1371/journal.pone.0059366

Sebaihia, M., Bocsanczy, A. M., Biehl, B. S., Quail, M. A., Perna, N. T., Glasner, J. D., & Beer, S. V. (2010). Complete genome sequence of the plant pathogen *Erwinia amylovora* strain ATCC 49946. *Journal of Bacteriology, 192*(7), 2020–2021. https://doi.org/10.1128/jb.00022-10

Singh, A., & Bhatia, P. (2020). Comparative sequencing data analysis of Ion Torrent and MinION sequencing platforms using a clinical diagnostic haematology panel. *International Journal of Laboratory Hematology, 42*(6), 833–841. https://doi.org/10.1111/ijlh.13286

Slatko, B. E., Gardner, A. F., & Ausubel, F. M. (2018). Overview of next-generation sequencing technologies. *Current Protocols in Molecular Biology, 122*(1), e59. https://doi.org/10.1002/cpmb.59

Stapleton, J. A., Kim, J., Hamilton, J. P., Wu, M., Irber, L. C., Maddamsetti, R., & Whitehead, T. A. (2016). Haplotype-phased synthetic long reads from short-read sequencing. *PLoS ONE, 11*(1), e0147229. https://doi.org/10.1371/journal.pone.0147229

Tamaki, H., Wright, C. L., Li, X., Lin, Q., Hwang, C., Wang, S., & Liu, W.-T. (2011). Analysis of 16S rRNA amplicon sequencing options on the Roche/454 next-generation titanium sequencing platform. *PLoS ONE, 6*(9), e25263. https://doi.org/10.1371/journal.pone.0025263

Thermo Fisher Scientific. (n.d.). *SeqStudio genetic analyzer*. Thermo Fisher Scientific. https://www.thermofisher.com/my/en/home/life-science/sequencing/sanger-sequencing/genetic-analyzers/models/seqstudio-flex.html

Thompson, J. F., & Milos, P. M. (2011). The properties and applications of single-molecule DNA sequencing. *Genome Biology, 12*, 217. https://doi.org/10.1186/gb-2011-12-2-217

Walker, B. J., Abeel, T., Shea, T., Priest, M., Abouelliel, A., Sakthikumar, S., & Earl, A. M. (2014). Pilon: An integrated tool for comprehensive microbial variant detection and genome assembly improvement. *PLoS ONE, 9*(11), e112963. https://doi.org/10.1371/journal.pone.0112963

Weng, L. W., Lin, Y. C., Su, C. C., Huang, C. T., Cho, S-. T., Chen, A. P., Chou, S. J., Tsai, C. W., & Kuo, C. H. (2021). Complete genome sequence of *Xylella taiwanensis* and comparative analysis of virulence gene content with *Xylella fastidiosa*. *Frontiers in Microbiology, 12*, 684092. https://doi.org/10.3389/fmicb.2021.684092

Xiao, T., & Zhou, W. (2020). The third generation sequencing: the advanced approach to genetic diseases. *Translational Pediatrics, 9*(2), 163–173. https://doi.org/10.21037/tp.2020.03.06

Xu, R., Adam, L., Chapados, J., Soliman, A., Daayfm, F., & Tambong, J. T. (2021). MinION Nanopore-based detection of *Clavibacter nebraskensis*, the corn Goss's wilt pathogen, and bacteriomic profiling of necrotic lesions of naturally-infected leaf samples. *PLoS ONE, 16*(1), e024533. https://doi.org/10.1371/journal.pone.0245333

Xu, Y., Lin, Z., Tang, C., Tang, Y., Cai, Y., Zhong, H., Wang, X., Zhang, W., Xu, C., Wang, J., Wang, J., Yang, H., Yang, L., & Gao, Q. (2019). A new massively parallel nanoball sequencing platform for whole exome research. *BMC Bioinformatics, 20*(1), 153. https://doi.org/10.1186/s12859-019-2751-3

Yang, R., Santos Garcia, D., Pérez Montaño, F., da Silva, G. M., Zhao, M., Jiménez Guerrero, I., & Burdman, S. (2019). Complete assembly of the genome of an *Acidovorax citrulli* strain reveals a naturally occurring plasmid in this species. *Frontiers in Microbiology, 10*, 1400. https://doi.org/10.3389/fmicb.2019.01400

Zhou, Q., Sun, W., Liu, X., Wang, X., Xiao, Y., Bi, D., & Shi, D. (2016). Third-generation sequencing and analysis of four complete pig liver esterase gene sequences in clones identified by screening BAC library. *PLoS ONE, 11*(10), e0163295. https://doi.org/10.1371/journal.pone.0163295

Utilization of Biodiversity for Sustainable Plant Disease Management

Mui-Yun Wong, Yee-Min Kwan, and H. Sathyapriya

Abstract Modern crop production systems that have focused on maximizing yield through the use of synthetic fertilizers and pesticides have led to negative impacts on human health and ecosystems. A major component in any crop production system is crop protection, where chemically synthesized pesticides are routinely and injudiciously applied. More sustainable crop protection systems are desirable. Mother nature has provided nature-based solutions to these problems. There are many documented plant and microbial species currently known to science that we can use for increasing sustainability in crop protection. Many plant species exhibiting potential for disease control are found in the five largest flowering plant families: Asteraceae, Orchidaceae, Fabaceae, Rubiaceae, and Poaceae. Various bacterial members from the *Frankia*, *Bacillus*, *Pseudomonas*, *Azospirillum*, and *Azotobacter* genera and fungal members from the *Trichoderma*, *Coniothyrium*, *Gliocladium*, *Purpureocillium*, and *Phlebiopsis* genera have been studied for their roles in plant growth promoting and disease resistance. This chapter will explore the potential of utilizing botanical extracts from various plant species and beneficial microorganisms for sustainable crop protection against plant pathogens.

Keywords Plant species diversity · Antimicrobial compounds · Plant microbiota · Microbiome · Biopesticides

---

M.-Y. Wong (✉)
Department of Plant Protection, Faculty of Agriculture, Universiti Putra Malaysia, 43400 Serdang, Selangor, Malaysia
e-mail: muiyun@upm.edu.my

Institute of Plantation Studies, Universiti Putra Malaysia, 43400 Serdang, Selangor, Malaysia

Y.-M. Kwan
Department of Crop Science, Faculty of Agricultural Science and Forestry, Universiti Putra Malaysia, Bintulu Sarawak Campus, Bintulu, Malaysia
e-mail: yeemink@upm.edu.my

H. Sathyapriya
Institute of Tropical Agriculture and Food Security, Universiti Putra Malaysia, 43400 Serdang, Selangor, Malaysia
e-mail: sathyapriyahamid@gmail.com

M. Wong (ed.), *Advances in Tropical Crop Protection*,
https://doi.org/10.1007/978-3-031-59268-3_12

1 Introduction

A rise in intense and unsustainable farming besides other factors such as increased demand for animal protein, the increased use and exploitation of wildlife, and the climate crisis, are all driving the increased emergence of zoonotic diseases—diseases transmitted from wildlife to people—like COVID-19 (United Nations, Sustainable Development Goals 15). The COVID-19 pandemic, which started in November 2019, has impacted all aspects of our lives on Earth and has highlighted the power of nature. There are 2.6 billion people who depend directly on agriculture, but 52% of the land used for agriculture is moderately or severely affected by soil degradation. Of the over 80,000 tree species, less than 1% have been studied for potential use, and over 80% of the human diet is provided by plants, where three cereal crops (rice, maize, and wheat) provide 60% of energy intake. Microorganisms and invertebrates are essential to ecosystem services, but their contributions are still poorly known and rarely acknowledged (United Nations, Sustainable Development Goals).

Modern crop production systems utilize the intensification model involving intensive use of farm inputs such as water, fertilizers, and pesticides, causing negative impacts on the environment, leading to pollution and reduction of biodiversity. This contradicts the United Nations Sustainable Development Goals, especially Goal 15, Life on Land. Its goal is to manage forests sustainably, prevent desertification and land degradation, and stop to the loss of biodiversity (United Nations, Sustainable Development Goals). Using environmentally friendly and renewable input sources such as biofertilizers and biopesticides could help to prevent soil degradation and loss of biodiversity.

A major component of crop production is pest and disease management, where chemical pesticides are commonly used injudiciously to reduce yield losses caused by pests and diseases. These chemicals are used due to their immediate and indiscriminate effects, relatively low cost, and easy availability in the market. High usage of chemical pesticides is an issue of grave concern among agriculture experts because of its negative impact on humans, animals, plants, and the environment. In 2020, global pesticide consumption in agriculture was 2.7 million metric tons, with a global application rate of 1.8 kg/ha. However, average pesticide use such as bactericides, fungicides, herbicides, and insecticides has increased by nearly 50% since the 1990s. Furthermore, the pesticide trade reached 7.2 million metric tons of formulated products worth USD 41.1 billion in 2020 (FAO, 2022). While pesticide use is necessary to increase yield to feed a growing population, it is considered the last option in the integrated pest management approach. Non-chemical strategies for crop protection that are safe for living organisms and the environment are available, including biological control agents, biopesticides, botanical extracts, RNA molecules, and ecological engineering. Conservation of natural enemies and high species diversity in the agroecosystems often reduce the risk of pests and diseases in general.

Recent research efforts have demonstrated the potential of various plant species and beneficial microbes for effective disease control of many economically important crops. According to Qian et al. (2022), 369,054 vascular plant species are currently known to science, with 354,072 species (95.9%) belonging to angiosperms. Nearly 2,500 new plant species are published annually (RBG Kew, 2023). Australia, Brazil, and China are the top three sources for many new species discovered every year since the 1990s. Brazil is home to more seed plants than any other country in the world, and the knowledge of its flora is growing at a "record-breaking" pace (Cheek et al., 2020). However, only about 31,000 of these species have at least one documented use, including uses for food, medicine, recreation, genes, poisons, animal feed, and building material (RBG Kew, 2016). The five largest families of flowering plants, with their species numbers rounded to the nearest 1000, are Asteraceae (daisies) with 32,000 species, Orchidaceae (orchids) with 30,000 species, Fabaceae (legumes) with 22,000 species, Rubiaceae (coffees and bedstraws) with 14,000 species and Poaceae (grasses) with 12,000 species (Cheek et al., 2020).

Plant diseases are caused by microorganisms such as oomycetes, fungi, bacteria, mollicutes, viruses, viroids, and nematodes. They can potentially cause tremendous yield losses, approximately 20–30% or 100%, if they are not managed well. Impacts of control methods pose another challenge to sustainability in crop protection, where pesticide usage is the main method of control. There is an urgent need to find a safer and sustainable alternative to crop protection. This chapter will focus on the role of plant and microbial diversity in providing sustainable disease protection of crops, covering the biodiversity of plant species and their bioactive compounds, efficacy in the laboratory and field, the biodiversity of beneficial microorganisms, the interactive relationships between host plants and their associated organisms, and microbial functional potentials in crop protection.

2 Biodiversity of Plant Species with Potential for Sustainable Disease Management

A substantial amount of research was done demonstrating the potential of utilizing bioactive compounds extracted from various plant species to effectively control plant diseases in the field, thanks to the high biodiversity of plant species that Mother Nature has given us. Plant species reported to have antimicrobial potentials against plant pathogens and nematodes *in vitro*, *in vivo*, and in the field come from various orders and families and geographical regions of the world. They include orders from Solanales, Liliales, Rutales, Myrtales, Lamiales, Asterales, Poales, Brassicales, Malpighiales, Rosales, Piperales, Malvales, Fabales, Ephedrales, Asparagales; and families from Solanaceae, Liliaceae, Meliaceae, Myrtaceae, Verbenaceae, Acanthaceae, Lamiaceae, Asteraceae, Poaceae, Caricaceae, Brassicaceae, Euphorbiaceae, Moraceae, Urticaceae, Piperaceae, Thymelaeaceae, Fabaceae, Ephedraceae, Asphodelaceae; and this list is non-exhaustive. These plant species are found all over the world, from temperate to subtropical to tropical regions.

Due to different genetic makeup and environmental conditions around the world, the type of bioactive component and amount of each component found in each plant species also vary. For example, essential oils from Eucalyptus are widely used in the medicinal, pharmaceutical, cosmetic, and food industries (Moghaddam et al., 2020). They are also known to have antibacterial, antifungal and antiseptic properties (Brooker & Kleini, 2006). A comparative analysis of the essential oil from *Eucalyptus camaldulensis* in three regions, Taiwan (Cheng et al., 2009), Morocco (Farah et al., 2001), and Iran (Moghaddam et al., 2020), revealed variations in their two major components, 1,8-cineol and α-pinene. The essential oil from Iran had the highest content of 1,8-cineol at 66.7%, followed by Morocco at 42.3% and Taiwan at 9.4%. Conversely, the concentration of α-pinene in Moroccan Eucalyptus oil was the highest at 28.3%, followed by Taiwanese Eucalyptus oil at 22.5% and Iranian Eucalyptus oil at 17.9%. Similarly, essential oils isolated from the buds of *Capparis spinosa*, a medicinal plant with antimicrobial properties (Benzidane et al., 2020), contained hexadecanoic acid as its primary component. The amounts of hexadecanoic acid varied in Moroccan, Turkish, and Italian samples, with Moroccan samples having the highest content at 36.67%, followed by Turkish samples at 25.03% and Italian samples at 13.9% (Stefanucci et al., 2018). Therefore, understanding the variations in bioactive components is crucial for determining each plant species' efficacy and potential applications in different regions.

Plants can potentially be used as biopesticides for disease management in crop production and postharvest applications as they contain numerous antimicrobial compounds and plant defense signaling molecules that are involved in defense against plant pathogens and nematodes. Plants also contain secondary metabolites that cause inhibition of pathogens by functional disruption of mitochondria, cell membrane, and cell wall; suppression of cell division and disorder of DNA, RNA, and protein syntheses (Freiesleben & Jagar, 2014; Tanifuji et al., 2019). Plant defensins are the abundant category of all plant antimicrobial peptides (AMPs) and can be found in all monocots and dicots (Cools et al., 2017). They are generally 45–55 amino acids with 4–5 disulfide bonds belonging to the cationic group. Previous studies have demonstrated the effectiveness of plant defensins as AMPs in various plant species. For instance, defensins from radish and petunia have shown antifungal activity against the fungal pathogens *Rhizoctonia solani* and *Magnaporthe oryzae* (Jha & Chattoo, 2010), as well as *Fusarium oxysporum* f. sp. *cubense* (Ghag et al., 2012), respectively. Additionally, MtDef4 and its derived peptides from *Medicago truncata* have exhibited intense antifungal action against multiple pathogens (Sagaram et al., 2011; Sathoff et al., 2019; Tetorya et al., 2023). However, plant defensins' antibacterial activity remains less common than their antifungal activity (Parisi et al., 2019). AsDef1 from oat and BcDef1 from *Brugmansia* x *candida* were identified as efficient antibacterial defensins against both Gram-positive and Gram-negative bacterial strains (Emamifar et al., 2021; Kaewklom et al., 2018). Recent research has also explored tobacco defensin, NAD1, in controlling oomycete diseases (Khatun, 2023). The antimicrobial properties and biotic stress responses that aid in plant growth and development of plant defensins (Cools et al., 2017; Tam et al., 2015) make them ideal candidates for biopesticide development.

Some of the volatile aromatic components of significant importance for providing protection against pathogens during the postharvest stage include aldehydes such as acetaldehyde, benzaldehyde, and hexane, while acetic acid is used as a fumigant for postharvest treatment. Aldehydes were shown to suppress the fungal spores and mycelial growth of *Rhizopus stolonifer*, *Penicillium digitatum*, *Colletotrichum* spp. during laboratory trials. These compounds can potentially be used effectively to control diseases of perishables (Chitranshi et al., 2020). Additionally, cinnamaldehyde from the genus *Cinnamomum* has also been identified as a potential biofungicide against economically important plant pathogens such as *Colletotrichum lindemuthianum* (Pereira et al., 2022) and *Phytophthora capsici* (Wang et al., 2021), as well as the citrus postharvest pathogen *Geotrichum citri-aurantii* (OuYang et al., 2019). The primary antifungal mechanisms of the aldehyde include fungal cell wall weakening and metabolic homeostasis disruption, which lead to structural damage and cell death (OuYang et al., 2019; Wang et al., 2021). More recently, the essential oil of Jenipapo (*Genipa americana* L.), which contains the active ingredient aldehyde (E)-2-decenal as its main constituent, was regarded as an eco-friendly and safer alternative in *Curvularia* leaf spot disease management because it controlled the pathogen infection and enhanced corn defense responses without negatively impacting the beneficial *Trichoderma asperellum* or ladybeetles (Guilherme et al., 2023).

Plant defense signaling molecules include hormones such as jasmonates [methyl jasmonate (MJ) and jasmonic acid (JA)] are involved in biotic stress responses in plants and have the potential to increase natural resistance against fruit decay during the postharvest stage (Chitranshi et al., 2020). The biosynthesis, regulation, and signaling systems of jasmonates in plants in response to biotic stressors have been extensively reviewed recently (Sood, 2023). Apart from that, the exogenous application of jasmonates has also been proven effective in improving host resistance in various plant pathosystems, including *Phaseolus vulgaris-Sclerotinia sclerotiorum* (Oliveira et al., 2015) and potato-*Phytophthora infestans* (Yang et al., 2022), through the induction of host immune responses. Naturally occurring compounds in the plants, such as glucosinolates (GSLs) and one of their hydrolysis products called allyl-isothiocyanate, were also shown to have antifungal properties during postharvest treatments (Fenwick et al., 1983; Mari et al., 2002). Moreover, the literature analysis on different GSLs and their hydrolysis products has revealed that the metabolites have a natural pesticidal effect on a variety of fungal and oomycete pathogens, *in vitro*, *in planta*, and in the field (Poveda et al., 2020). This suggests that these metabolites can effectively be used as natural biocides to control various fungal and oomycete diseases in agriculture. Plants are rich in secondary metabolites with antimicrobial properties, and these compounds can be extracted from all parts of the plants, such as leaf, flower, fruit, stem, and root. The samples are dried before extraction to preserve the integrity of the secondary metabolites and ensure their stability during the extraction process. Some of the most common drying techniques are sun drying, shadow drying, freeze drying, and hot air drying. The choice of drying technique for each plant species can significantly impact the total essential oil output and the bioactive compounds extracted (Nurhaslina et al., 2022). Plants produce essential oils as natural aromatic complex mixtures of compounds mainly involving two chemical categories: terpenoids and phenylpropanoids (Garay et al., 2020). The choice of solvent for extracting compounds from plants depends on the chemical class of the target compounds. For example, water was used to extract phenolic acids, alkaloids, flavonoids, lectins, glycosides, and phenylpropanoids due to their solubility in aqueous solutions (Chikezie et al., 2015). Methanol was preferred for extracting saponins, tannins, polyphenols, terpenoids, flavones, and anthocyanins because of its ability to extract a wide range of polar and nonpolar compounds, while ethanol was suitable for extracting tannins, polyphenols, flavanols, terpenoids, and alkaloids due to its ability to solubilize a diverse range of compounds (Azmir et al., 2013). Acetone was the solvent of choice for extracting alkaloids, sugars, tannins, and quinones because of its ability to dissolve both polar and nonpolar compounds (Neelamathi & Kannan, 2016), while chloroform was effective in extracting anthocyanins, flavonoids, flavones, and polyphenols due to its ability to selectively extract lipophilic compounds (Selvamuthukumaran and Shi 2017). The choice of solvent is crucial in ensuring the efficient extraction of the desired compounds.

Many studies have explored the effectiveness of various botanical extracts in inhibiting the growth of major phytopathogens to manage devastating plant diseases that cause significant yield losses. For example, aqueous extracts of neem leaves (*Azadiracta indica*) along with seven Egyptian medicinal plant species namely, lantana leaves and fruits (*Lantana camara*), chili fruits (*Capsicum frutescens*), chamomile (*Matrecaria chamomilla*), sage (*Salvia officinalis*), garlic bulbs (*Allium sativum*), lemon grass leaves (*Cymbopogon citratus*), and onion seeds (*Allium cepa*), demonstrated effectiveness against *Alternaria solani* both *in vitro* and in open field conditions at varying levels, with notable reductions in fungal mycelial growth, inhibition of spore germination and disease severity (Afifi & Sahar, 2009). Furthermore, leaf extracts of *Datura stramonium*, *A. sativum*, and *A. indica*, each with a 5% concentration, displayed the maximum reduction in mycelial growth of *A. solani* in vitro bioassays. In greenhouse experiments, *D. stramonium* and *A. sativum* extracts at the same concentration significantly increased tomato fruit yield (Nashwa & Abo-Elyousr, 2012). Also, the leaf extracts of *Chromolaena odorata*, *Euphorbia heterophylla*, *Tithonia diversifolia*, *A. indica*, and *Carica papaya* demonstrated significant efficacy, notably reducing early blight severity, increasing fruit production, improving fruit length, and enhancing fruit weight. Among the botanicals tested, *C. papaya* leaf extracts exhibited the highest effectiveness in mitigating early blight, leading to an increased tomato plant yield compared to the control (Joseph et al., 2017). *Eucalyptus camaldulensis* leaf extracts were proven effective against the early blight of tomato (Chourasiya et al., 2013). Under *in vitro* conditions, extracts from *D. stramonium*, *A. sativum*, *A. indica*, *E. globulus*, and *L. camera* at concentrations of 5, 10, 15, and 20% were also notably effective in suppressing the conidial germination of *A. solani* when compared to the control (Kumar et al., 2023). These studies suggest that tomato early blight disease, one of the most significant and prevalent crop diseases in the world (Song et al., 2011), can be managed by using botanical extracts.

Backhousia citriodora, *Clinacanthus nutans*, *Ficus deltoidea*, *Phaleria macrocarpa*, and *Piper betle* extracts demonstrated varying levels of antifungal efficacy when assessed *in vitro* against three economically significant fungal pathogens.

Notably, *P. macrocarpa* emerged as the most promising candidate for developing biofungicides, particularly against *Fusarium oxysporum* f.sp. *cubense* Tropical Race 4 (FocTR4), *Ganoderma boninense*, and *Rhizoctonia solani*, followed closely by *P. betle*. Additionally, *B. citriodora* and *C. nutans* leaf extracts were found to be valuable in managing Fusarium wilt disease caused by FocTR4 in banana and controlling *R. solani* sheath blight in rice, respectively (Wong et al., 2020). On the other hand, the ethanol extracts of *Paris polyphylla*, *Phellodendron chinense*, and *Menispermum dauricum* medicinal plants inhibited *Magnaporthe oryzae* spore germination by 100%, 70%, and 50%, respectively, with *P. polyphylla* pericarp extract significantly improving blast resistance in rice seedlings, indicating its potential as an effective antifungal agent for rice blast disease control (Yanping et al., 2023). These findings highlight the potential of these plant extracts in developing biofungicides.

The microbicidal effects of various botanical extracts have also been explored to manage various bacterial and viral plant diseases. The emulsion formulations of *Piper sarmentosum* extracts, for instance, effectively controlled sheath brown rot caused by *Pseudomonas fuscovaginae* and bacterial leaf blight (BLB) caused by *Xanthomonas oryzae* pv. *oryzae* when tested *in vitro* and *in planta*, indicating their potential as promising alternatives for disease management in rice (Syed-Ab-Rahman et al., 2020). The ethyl acetate and chloroform fractions of Indian siris (*Albizia lebbeck*) bark extracts exhibited remarkable antibacterial activity against *P. syringae* and *Xanthomonas campestris in vitro*, and their potential for managing bacterial speck disease caused by *P. syringae* in tomato crops was highlighted by their ability to reduce disease incidence under greenhouse conditions (Arasu et al., 2022). More recently, a study on *Hedychium coronarium* methanol and aqueous extracts found that priming rice seedlings with the extracts can improve rice resistance to BLB disease by stimulating plant defense responses and improving photosynthetic efficiency (Dubey et al., 2023). On the other hand, the antimicrobial activity of aqueous rosemary extract (ARE) against various plant pathogens, including citrus canker causative agent *Xanthomonas citri* subsp. *citri* and tobacco necrosis virus strain A demonstrated ARE's potential use as a bioprotective agent in plant disease management (Martin et al., 2023). Moreover, exogenous application of plant-based compounds such as eugenol and anise oil was also proven to increase the resistance of tomato plants to tomato yellow leaf curl Thailand virus by activating SA-dependent signaling pathways (Tsai et al., 2019). Exploring botanical extracts and their natural compounds for their bactericidal and antiviral activities presents promising solutions for managing various bacterial and viral plant diseases.

Foliar sprays of aqueous leaf extracts from three sun-dried medicinal plants, namely, *Podophyllum hexandrum*, *Withania somnifera*, and *Xanthium strumarium*, at a concentration of 25% (w/v) applied at three-day intervals significantly decreased the severity of potato late blight, resulting in higher tuber and biological yields compared to the control. Notably, among the tested plant extracts, *P. hexandrum* exhibited the highest effectiveness in reducing disease incidence, and it produced superior biological yield per plant (24.92 g) and tuber yield (14.93 t/ha) compared to the other treatments (Majeed et al., 2011). In a pot experiment, neem leaf extract proved highly effective in controlling damping-off disease caused by the oomycete

Pythium ultimum in brinjal, with the highest seed germination rate (69%) compared to the control (27%) and other tested leaf extracts including garlic clove (63%), Datura (57%), onion bulb (53%), turmeric (50%), ginger (47%), and Tulsi (43%) extracts (Choudhary et al., 2017). Application of five plant extracts, namely henna (*Lawsonia inermis*), acalypha (*Acalypha wilkesiana*), chinaberry (*Melia azedarach*), pomegranate (*Punica granatum*), and lantana (*L. camara*), demonstrated significant effectiveness in reducing leaf rust infection caused by *Puccinia triticina* Eriks, with lantana extract being the most effective. All treatments also significantly improved wheat yield and increased chlorophyll content, total phenolics, and oxidative enzyme activities, indicating that these plant extracts have the potential to induce resistance in wheat against leaf rust (Draz et al., 2019). Licorice (*Glycyrrhiza glabra*) cold water extract demonstrated strong antifungal activity, particularly against *P. aphanidermatum*, the causal agent of damping-off in *Capsicum annuum*, with a remarkable 89.9% antifungal efficacy at a concentration of 200 mg/ml, highlighting its significant potential for disease control in chili plants (Arora et al., 2023). These findings highlight the valuable role of plant extracts in managing diseases caused by oomycetes in various crops.

Plant extracts were also studied for their efficacy in controlling plant diseases caused by nematodes. *A. sativum*, *Urtica dioica*, *Sophora mollis*, *Ephedra intermedia*, and *Tanacetum baltistanicum* showed excellent results against root–knot nematodes, with a total mortality rate of 75–95% at concentrations of 0.125–1% after 72 h of treatments (Ismail et al., 2020). In another study, rape seed (*Brassica napus*), lantana (*L. camara*), African marigold (*Tagetes erecta*) and neem (*A. indica*) at 5% and 10% concentrations and *T. harzianum* at 5% inhibited nematode egg hatching and resulted in 84.67–100% mortality of J2 juveniles of *Meloidogyne incognita* in vitro at 10% concentration after 72 h of exposure time. *L. camara* extract at 5% combined with *T. harzianum* caused a significant reduction in root–knot nematode population, nematode reduction rate, number of galls, and egg masses per plant (Feyisa et al., 2015). In addition, extracts of tobacco (*Nicotiana tabacum*), clove (*Syzygium aromaticum*), and betel vine (*Piper betle*) were highly toxic to *M. incognita* at 5 mg extract/ml water, killing more than 80% of the nematodes and with an EC_{50} that was 5–10 times lower than synthetic pesticides chlorpyrifos, carbosulfan, and deltamethrin (Wiratno et al., 2009).

A study on the nematicidal activity of a *Moringa oleifera* ethyl acetate leaf extract against the eggs and larvae of two nematodes, *Haemonchus contortus*, and *Nacobbus aberrans*, also demonstrated its potential for nematode control. The extract inhibited egg hatching and caused larval mortality, with major compounds identified such as 9,12,15-octadecatrienoic acid and n-hexadecanoic acid, highlighting *M. oleifera*'s promising role as a potential agent for agricultural nematode management (Páez-León et al., 2022). More recently, the extracts of *Piper cumbricola* and *P. eriopodon* were found to have significant antioxidant activity and nematicidal potential against *Radopholus* sp. and *Meloidogyne* spp. that affect *Musa paradisiaca*. Extracts from both species exhibited high radical-scavenging activities, and specific compounds, such as 2,4-dimethoxy-5,6-methylenedioxy-1-(2-propenyl) benzene in *P. cumbricola* and gibbilimbol B (4-[3É)-decenyl] phenol) in *P. eriopodon* displayed notable biocidal effects, making them promising candidates for controlling these endoparasites in *Musa paradisiaca* (Vanegas et al., 2023). Interestingly, the use of Brassicaceae plant tissues, such as chopped cabbage and cauliflower, as bio-fumigants has also been shown to be successful in reducing the populations of root–knot, lance, spiral, and stunt nematodes in okra (*Abelmoschus esculentus*) by almost 80% (Behera et al., 2020).

3 Plant Holobiont

Plants engage in diverse ecological interactions, including competition, mutualism, commensalism, and parasitism, with various microorganisms within their natural ecosystem to achieve environmental sustainability. Therefore, plant–microbe interaction has a direct impact on plant growth and survival. The beneficial interaction between plants and microbes can confer advantages such as disease suppression, mitigation of abiotic stresses, and enhanced nutrient uptake. The symbiotic interaction between a host plant and its associated microorganisms is called the plant holobiont. The concept of the "holobiont" was first conceptualized in 1943 by a German theoretical biologist, Adolf Mayer, to characterize a host and symbiont system that exhibited a high degree of independence and contributed to evolutionary modifications. This concept was further expounded by an American biologist, Lynn Margulis, in the 1990s on the evolutionary endosymbiotic theory (Baedke et al., 2020). Margulis' idea of endosymbiosis described the co-evolution of the symbiotic-associated host plant and microbes through genome fusion (Margulis, 1991). These mutational changes occur between the host plant and microbes, leading to new phylogenic forms. Microbes likely provide additional fitness-related genes to improve host plants' adaptation to changing environmental conditions (Bulgarelli et al., 2013; Vandenkoornhuyse et al., 2015). The current advancement of omics technologies has significantly enhanced the investigation of the impact of microbiomes on plant stress responses and their adaptation.

In the plant holobiont system, the microbiomes (or plant microbiota) consist of various microorganisms, including archaea, bacteria, fungi, and protists. Bacteria and fungi represent the predominant groups of plant microbiota. The major bacterial phyla associated with plants are Proteobacteria, Actinobacteria, and Bacteroidetes. In contrast, the most prevalent fungal communities associated with plants are comprised of members of the Ascomycota and Basidiomycota phyla. (Hassani et al., 2018). Plant microbiota proliferates in the rhizosphere (soil around the roots), phyllosphere (aerial plant parts in contact with the external environment), and endosphere (external environment) (Compant et al., 2019). They can be inherited (vertical transmission) by subsequent generations via seeds. The seed-associated endophytic microbiota comprises a limited range of archaea, bacteria, and fungi that can survive desiccation and seed storage conditions (Truyens et al., 2014). They can be found in various seed tissues such as seed coat, embryo, cotyledon, and endosperm. They are particularly useful in facilitating rapid seed germination, seedling survival, plant growth, and productivity (Nelson et al., 2018). Seed microbiota acts as the early colonizers of plant tissues and can avoid direct competition with soil microbiota and protect plants against invading pathogenic microbes.

Alternatively, plants can actively recruit (horizontal acquisition) microbes from the surrounding environment. This ability is a necessary adaptation that helps to improve nutrient uptake, protection from disease, and increase tolerance to environmental stresses. The enlistment of microbiota from the rhizosphere is facilitated by the secretion of root exudates, which contain soluble carbon-based metabolites, including organic acids, amino acids, and sugars (Canarini et al., 2019). These molecules are unloaded as rhizodeposits at the actively growing root tips through several specific efflux carriers and channels, such as ALMT transporters (for organic acids), CAT transporters (for amino acids), and SWEET transporters (for sugars) (Ross-Elliott et al., 2017). Root exudates serve as a valuable carbonaceous substrate to the soil microbiota, where their growth is often limited by labile carbon sources (Bais et al., 2006). However, these growth substrates can also serve as chemical signals to attract pathogenic microbiota. Thus, plants have evolved specific microbe-associated molecular patterns (MAMPs) and damage-associated molecular patterns (DAMPs) recognition mechanisms to discriminate the pathogenic microbiota. Examples of MAMPs from bacteria are bacterial flagellin, elongation factor EF-Tu, superoxide dismutase, 23-amino acid peptides, lipoligosaccharides, and peptidoglycans. These molecules can be detected by extracellular pattern recognition receptors (PRRs) at millimolar to subnanomolar concentrations. Ergosterol, arachidonic acids, oligosaccharides, chitin, glucan, and xylan are examples of fungal MAMPs (Köhl et al., 2019). Besides, the plant microbiota has also evolved mechanisms for host recognition and avoiding triggered innate plant immunity. Therefore, specific metabolites exuded by the host roots can attract specific rhizosphere microbes. In such cases, particular microorganisms strongly associated with a specific plant species independent of environmental variables are referred to as the core microbiome (Toju et al., 2018). In addition to root exudates, external environmental factors such as climate, soil type, soil structure, soil pH, salinity, mineral nutrients, moisture, organic matter, and agronomic practices play essential roles in shaping the plant microbiota (Compant et al., 2019).

4 Functional Potentials of Plant Microbiota

The presence of plant microbiota in diverse settings enhances the host plant's adaptability to biotic and abiotic environmental factors by providing supplementary resources. This underscores the functional potential of plant microbiota. Moreover, plant microbiota-mediated disease protection highlights the interconnectedness of plant–microbe and microbe–microbe interactions in the natural environment. The beneficial effects of plant microbiota partners in plant protection include: (1)

enhancing nutrient acquisition, (2) phytohormone manipulation, (3) direct or indirect antagonism of pathogenic microbial strains, and (4) activation of plant innate immunity (Hartman & Tringe, 2019; Jacoby et al., 2017).

4.1 Nutrition Acquisition

Microbes are natural decomposers. They play an active role in depolymerizing and mineralizing soil organic matter, thereby improving the bioavailability of nutrients for plants. In the case of an organic farming system, using green manure as fertilizers and microbial inoculants as decomposers has the advantage of gradual nutrient release, which helps to reduce nutrient runoff caused by leaching and volatilization (Reganold & Wachter, 2016). Additionally, microbes find application in the phytoremediation of nutrient-poor or contaminated soils (Abhilash et al., 2012). Under nutrient scarcity conditions, the release of root exudates can alter the soil microbial population, thereby improving nutrient acquisition and plant growth. Rhizobia and mycorrhiza are beneficial microbes that have been utilized as biofertilizers and biopesticides. The specific symbiotic association between *Rhizobium* and legume plants is triggered by nitrogen deprivation experienced by the host plant. This symbiotic association facilitates atmospheric nitrogen fixation and thus improves plant growth. The mycorrhizae–root association operates in a similar way to enhance phosphorus uptake. Ectomycorrhizal (ECM) and arbuscular endomycorrhiza (AM) fungi are the two types of mycorrhizal fungi. The formation of a mycelial mantle and a Hartig net on the root surface is ECM fungi's characteristics. In contrast, AM fungi penetrate the root epidermal and cortical cells to form nutrient exchange sites known as arbuscules (Sui et al., 2019).

4.2 Phytohormone Manipulation

Aside from their role in nutrient acquisition, Rhizobia and mycorrhiza can modulate plant responses via lipo-chito-oligosaccharide signaling and hormone manipulation (Buhian & Bensmihen, 2018). For example, indole-3-acetic acid (auxin) promotes the growth of lateral roots and root hairs to facilitate mycorrhizal colonization and nodulation in legume-Rhizobial symbiosis (Boivin et al., 2016). Additionally, Rhizobia and mycorrhiza provide disease protection to host plants through the production of phytohormones such as auxin, cytokinin, gibberellic acid, and 1-aminocyclopropane-1-carboxylic acid (ACC) deaminase to counteract the detrimental effects of ethylene and abscisic acid induction caused by pathogens (Friesen et al., 2011; Sui et al., 2019). The interplay between antagonistic phytohormones (e.g., auxin–cytokinin and abscisic acid-gibberellins) and their interactions with jasmonic acid (JA), salicylic acid (SA), and brassinosteroids (BR) can significantly affect plant immunity. JA, SA, and BR are stress-specific signaling molecules that closely interact with phytohormones in plant biotic stress mitigation. Auxin and JA syntheses are induced in the early stages of Rhizobia and mycorrhiza root colonization to avoid triggering the SA-mediated defense responses (Gutjahr & Paszkowski, 2009). Various bacterial members from the *Frankia*, *Bacillus*, *Pseudomonas*, *Azospirillum*, and *Azotobacter* genera and fungal members from the *Trichoderma*, *Ganoderma*, and *Saccharomyces* genera have been studied for similar plant growth promoting and disease resistance effects (Tian et al., 2020). These microbes are known to promote plant growth and induce systemic resistance by producing growth-promoting hormones and antagonistic molecules.

4.3 Antagonistic Mechanisms

The mutualistic, endosymbiotic, competitive, antagonistic, parasitic, and predatory interactions between plant microbiota may, directly and indirectly, affect host plants (Braga et al., 2016). Through plant microbiota interaction, the population can collaboratively reduce the offensive qualities of an invading pathogenic microbe to achieve satisfactory plant protection outcomes. Plants have capitalized on specific groups of microorganisms with plant-beneficial properties to suppress plant diseases. In such cases, applying a high density of selected antagonistic beneficial microorganisms to induce pathogen suppression is known as augmentative biological control (van Lenteren, 2012). Microbial biological control agents (MBCAs) employ different mechanisms, such as competition and antagonistic activities (hyperparasitism and production of antimicrobial metabolites) to suppress the pathogen population.

Competitive interaction occurs between microbes under limited nutrient resources and space. Rapid growth and efficient nutrient sequestration and utilization are the competitive mechanisms employed by successful MBCAs to out-compete opposing microbes in a resource-limited environment. The efficacy of MBCAs in the competition depends on nutrient availability, pathogen nutritional requirement and acquisition strategies (which may vary across life stages), and MBCAs distribution at the point of interaction (Köhl et al., 2019). Pathogen inoculum depends on exogenous nutrients to initiate germination and growth, aiming to attain an effective population size for host plant invasion. Effective biological control is achieved when an established MBCA population gains a competitive advantage against the pathogen. For example, iron is required explicitly as a cofactor for many enzymes involved in electron transfer and oxidative stress regulation (Crichton et al., 2002). Under a limited iron condition, the production of siderophores, a group of low molecular weight peptides with high affinity for ferric iron by certain beneficial bacterial and fungi (e.g., *Pseudomonas* spp. and *Trichoderma* spp.), is a competitive iron acquisition strategy (Aude & Alia, 2015; Segarra et al., 2010). The active iron sequestration by endophytic MBCAs will lead to enzyme inactivation in pathogens and virulent expression and indirectly promoting plant survival.

Antagonism is a contact-dependent competition achieved via hyperparasitism and the production of antimicrobial metabolites to regulate the growth of pathogens, such as interfering with signaling molecules required for quorum sensing among pathogenic bacterial cells (Friesen et al., 2011). *Trichoderma*, *Pythium*, and *Clonostachys* are recognized for their ability to parasitize other microbes. Hyperparasitism is a direct competition between two organisms where the host is also a parasite. This situation is most common in the fungal community. The hyperparasitism mechanisms employed by the biotrophic fungal parasite are the production of attachment and infection structures (e.g., haustoria) to invade fungal spores or mycelia, whereas the strategies employed by necrotrophic fungal parasites typically involve a combination of antimicrobial metabolites, cell wall degrading enzymes (CWDEs), and proteases (Köhl et al., 2019). The activities of CWDEs are highly specific, and their production is triggered by signaling associated with the potential host.

Quorum sensing is a mechanism utilized by various bacterial species, whereby the production of specific signaling molecules such as acyl-homoserine-lactones (AHLs) in proteobacteria, gamma-butyrolactones in *Streptomyces*, cis-11-methyl-2-dodecanoic acid in *Xanthomonas* and *Xylella*, and oligopeptides in gram-positive bacteria to enable the bacteria population to coordinate gene expression and biofilm formation (Danhorn & Fuqua, 2007). Many phytopathogenic bacteria rely on this communication system to regulate virulent gene expression for host infection. Therefore, inhibiting quorum sensing and biofilm formation offers a promising mechanism to fight plant diseases. The mechanisms of quorum sensing interference include inhibiting the biosynthesis of enzymes involved in signal generation, destroying the signaling molecules in the medium, or inhibiting the activation of receptors (Helman & Chernin, 2015). For example, beneficial bacteria (*Bacillus*, *Arthrobacter*, and *Klebsiella*) produce AHL-degrading lactonase to reduce the virulent gene expression in pathogenic bacteria (Reading & Sperandio, 2006). In addition to quorum sensing, bacteria (*Streptomyces*, *Pseudomonas*, *Stenotrophomonas*, and *Serratia*), fungi (*Trichoderma*), and oomycetes (*Pythium*) can also produce volatile organic compounds (VOCs) to mediate intra- and inter-kingdom interactions (Cordovez et al., 2015; Hassani et al., 2018). VOCs are low molecular weight carbon-containing compounds (<300 Da) with low boiling points and high-vapor pressure that can readily evaporate and diffuse through the rhizosphere environment. Examples of microbial VOCs include alkenes, ketones, terpenes, benzenoids, aldehydes, pyrazines, esters, and sulfur-containing compounds (Effmert et al., 2012). These species-specific signaling molecules can have bacteriostatic, fungistatic, and robust inhibitory effects on competing microbes. They can also promote the growth of neighboring bacteria in the rhizosphere to compete against their competitors or plant pathogens (Schulz-Bohm et al., 2017). Microbes that produce VOCs are good biocontrol candidates due to the volatile properties of VOCs to contact and manipulate the metabolic pathways of plant pathogens.

4.4 Microbial-mediated Disease Resistance

Besides antagonistic mechanisms, the plant's immune system could be strengthened against pathogen infections through the priming effect of plant microbiota, even without direct interaction with the plant pathogen. Without mobile immune cells and a somatic adaptive immune system, plants have developed various adaptive defense mechanisms designated to detect and suppress microbial pathogens. Plants mainly activate two interrelated pattern-triggered immunity (PTI) and effector-triggered immunity (ETI) defense mechanisms in response to pathogen invasions (Han, 2019). MAMPs are specific molecular signatures derived from the interacting microbes, whereas DAMPs are self-derived peptides/polypeptides released by wound-induced or dying cells (Choi & Klessig, 2016). However, the conserved nature of MAMPs often makes it difficult for the membrane-localized pattern recognition receptors (PRRs) to efficiently discriminate them from the DAMPs. Therefore, MAMPs and DAMPs can activate the immune system to enhance plant defense capability. Plant-beneficial microbes can trigger the induced systemic resistance (ISR) response. The priming effects of the beneficial plant microbes will lead to accelerated activation of defense response at a later post-challenge or transgenerational stage (Conrath et al., 2015). These enhanced plant defense responses are achieved through the precondition of plant defense mechanisms before pathogen infection via the ISR-mediated defense mechanisms, including active production of various phytohormones and signaling molecules such as ethylene, jasmonic acid, reactive oxygen species (ROS), phenolic compound, phytoalexins, PR proteins, and the modification of cuticle and cell wall (e.g., lignin deposition) (Romera et al., 2019; Wiesel et al., 2014). In addition, beneficial microbes can induce the production of peroxidase, polyphenol oxidase, and CWDEs, such as chitinases and β-1,3-glucanases in host plants to confer resistance against pathogenic microbes (Baiyee et al., 2019).

4.5 Microbial Biopesticides

Biopesticides are a broad group of biologically based agents derived from animals, plants, fungi, and microbes for agricultural pest control. The global biopesticide market is projected to reach USD10.19 billion in 2025, based on a compound annual growth rate of 14.93% (Fortune Business Insights, 2019). In 2017, 101 microbial biological control agents (MBCAs) were registered in Australia, Brazil, Canada, the European Union, Japan, New Zealand, and the USA for augmentative biological control in plant pest and disease management (van Lenteren et al., 2018). The mechanisms that lead to pathogen suppression include competition, antibiosis, parasitic activity, and strengthening host plant resistance (see Table 12.1). Bacteria and fungi account for the largest share of biopesticide products for all crop types.

Table 12.1 List of major groups of commercialized microbial biological control agents and their mechanisms of action

Microbes	Mechanism of action
Bacteria	
Bacillus spp.	Competition, antibiosis, and induced resistance
Pseudomonas spp.	Competition, antibiotics, volatiles, and growth-promoting substances
Streptomyces spp.	Competition, antibiotics, and cell wall degrading enzymes (CWDEs)
Fungi	
Coniothyrium minitans	Hyperparasitism
Gliocladium catenulatum	Antibiosis
Purpureocillium lilacinum	Hyperparasitism
Phlebiopsis gigantea	Competition
Trichoderma spp.	Competition, hyperparasitism, antibiotics, and CWDEs
Oomycetes	
Pythium oligandrum	Hyperparasitism

Source Pertot et al. (2015)

5 Perspectives

Next-generation crop production systems should focus on enhancing soil health and minimizing synthetic input of fertilizers and pesticides. There should be a paradigm shift in the current crop protection systems where synthetic pesticides are used indispensably. Botanical extracts and microbial biological control agents (MBCAs) provide a safe and sustainable alternative to crop protection. These are renewable resources that nature offers. However, there are some challenges in the adoption of botanical pesticides. Despite proof of the efficacy of botanical pesticides against a wide range of crop diseases, they are not easily found in the pesticide market. Issues like the availability of source plants in large quantities, storage and processing facilities, competition from synthetic pesticides, stringent regulatory procedures similar to synthetic products leading to high production costs, and low awareness of farmers on the use of botanical pesticides have to be addressed to facilitate this paradigm shift.

The robustness of plant microbiota-mediated disease protection is greatly improved by the diversity of the microbial community. Crop diversification and natural farming system that promotes sustainable agricultural practices have resulted in higher microbiota richness. In addition, the redundant (expressed by multiple microbes in the community) and dominant expression of the desirable functional traits (plant growth and survival) can lead to a synergistic effect in crop protection. In addition to the environmental influence on the microbiota community, the specific relation between crop genotypes and core microbiota is potential plant–microbe interaction and breeding research focus. It is, therefore, possible to include the ability to attract beneficial rhizosphere microbiota as a selection criterion for future crop breeding programs.

It is recommended that governments lead the paradigm shift by adopting policies that enhance the usage of bio-based products such as botanical pesticides and MBCAs. Similarly, agrochemical companies may play their role by taking up the responsibility of producing sufficient botanical pesticides for the agriculture industry. Consumers should play a critical role by pushing for chemical-free agricultural produce and products. Hence, plant and microbial diversity play an essential role in sustainable crop protection.

References

Abhilash, P. C., Powell, J. R., Singh, H. B., & Singh, B. K. (2012). Plant-microbe interactions: Novel applications for exploitation in multipurpose remediation technologies. *Trends in Biotechnology, 30*(8), 461–420. https://doi.org/10.1016/j.tibtech.2012.04.004

Afifi, M. A., & Sahar, A. M. Z. (2009). Antifungal effect of neem and some medicinal plant extracts against (*Alternaria solani*) the causes of tomato early blight. *Aspects of Applied Biology, 96*, 251–257.

Arasu, M. V., Al-Dhabi, N. A., Choi, K. C., Bensy, A. D. V., & Rajaselvam, J. (2022). Bioactive potential of *Albizia lebbeck* extract against phytopathogens and protective properties on tomato plant against speck disease in greenhouse. *Physiological and Molecular Plant Pathology, 117*, 101750. https://doi.org/10.1016/j.pmpp.2021.101750

Arora, H., Sharma, A., & Sharma, S. (2023). Thyme essential oil fostering the efficacy of aqueous extract of licorice against fungal phytopathogens of *Capsicum annuum* L. *Journal of Bioscience and Bioengineering, 135*(6), 466–473. https://doi.org/10.1016/j.jbiosc.2023.03.003

Aude, A., & Alia, D. (2015). New insights into the role of siderophores as triggers of plant immunity: What can we learn from animals? *Journal of Experimental Botany, 66*(11), 3001–3010. https://doi.org/10.1093/jxb/erv155

Azmir, J., Zaidul, I. S. M., Rahman, M. M., Sharif, K. M., Mohamed, A., Sahena, F., Jahurul, M. H. A., Ghafoor, K., Norulaini, N. A. N., & Omar, A. K. M. (2013). Techniques for extraction of bioactive compounds from plant materials: A review. *Journal of Food Engineering, 117*, 426–436. https://doi.org/10.1016/j.jfoodeng.2013.01.014

Baedke, J., Fábregas-Tejeda, A., & Delgado, A. N. (2020). The holobiont concept before Margulis. *Molecular and Developmental Evolution, 334*(3), 149–155. https://doi.org/10.1002/jez.b.22931

Bais, H. P., Weir, T. L., Perry, L. G., Gilroy, S., & Vivanco, J. M. (2006). The role of root exudates in rhizosphere interactions with plants and other organisms. *Annual Review of Plant Biology, 57*, 233–266. https://doi.org/10.1146/annurev.arplant.57.032905.105159

Baiyee, B., Ito, S. I., & Sunpapao, A. (2019). *Trichoderma asperellum* T1 mediated antifungal activity and induced defense response against leaf spot fungi in lettuce (*Lactuca sativa* L.). *Physiological and Molecular Plant Pathology, 106*, 96–101. https://doi.org/10.1016/j.pmpp.2018.12.009

Behera, S. K., Sahu, A., Das, N., & Kumari, A. (2020). Effect of bio-fumigation on nematode population and nutrient status of soil in okra. *Journal of Entomology and Zoology Studies, 8*(2), 394–397.

Benzidane, N., Aichour, R., Guettaf, S., Laadel, N., Khennouf, S., Baghiani, A., & Arrar, L. (2020). Chemical investigation, the antibacterial and antifungal activity of different parts of *Capparis spinosa* extracts. *Journal of Drug Delivery and Therapeutics, 10*(5), 118–125. https://doi.org/10.22270/jddt.v10i5.4388

Boivin, S., Fonouni-Farde, C., & Frugier, F. (2016). How auxin and cytokinin phytohormones modulate root microbe interactions. *Frontiers in Plant Science, 7*, 1240. https://doi.org/10.3389/fpls.2016.01240

Braga, R. M., Dourado, M. N., & Araujo, W. L. (2016). Microbial interactions: Ecology in a molecular perspective. *Environmental Biology, 47*, 86–89. https://doi.org/10.1016/j.bjm.2016.10.005

Brooker, M. I. H., & Kleini, D. A. (2006). *Field guide to eucalyptus* (Vol. 1, South-eastern Australia, 3rd ed.). Bloomings.

Buhian, W. P., & Bensmihen, S. (2018). Mini-review: Nod factor regulation of phytohormone signaling and homeostasis during Rhizobia-legume symbiosis. *Frontiers in Plant Science, 9*, 1247. https://doi.org/10.3389/fpls.2018.01247

Bulgarelli, D., Schlaeppi, K., Spaepen, S., Ver Loren van Themaat, E., & Schulze-Lefert, P. (2013). Structure and functions of the bacterial microbiota of plants. *Annual Review of Plant Biology, 64*, 807–838. https://doi.org/10.1146/annurev-arplant-050312-120106

Canarini, A., Kaiser, C., Merchant, A., Richter, A., & Wanek, W. (2019). Root exudation of primary metabolites: Mechanisms and their roles in plant responses to environmental stimuli. *Frontiers in Plant Science, 10*, 157. https://doi.org/10.3389/fpls.2019.00157

Cheng, S. S., Huang, C. G., Chen, Y. J., Yu, J. J., Chen, W. J., & Chang, S. T. (2009). Chemical compositions and larvicidal activities of leaf essential oils from two eucalyptus species. *Bioresource Technology, 100*(1), 452–456. https://doi.org/10.1016/j.biortech.2008.02.038

Cheek, M., Nic Lughadha, E., Kirk, P., Lindon, H., Carretero, J., Looney, B., Douglas, B., Haelewaters, D., Gaya, E., Llewellyn, T., Ainsworth, A. M., Gafforov, Y., Hyde, K., Crous, P., Hughes, M., Walker, B. E., Forzza, R. C., Wong, K. M., & Niskanen, T. (2020). New scientific discoveries: Plants and fungi. *Plants, People, Planet, 2*(5), 371–388. https://doi.org/10.1002/ppp3.10148

Chikezie, P., Ojiako, O., & Nwufo, C. (2015). Diabetes and metabolism overview of anti-diabetic medicinal plants: the Nigerian research experience. *Journal of Diabetes and Metabolism, 6*, 6. https://doi.org/10.4172/2155-6156.1000546

Chitranshi, S., Dubey, N., & Sajjad, M. (2020). Sustainable botanical products for safe post-harvest management of perishable produce: A review. *Journal of Horticulture and Postharvest Research, 3*(1), 125–140.

Choi, H. W., & Klessig, D. F. (2016). DAMPs, MAMPs, and NAMPs in plant innate immunity. *BMC Plant Biology, 16*, 232. https://doi.org/10.1186/s12870-016-0921-2

Choudhary, R., Kakraliya, S. S., Sheshma, M. K., & Bajiya, M. R. (2017). Bio-efficacy of few plant extracts/botanicals against damping off of brinjal (*Pythium ultimum*). *International Journal of Chemical Science, 1*(1), 1–3.

Chourasiya, S. S., Sreedhar, E., Babu, K. S., Shankaraiah, N., Nayak, V. L., Ramakrishna, S., Sravani, S., & Rao, M. V. B. Isolation, synthesis and biological evaluation of phenylpropanoids from the rhizomes of *Alpania galanga. Natural Product Communications, 8*(12), 1741–1746.

Compant, S., Samad, A., Faist, H., & Sessitsch, A. (2019). A review on the plant microbiome: Ecology, functions, and emerging trends in microbial application. *Journal of Advanced Research, 19*, 29–37. https://doi.org/10.1016/j.jare.2019.03.004

Conrath, U., Beckers, G. J., Langenbach, C. J., & Jaskiewicz, M. R. (2015). Priming for enhanced defense. *Annual Review of Phytopathology, 53*, 97–119.

Cools, T. L., Struyfs, C., Cammue, B. P., & Thevissen, K. (2017). Antifungal plant defensins: Increased insight in their mode of action as a basis for their use to combat fungal infections. *Future Microbiology, 12*(5), 441–454. https://doi.org/10.2217/fmb-2016-0181

Cordovez, V., Carrion, V. J., Etalo, D. W., Mumm, R., Zhu, H., van Wezel, G. P., & Raaijmakers, J. M. (2015). Diversity and functions of volatile organic compounds produced by Streptomyces from a disease-suppressive soil. *Frontiers in Microbiology, 6*, 1081. https://doi.org/10.3389/fmicb.2015.01081

Crichton, R. R., Wilmet, S., Legssyer, R., & Ward, R. J. (2002). Molecular and cellular mechanisms of iron homeostasis and toxicity in mammalian cells. *Journal of Inorganic Biochemistry, 91*(1), 9–18. https://doi.org/10.1016/s0162-0134(02)00461-0

Danhorn, T., & Fuqua, C. (2007). Biofilm formation by plant-associated bacteria. *Annual Review of Microbiology, 61*, 401–422. https://doi.org/10.1146/annurev.micro.61.080706.093316

Draz, I. S., Elkhwaga, A. A., Elzaawely, A. A., El-Zahaby, H. M., & Ismail, A. W. A. (2019). Application of plant extracts as inducers to challenge leaf rust of wheat. *Egyptian Journal of Biological Pest Control, 29*(1), 1–8. https://doi.org/10.1186/s41938-019-0109-9

Dubey, S. K., Mansoori, A., Mohan, M., Sharma, K., & Kumar, A. (2023). *Hedychium coronarium* flower extract modulates rice plant photosystem II and antioxidant enzymes activity to induce resistance against bacterial blight. *Physiological and Molecular Plant Pathology, 127*, 102114.

Effmert, U., Kalderás, J., Warnke, R., & Piechulla, B. (2012). Volatile mediated interactions between bacteria and fungi in the soil. *Journal of Chemical Ecology, 38*, 665–703. https://doi.org/10.1007/s10886-012-0135-5

Emamifar, S., Abolmaali, S., Sohrabi, S. M., Mohammadi, M., & Shahmohammadi, M. (2021). Molecular characterization and evaluation of the antibacterial activity of a plant defensin peptide derived from a gene of oat (*Avena sativa* L.). *Phytochemistry, 181*, 112586. https://doi.org/10.1016/j.phytochem.2020.112586

Farah, A., Satrani, B., Fechtal, M., Chaouch, A., & Talbi, M. (2001). Chemical composition and antimicrobial activity of essential oils from leaves of *Eucalyptus camaldulensis* and its natural hybrid (clone 583). *Acta Botanical Gallica, 148*, 183–190.

FAO. (2022). Pesticides use, pesticides trade and pesticides indicators—Global, regional and country trends, 1990–2020. In *FAOSTAT Analytical Briefs* (No. 46), Rome. https://doi.org/10.4060/cc0918en

Fenwick, G. R., Heaney, R. K., & Mullin, W. J. (1983). Glucosinolates and their break-down products in food and food plants. *Critical Reviews in Food Science and Nutrition, 18*, 123–201. https://doi.org/10.1080/10408398209527361

Feyisa, B., Lencho, A., Selvaraj, T., & Getaneh, G. (2015). Evaluation of some botanicals and *Trichoderma harzianum* for the management of tomato root-knot nematode (*Meloidogyne incognita* (Kofoid and White) Chit Wood). *Advances in Crop Science and Technology, 4*(1), 1–10. https://doi.org/10.4172/2329-8863.1000201

Fortune Business Insights. (2019). https://www.fortunebusinessinsights.com/industry-reports/biopesticides-market-100073. Accessed at August 21, 2020

Freiesleben, S. H., & Jagar, A. K. (2014). Correlation between plant secondary metabolites and their antifungal mechanisms—a review. *Medicinal and Aromatic Plant, 3*, 154. https://doi.org/10.4172/2167-0412.1000154

Friesen, M. L., Porter, S. S., Stark, S. C., von Wettberg, E. J., Sachs, J. L., & Martinez-Romere, E. (2011). Microbially mediated plant functional traits. *Annual Review of Ecology, Evolution, and Systematics, 42*, 23–46.

Garay J., Brennan T., & Bon, D. (2020). Essential oils a viable pest control alternative. *International Journal of Ecotoxicology and Ecobiology, 5*(2), 13–22. https://doi.org/10.11648/j.ijee.20200502.11

Ghag, S. B., Shekhawat, U. K. S., & Ganapathi, T. R. (2012). Petunia floral defensins with unique prodomains as novel candidates for development of Fusarium wilt resistance in transgenic banana plants. *PLoS ONE, 7*(6), e39557. https://doi.org/10.1371/journal.pone.0039557

Guilherme, E. O., Giongo, M. V., Araujo, S. H., Ferreira, T. P., Moraes, C. B., Moura, W. S., Jumbo, L. O. V, Svacina, T., Oliveira, A. C. S. S., Aguiar, R. W. A., Oliveira, E. E., & Santos, G. R. (2023). Jenipapo, *Genipa americana* L., essential oil and *Curvularia lunata* control: Potential mode-of-action; plant immune responses and selectivity against beneficial non-target organisms. *Industrial Crops and Products, 199*, 116708. https://doi.org/10.1016/j.indcrop.2023.116708

Gutjahr, C., & Paszkowski, U. (2009). Weights in the balance: Jasmonic acid and salicylic acid signaling in root-biotroph interactions. *Molecular Plant-Microbe Interactions, 22*, 763–772. https://doi.org/10.1094/MPMI-22-7-0763

Han, G. (2019). Origin and evolution of the plant immune system. *New Phytologist, 222*(1), 70–83. https://doi.org/10.1111/nph.15596

Hartman, K., & Tringe, S. G. (2019). Interactions between plants and soil shaping the root microbiome under abiotic stress. *The Biochemical Journal, 476*(19), 2705–2724. https://doi.org/10.1042/BCJ20180615

Hassani, M. A., Durán, P., Hacquard, S. (2018). Microbial interactions within the plant holobiont. *Microbiome, 6*(1), 58. https://doi.org/10.1186/s40168-018-0445-0

Helman, Y., Chernin, L. (2015). Silencing the mob: Disrupting quorum sensing as a means to fight plant disease. *Molecular Plant Pathology, 16*(3), 316–329. https://doi.org/10.1111/mpp.12180

Ismail, N. Z., Md, T. Z., Muhamad, M., Nik Mohamed Kamal, N., Mohamad Zai, N. N., & Arsad, H. (2020). Antioxidant effects, antiproliferative effects, and molecular docking of *Clinacanthus nutans* leaf extracts. *Molecules, 25*(9), 2067. https://doi.org/10.3390/molecules25092067

Jacoby, R., Peukert, M., Succurro, A., Koprivova, A., & Kopriva, S. (2017). The role of soil microorganisms in plant mineral nutrition-current knowledge and future directions. *Frontiers in Plant Science, 8*, 1617. https://doi.org/10.3389/fpls.2017.01617

Jha, S., & Chattoo, B. B. (2010). Expression of a plant defensin in rice confers resistance to fungal phytopathogens. *Transgenic Research, 19*, 373–384. https://doi.org/10.1007/s11248-009-9315-7

Joseph, A., Akere Ese, E. I., Ademiluyi, B. O., & Aluko, P. A. (2017). Efficacy of selected plant extracts in the management of tomato early blight disease caused by *Alternaria solani. Asian Journal of Plant Pathology, 11*(1), 48–52. https://doi.org/10.3923/ajppaj.2017.48.52

Kaewklom, S., Wongchai, M., Petvises, S., Hanpithakphong, W., & Aunpad, R. (2018). Structural and biological features of a novel plant defensin from *Brugmansia* x *candida. PLoS ONE, 13*(8), e0201668. https://doi.org/10.1371/journal.pone.0201668

Khatun, A. (2023). *Targeting cell wall formation in the Oomycete Phytophthora cinnamomi for disease control* (Doctoral dissertation). https://digital.library.adelaide.edu.au/dspace/handle/2440/139673

Köhl, J., Kolnaar, R., Ravensberg, W. J. (2019). Mode of action of microbial biological control agents against plant diseases: relevance beyond efficacy. *Frontiers in Plant Science, 10*, 845. https://doi.org/10.3389/fpls.2019.00845

Kumar, S., Phogat, M., Kumar, N., & Kumar, B. (2023). Evaluating the effectiveness of bio-control agents in suppressing early blight (*Alternaria solani*) in tomato plants. *The Pharma Innovation Journal, 12*(7), 3585–3588.

Majeed, A., Ahmad, H., Chaudhry, Z., Jan, G., Alam, J., & Muhammad, Z. (2011). Assessment of leaf extracts of three medicinal plants against late blight of potato in Kaghan Valley, Pakistan. *Journal of Agricultural Technology, 7*(4), 1155–1161.

Margulis, L. (1991). Symbiogenesis and symbionticism. In L. Margulis & R. Fester (Eds.), *Symbiosis as a source of evolutionary innovation: Speciation and morphogenesis* (pp. 1–14). Cambridge University Press.

Mari, M., Leoni, O., Lori, R., & Cembali, T. (2002). Antifungal vapour-phase activity of allyl isothyocyanates against *Penicillium expansum* on pears. *Plant Pathology, 51*, 231–236. https://doi.org/10.1046/j.1365-3059.2002.00667.x

Martin, A. P., Martínez, M. F., Chiesa, M. A., Garcia, L., Gerhardt, N., Uviedo, F., Torres, P. S., & Marano, M. R. (2023). Priming crop plants with rosemary (*Salvia rosmarinus* Spenn, syn *Rosmarinus officinalis* L.) extract triggers protective defense response against pathogens. *Plant Physiology and Biochemistry, 197*, 107644. https://doi.org/10.1016/j.plaphy.2023.107644

Moghaddam, E. H., Karamian, R., Shaaban, M., & Mohammadian, A. (2020). Identification and comparison of some essential oils components in seven *Eucalyptus* species cultivated in Khoramabad. *Journal of Medicinal Plants and by-Products, 1*, 59–66.

Nashwa, S. M. A., & Abo-elyousr, K. A. M. (2012). Evaluation of various plant extracts against the early blight disease of tomato plants under greenhouse and field conditions. *Plant Protection Science, 48*, 74–79.

Neelamathi, E., & Kannan, R. (2016). Screening and characterization of bioactive compounds of *Turbinaria ornata* from the Gulf of Mannar, India. *International Advanced Research Journal in Science, Engineering and Technology, 2*, 49–54.

Nelson, E. B., Simoneau, P., Barret, M., Mitter, B., & Compant, S. (2018). Editorial special issue: The soil, the seed, the microbes and the plant. *Plant and Soil, 422*, 1–5. https://doi.org/10.1007/s11104-018-3576-y

Nurhaslina, C. R., Bacho, S. A., & Mustapa, A. N. (2022). Review on drying methods for herbal plants. *Materials Today Proceedings, 63*(4), 122–139. https://doi.org/10.1016/j.matpr.2022.02.052

Oliveira, M. B., Junior, M. L., Grossi-de-Sá, M. F., & Petrofeza, S. (2015). Exogenous application of methyl jasmonate induces a defense response and resistance against *Sclerotinia sclerotiorum* in dry bean plants. *Journal of Plant Physiology, 182*, 13–22. https://doi.org/10.1016/j.jplph.2015.04.006

OuYang, Q., Duan, X., Li, L., & Tao, N. (2019). Cinnamaldehyde exerts its antifungal activity by disrupting the cell wall integrity of *Geotrichum citri-aurantii*. *Frontiers in Microbiology, 10*, 55. https://doi.org/10.3389/fmicb.2019.00055

Páez-León, S. Y., Carrillo-Morales, M., Gómez-Rodríguez, O., López-Guillén, G., Castañeda-Ramírez, G. S., Hernández-Núñez, E., Wong-Villarreal, A., & Aguilar-Marcelino, L. (2022). Nematicidal activity of leaf extract of *Moringa oleifera* Lam. against *Haemonchus contortus* and *Nacobbus aberrans*. *Journal of Helminthology, 96*, e13. https://doi.org/10.1017/S0022149X22000025

Parisi, K., Shafee, T. M. A., Quimbar, P., van der Weerden, N. L., Bleackley, M. R., & Anderson, M. A. (2019). The evolution, function and mechanisms of action for plant defensins. *Seminars in Cell and Developmental Biology, 88*, 107–118. https://doi.org/10.1016/j.semcdb.2018.02.004

Pereira, F. A. C., Silva, A. A., de Souza, E. A., de Oliveira, D. F., Macedo, W. R., & Silva, G. H. (2022). *Cinnamomum cassia* essential oil and (E)-cinnamaldehyde as control agents of anthracnose on common bean seeds. *Journal of Phytopathology, 170*(6), 414–421. https://doi.org/10.21203/rs.3.rs-845460/v1

Pertot, I., Alabouvette, C., Esteve, E. H., & Soraya, F. (2015). *Mini-paper—The use of microbial biocontrol agents against soil-borne diseases*. https://ec.europa.eu/eip/agriculture/sites/agri-eip/files/8_eip_sbd_mp_biocontrol_final.pdf

Poveda, J., Eugui, D., & Velasco, P. (2020). Natural control of plant pathogens through glucosinolates: An effective strategy against fungi and oomycetes. *Phytochemistry Reviews, 19*(4), 1045–1059. https://doi.org/10.1007/s11101-020-09699-0

Qian, H., Zhang, J., & Zhao, J. (2022). How many known vascular plant species are there in the world? An integration of multiple global plant databases. *Biodiversity Science, 30*(7), 22254. https://doi.org/10.17520/biods.2022254

RBG Kew. (2016). *The State of the World's Plants Report—2016*. Royal Botanic Gardens, Kew.

RBG Kew. (2023). *The State of the World's Plants Report—2023*. Royal Botanic Gardens, Kew.

Reading, N. C., & Sperandio, V. (2006). Quorum sensing: The many languages of bacteria. *FEMS Microbiology Letters, 254*(1), 1–11. https://doi.org/10.1111/j.1574-6968.2005.00001.x

Reganold, J. P., & Wachter, J. M. (2016). Organic agriculture in the twenty-first century. *Nature Plants, 2*, 15221. https://doi.org/10.1038/nplants.2015.221

Romera, F. J., García, M. J., Lucena, C., Martínez-Medina, A., Aparicio, M. A., Ramos, J., Alcántara, E., Angulo, M., & Pérez-Vicente, R. (2019). Induced systemic resistance (ISR) and Fe deficiency responses in dicot plants. *Frontiers in Plant Science, 10*, 287. https://doi.org/10.3389/fpls.2019.00287

Ross-Elliott, T. J., Jensen, K. H., Haaning, K. S., Wager, B. M., Knoblauch, J., Howell, A. H., Mullendore, D. L., Monteith, A. G., Paultre, D., Yan, D., Otero, S., Bourdon, M., Sager, R., Lee, J. Y., Helariutta, Y., Knoblauch, M., & Oparka, K. J. (2017). Phloem unloading in *Arabidopsis* roots is convective and regulated by the phloem-pole pericycle. *eLife, 6*, e24125. https://doi.org/10.7554/eLife.24125

Sagaram, U. S., Pandurangi, R., Kaur, J., Smith, T. J., & Shah, D. M. (2011). Structure-activity determinants in antifungal plant defensins MsDef1 and MtDef4 with different modes of action against *Fusarium graminearum*. *PLoS ONE, 6*, e18550. https://doi.org/10.1371/journal.pone.0018550

Sathoff, A. E., Velivelli, S., Shah, D. M., & Samac, D. A. (2019). Plant defensin peptides have antifungal and antibacterial activity against human and plant pathogens. *Phytopathology, 109*, 402–408. https://doi.org/10.1094/PHYTO-09-18-0331-R

Schulz-Bohm, K., Martín-Sánchez, L., & Garbeva, P. (2017). Microbial volatiles: Small molecules with an important role in intra- and inter-kingdom interactions. *Frontiers in Plant Science, 8*, 2484. https://doi.org/10.3389/fmicb.2017.02484

Segarra, G., Casanova, E., Avilés, M., & Trillas, I. (2010). *Trichoderma asperellum* strain T34 controls *Fusarium* wilt disease in tomato plants in soilless culture through competition for iron. *Microbial Ecology, 59*(1), 141–149. https://doi.org/10.1007/s00248-009-9545-5

Selvamuthukumaran, M., & Shi, J. (2017). Recent advances in extraction of antioxidants from plant by-products processing industries. *Food Quality and Safety, 1*(1), 61–81.

Song, W., Ma, X., Tan, H., & Zhou, J. (2011). Abscisic acid enhances resistance to *Alternaria solani* in tomato seedlings. *Plant Physiology and Biochemistry, 49*(7), 693–700. https://doi.org/10.1016/j.plaphy.2011.03.018

Sood, M. (2023). Jasmonates: "The Master Switch" for regulation of developmental and stress responses in plants. *Journal of Plant Growth Regulation, 8*, 5247–5265. https://doi.org/10.1007/s00344-023-11047-3

Stefanucci, A., Zengin, G., Locatelli, M., Macedonio, G., Wang, C. K., Novellino, E., Mahomoodally, M. F., & Mollica, A. (2018). Impact of different geographical locations on varying profile of bioactives and associated functionalities of caper (*Capparis spinosa* L.). *Food and Chemical Toxicology, 118*, 181–189. https://doi.org/10.1016/j.fct.2018.05.003

Sui, X. L., Zhang, T., Tian, Y. Q., Xue, R. J., & Li, A. R. (2019). A neglected alliance in battles against parasitic plants: Arbuscular mycorrhizal and rhizobial symbioses alleviate damage to a legume host by root hemiparasitic Pedicularis species. *New Phytologist, 221*, 470–481. https://doi.org/10.1111/nph.15379

Syed-Ab-Rahman, S. F., Carvalhais, L. C., & Omar, D. (2020). Development of plant-based emulsion formulations to control bacterial leaf blight and sheath brown rot of rice. *Heliyon, 6*(1), e03151. https://doi.org/10.1016/j.heliyon.2019.e03151

Tam, J. P., Wang, S., Wong, K. H., & Tan, W. L. (2015). Antimicrobial peptides from plants. *Pharmaceuticals, 8*(4), 711–757. https://doi.org/10.3390/ph8040711

Tetorya, M., Li, H., Djami-Tchatchou, A. T., Buchko, G. W., Czymmek, K. J., Shah, D. M. (2023). Plant defensin MtDef 4-derived antifungal peptide with multiple modes of action and potential as a bio-inspired fungicide. *Molecular Plant Pathology, 24*(8), 896–913. https://doi.org/10.1111/mpp.13336

Tanifuji, R., Tsukakoshi, K., Ikebukuro, K., Oikawa, H., & Oguri, H. (2019). Generation of C5-desoxy analogs of tetrahydroisoquinoline alkaloids exhibiting potent DNA alkylating ability. *Bioorganic & Medicinal Chemistry Letters, 29*(14), 1807–1811. https://doi.org/10.1016/j.bmcl.2019.05.009

Tian, L., Li, X., Tian, J., Ji, L., Chen, Y., Tran, L. P., & Tian, C. (2020). Research advances of beneficial microbiota associated with crop plants. *International Journal of Molecular Sciences, 21*(5), 1792. https://doi.org/10.3390/ijms21051792

Toju, H., Peay, K. G., Yamamichi, M., Narisawa, K., Hiruma, K., Naito, K., Fukuda, S., Ushio, M., Nakaoka, S., Onoda, Y., Yoshida, K., Schlaeppi, K., Bai, Y., Sugiura, R., Ichihashi, Y., Minamisawa, K., & Kiers, E. T. (2018). Core microbiomes for sustainable agroecosystems. *Nature Plants, 4*, 247–257. https://doi.org/10.1038/s41477-018-0139-4

Truyens, S., Jambon, I., Croes, S., Janssen, J., Weyens, N., Mench, M., Carleer, R., Cuypers, A., & Vangronsveld, J. (2014). The effect of long-term Cd and Ni exposure on seed endophytes of Agrostis capillaris and their potential application in phytoremediation of metal-contaminated soils. *International Journal of Phytoremediation, 16*, 643–659. https://doi.org/10.1080/15226514.2013.837027

Tsai, W. A., Weng, S. H., Chen, M. C., Lin, J. S., & Tsai, W. S. (2019). Priming of plant resistance to heat stress and tomato yellow leaf curl Thailand virus with plant-derived materials. *Frontiers in Plant Science, 10*, 906. https://doi.org/10.3389/fpls.2019.00906

United Nations, Sustainable Development Goals. https://www.un.org/sustainabledevelopment/sustainable-development-goals/

van Lenteren, J. C. (2012). The state of commercial augmentative biological control: Plenty of natural enemies, but a frustrating lack of uptake. *BioControl, 57*, 1–20. https://doi.org/10.1007/s10526-011-9395-1

van Lenteren, J. C., Bolckmans, K., Köhl, J., Ravensberg, W. J., & Urbaneja, A. (2018). Biological control using invertebrates and microorganisms: Plenty of new opportunities. *BioControl, 63*, 39–59. https://doi.org/10.1007/s10526-017-9801-4

Vandenkoornhuyse, P., Quaiser, A., Duhamel, M., Le Van, A., & Dufresne, A. (2015). The importance of the microbiome of the plant holobiont. *The New Phytologist, 206*(4), 1196–1206. https://doi.org/10.1111/nph.13312

Vanegas, A. M. M., Arenas, J. W., Omar Ocampo Jiménez, O., & Zulma Monsalve Fonnegra, I. (2023). Nematicidal activity and *in vitro* radical scavenging from *Piper cumbricola* and *Piper eriopodon. Biocatalysis and Agricultural Biotechnology, 47*, 102595. https://doi.org/10.1016/j.bcab.2022.102595

Wang, Y., Wang, M., Li, M., Zhao, T., & Zhou, L. (2021). Cinnamaldehyde inhibits the growth of *Phytophthora capsici* through disturbing metabolic homoeostasis. *PeerJ, 9*, e11339. https://doi.org/10.7717/peerj.11339

Wiesel, L., Newton, A. C., Elliott, I., Booty, D., Gilroy, E. M., Birch, P. R., & Hein, I. (2014). Molecular effects of resistance elicitors from biological origin and their potential for crop protection. *Frontiers in Plant Science, 5*, 655. https://doi.org/10.3389/fpls.2014.00655

Wiratno, W., Taniwiryono, D., van den Berg, H., Riksen, J. A. G., Rietjens, I., Djiwanti, S. R., Kammenga, J. E., & Murk, A. T. (2009). Nematicidal activity of plant extracts against the root-knot nematode, *Meloidogyne incognita. The Open Natural Products Journal, 2*, 1

Wong, M., Hamid, S., Nor Afifah, I. S., Nur Maliya, R., Zainina, Z. K., & Nurul Husna, A. R. (2020). Antifungal potential of six herbal plants against selected plant fungal pathogens. *Pertanika Journal of Tropical Agricultural Science, 43*(2), 107–117.

Yang, Y., Yang, X., Guo, X., Hu, X., Dong, D., Li, G., & Xiong, X. (2022). Exogenously applied methyl Jasmonate induces early defense related genes in response to *Phytophthora infestans* infection in potato plants. *Horticultural Plant Journal, 8*(4), 511–526. https://doi.org/10.1016/j.hpj.2022.04.003

Yanping, T., Shiqi, D., Yonghua, Q., Xin, X., You, Y., Liu, C., Chuntai, W., Changjie, J., & Xinqiong, L. (2023). Evaluation of medicinal plant extracts for rice blast disease control. *Rice Science, 30*(1), 6. https://doi.org/10.1016/j.rsci.2022.03.003

Advances in Other Tropical Pests Management

Management of Vertebrates Pests

Wan Mohd Hafezul Wan Abdul Ghani

Abstract Vertebrate animals can also be nuisance, harmful, and menace toward human. They can be considered as a pest if they periodically or consistently threaten human life and well-being or significantly has conflicts human activities or interests. In general, vertebrate pests are a diverse group of animals, but only some of the groups are considered as the important pests such as rodent (small mammal) and starling (bird). The conflict between men and these animals have been recorded in many parts of the world such as the vectors for diseases and direct attacks on humans. However, the damage to agricultural crops is the most common factors of conflicts. In crop protection, this type of infestation is more difficult to cope, and the control program will consume a lot of time and money. Nevertheless, their infestation resulted on huge loss in agriculture-based economy especially the primary commodity of a country. For example, the rodent infestation in oil palm plantation contributed up to 80% of loss in immature palm and reduced up to 10% of total production. Management and control of their infestation is also facing many issues such as animal welfare concern, tolerance to method of killing, cost effectiveness, and controlling at an adequate scale. Therefore, this chapter will not only be pointing out the important vertebrate pests in line with tropical crop protection, but also will cover other issue which are related to the vertebrate pest management and control.

Keywords Vertebrate pests · Nuisance animals · Human–wildlife conflicts · Pest managements

1 Introduction

Vertebrate pests are defined as any vertebrate which are native or introduced, that consistently or indirectly become nuisance or harmful to human or become conflicts in some significant way with human activities or interests (Marsh & Salmon, 2010).

W. M. H. Wan Abdul Ghani (✉)
Department of Plant Protection, Faculty of Agriculture, Universiti Putra Malaysia, 43400 Serdang, Selangor, Malaysia
e-mail: wanhafezul@upm.edu.my

M. Wong (ed.), *Advances in Tropical Crop Protection*,
https://doi.org/10.1007/978-3-031-59268-3_13

They are a diverse group of animals which included mammal, birds, reptiles and amphibian. Although fish are also vertebrates, they are usually excluded in vertebrate pests discussion because the problem cause by this type of vertebrate in insignificant (Marsh & Salmon, 2010). However, the term 'pest' that is referred to vertebrate in sometime become highly emotional and controversial. For example, the presence of pigeons in urban areas which some people see the birds as an attraction and enjoy feeding pigeons in the public square, whereas property owners in the immediate vicinity may view them as nothing more than messy pests (Giunchi et al., 2012). Therefore, the term 'pest' is always in conjunction with use of different term such as 'nuisance' to balance between the different perceptions among publics.

The scope and magnitude of vertebrate pest problems affect a wide and varied range of activities and health issues. Furthermore, the vertebrate pest issues are different across the regions. For example, in India, the increase of the human population was shrinking and degrading the habitat of many of the 1,200 bird and 500 mammal species (Sridhara, 2006). This situation resulted on the competition between human and wild animal for the same resources, which later the wild animals were classified as pests. Meanwhile, in Malaysia, the most importance vertebrate pests are the animals that are disturbance both in agriculture and urban areas (Lim, 2016). However, beside the pest of agriculture, people are more familiar with the commensal pests that directly nuisance to human.

Due to the difference, the strategies of controlling the pests are depending to the situations. For example, the outbreak of rabies that occurred in 2015 in few states in Malaysia has led to the extensive vaccination and culling of the dogs and awareness campaign (Ahmed & Ibrahim, 2019). However, the presence of the stray dogs was ignored if there was no biohazard threat detected. Furthermore, the management and control of the stray dogs are always controversies, which many animal lover movements against the treatment toward the stray dog. In agriculture sectors, one of the important pests is rodent, which the attack of rodent in plantation caused the loss of million dollars. According to Seprido and Andriani (2022), rat's infestation in oil palm plantation contributed to 50% losses of total crude palm oil per hectare per year and 80% damage to immature crop. Meanwhile, according to Singleton (2003), smallholder rice farmers reported chronic yields losses of 20–30% per annum.

Therefore, this chapter is discussing the vertebrate pests in the tropic, especially in Malaysian perspective based on the awareness and perception of the people toward this issue. The objective of this chapter is to introduce the most importance species of vertebrate pests that the people always encounter and the approach that can be taken to resolve the issue.

2 Problems Caused by Vertebrate Pests

According to Conover (2001), human and animal interactions or conflicts that become problematic are numerous and highly varied; entire volumes are devoted to the subject. Therefore, vertebrate pest problems are categorized by the nature of the conflict. Table 13.1 showed the summary of the vertebrate pest problems which occur in urban and suburban environments, with examples of the implication of the issues.

3 Types of Vertebrate Pests

3.1 *Birds*

3.1.1 Crow

Species: *Corvus splenden* (Plate 13.1).

Characteristics: The body is black in color. *C. splendens* has grayish neck and breast which contrast with the glossy black body. The bird also has a long but slender and strong bill, and a pair of strong leg with hooks. Mostly scavenger and feed on rubbish and food waste. The bird usually resting near to the food source, but their nest can be up to 2 km from the roosting areas.

Control and management: In Malaysia, shooting the bird usually is one of the extreme approaches that is taken by the authorities, especially the city councils. However, this approach is criticized by many especially from the animal welfare movement, as inhuman. Furthermore, this technique only focus to reduce the crows population aggressively, without taking into consideration other factors that cause the unsuccessful in crows control and management. For example, affinity of this invasive birds toward open rubbish area, food scraps, anthropogenic waste, and urban food source, has contributed to their high population and turnover rate (Siti Aishah et al., 2022). Moreover, most of the factors are perennial especially in urban area, making the bird very successful and uneasy to control.

Therefore, the most suitable approach to control the birds are to educate the people to properly manage the rubbish and food waste, in order to reduce the food source toward the birds. However, for a short-term solution, a few approaches can be taken such as the use of avicide, nest detriment as well as to destroy the eggs. In their roosting area, several simple modifications can be made to disallow the bird from resting at the area.

3.1.2 Sparrows

Species (Plate 13.2; Table 13.2).

Symptoms: Sparrows are considered as nuisance in both urban and rural areas. In agriculture, these species of birds always correlated with the pest of field crop such as paddy and maize. In urban area, their presence near to human habitation is more tolerated compared to pigeon and crows. However, their nesting behavior which prefer to build nest at the buildings such as inside the attic, lighting fixture,

Table 13.1 Nature of select vertebrate pest problems occurring in urban and suburban environments, along with examples of the pests implicated and the issues (Marsh & Salmon, 2010)

Nature of problem	Examples
Animal-related health problems	
Disease potential	Pigeon droppings and histoplasmosis; bats and potential rabies; rodent and bird contamination in restaurants and food-processing facilities
Parasites and host-related arthropods	Pigeons and sparrows and mite infestations; deer mice and ticks; rats and fleas; raccoons and roundworms
Animal conflicts and injuries	Rat bites to children, elderly, and pets; coyote, mountain lion, and alligator attacks on children; coyotes and mountain lions feeding on pet cats and dogs
Vehicle collisions	Deer-caused automobile accidents; bird strikes contributing to aircraft accidents; deer and gulls making takeoffs unsafe
Phobias to animals	Fear of bats, rodents, or snakes in the home; fear of almost all mammals or birds
Product and material losses	
Food loss	Rats and house mice destroying human and animal food commodities in warehouses, residences, etc
Degradation or destruction of human-made structures and materials	Gnawing of doors, cabinets, electrical wiring by rats and house mice; woodpecker damage to houses; bird droppings defacing buildings; rodents destroying stored books, artwork, and manuscripts; raccoons ripping off shingles
Communication and power outages and fires	Power outages caused by squirrels nesting or birds roosting in power substations; fires caused by rats and mice gnawing on electrical wires; pocket gophers gnawing through underground communication cables
Vegetation destruction	Rabbit, pocket gopher, vole, and deer damage to landscape plantings, and flower and vegetable gardens
Environmental consequences	
Noise disturbances	Starlings roosting on building ledges; rats and bats in attics or in walls; raccoons in the chimney
Odor pollution	Skunks beneath houses; a dead rat within a wall; bird droppings in air-conditioning systems; roaming cats defecating and spraying in gardens

(continued)

Table 13.1 (continued)

Nature of problem	Examples
Unsightly conditions	Bird droppings on sidewalks, boats, and vehicles; garbage scattered by raccoons, opossums, feral dogs, and bears; frogs and voles fallen into swimming pools
Competition with pets	Rodents and raccoons stealing pet food and causing dogs to bark; birds diving at dogs and cats; rodents infesting outdoor aviaries

Plate 13.1 *Corvus splendens*. Photo by Pixabay

kitchen vents, louvers, rooftops, gutter, and crevices caused them as an unwanted guest in the urban area.

Control and Management: In many cases, control and management of these birds are not as serious as other avian pests such as pigeon. However, in area such as the airport, proper control and management are carried out due to the safety for the area operation. The control usually involved the use of mechanical approach such as repellent devices, installation of spikes at their resting areas, proper maintenance of grass and bush to discard their food source (insect) and other electronic devices that

Lonchura punctulata

Lonchura maja

Passer montanus

Plate 13.2 Few examples of sparrows. Photo by Pixabay

Table 13.2 Paddy bird pest species in Malaysia (Badrulhadza et al., 2022)

Species	Common name
Lonchura punctulata (Linnaeus)	Ciak/pipit pinang
Lonchura striata (Linnaeus)	Ciak/pipit tuli
Lonchura atricapila (Vieillot)	Ciak/pipit kepala hitam
Lonchura maja (Linnaeus)	Ciak/pipit uban
Lonchura mallaca (Linnaeus)	Ciak/pipit rawa
Lonchura leucogastroides (Moore)	Ciak/pipit jawa
Passer montanus (Linnaeus)	Ciak/pipit Eurasia

repel them from the area. It is very rare that chemical control will be used to control the birds.

3.1.3 Pigeon

Species: *Columba* spp. (Plate 13.3).

Characteristics: For adult, the size of the bird is around 11 inches in length. The bird nest near human habitation, similar to the sparrows. They lay between 1 and

Plate 13.3 *Columba* spp. Photo by Pixabay

2 eggs per clutch and the incubation period is 19 days. After four to six weeks, the fledglings will leave the nest. The bird feed on seeds, grain, livestock feed and manure and other discarded food.

Symptoms: The feral pigeon adapts well in man-made environment. Besides crows, pigeon is the most troublesome bird pest in urban areas. In the areas, food and water always in adequate supply due to human activities. Furthermore, in some area, feeding pigeon activities are allowed and become an attraction to the area.

Pigeon dropping deface and accelerate deterioration of buildings, houses, lawn, and equipment (vehicles). Their dropping area very easy to spot at their hotspot areas, which can contaminate the areas and exposed human to zoonotic disease. Pigeon is known to carry pathogen that can cause disease such as pigeon ornithosis, encephalitis, Newcastle disease, histoplasmosis, cryptococcosis, toxoplasmosis, pseudo-tuberculosis, pigeon coccidiosis, and salmonella. Beside pathogen, they can transmit ectoparasites such as chewing lice, fleas, ticks and mites. Due to this transmission of hazardous pathogen by pigeon, sometimes they are referred as 'rats with wings' (Giunchi et al., 2012).

Control and Management: A variety of control option area available to control and manage pigeon. It includes habitat modification, exclusion, nest destruction, and live trapping. In certain areas, pigeons are shoot. However, this approach always being questioned and criticized from animal lover movements. Therefore, it is important to educate the public about the seriousness of the bird infestation and threat that they are posing. Feeding the pigeons is one of the mistakes that need to be corrected.

3.2 *Mammals*

3.2.1 Rodent

Species (Table 13.3).

Table 13.3 Rats species in Malaysia

Species	Common name	Symptoms
Rattus norvegicus	Norway rat	Gnawing mark on household tools such as plastic container
Rattus rattus diardii	Roof rat	Gnawing mark on household tools such as electrical cable
Mus musculus	House rat	Gnawing mark on household tools such as electrical cable, plastic container, plastic tools
Rattus argentiventer	Ricefield rat	Gnawing at the paddy stalk and paddy fall down
Rattus tiomanicus	Bush rat	Gnawing mark on the oil palm fruit, leave ribs

Control and Management: As other pests, IPM can be used to control the infestation of rat, both in urban and agriculture areas. However, since the Economic Injury Level (EIL) of most rat infestation is low (< 20%), the most effective approach to control and manage the outbreak is the application of chemical control (Ariff Ateed et al., 2023).

The application of chemical pesticides (rodenticide) to control rat infestation started before 1940s with the heavy metals such as arsenic and thallium or poisons such as strychnine and red squill was used as the active ingredient. However, due to the safe toward other organism especially human, the use of these chemical was discontinued. Between 1940 and 1950s, the first kind of anticoagulant rodenticides were introduced and known as first-generation rodenticides.

Using rodenticides to control rats caused several unintended side effects. The most reported cases were the effect on the non-target organisms (Mowbray, 2022).

1. Chemical control

   (a) Non-anticoagulant rodenticide

   - Bromethalin
   - Zink phosphate

   (b) Anticoagulant rodenticide (see Table 13.4)

(c) Chemical Repellent

For commensal rats, it is always recommended to use rat repellent beside the chemical rodenticide. Since the rat area mingling around human habitation, the application of rodenticide is unsafe. Furthermore, it is also a potential threat to other vertebrate such as feral cats and stray dogs. However, the application of repellent is not as efficient as the rodenticide. It is applicable when the population of rats are low.

2. Biological control of rats

   Barn owl

   Species: *Tyto javanica javanica* (Plate 13.4).

Table 13.4 Examples of anticoagulant rodenticides

Class	Chemical name	(Trade name)
1st generation	Warfarin	KG 22®, TIKUMIN®
	Chlorophacinone	BUTIK S® CHECK BAIT®
	Coumatetralyl	RACUMIN®
2nd generation	Bromadiolone	EBOR®, BUTIK G2®
	Brodifacoum	ARATTA®, JASPER®, ECHO®, MATIKUS®
	Difenacoum	RATPUS®
	Flocoumafen	STORM®

Plate 13.4 *Tyto javanica javanica*. Photo by Pixabay

Control: Introduced species in Borneo (Sabah and Sarawak) to control rat populations in oil palm plantations. It has been used in agricultural fields since 1970s as a very efficient biocontrol against rats. In the nature, barn owls do not build their own nest, hence the provision of artificial nest boxes in the field are crucial to boost their population. Barn owls feed exclusively on rodent and 99% of their diet are rats. Therefore, the bird prefers the areas where the rats are abundant such as an oil palm plantations and rice fields. The owl requires about 90 g of food a day which is equivalent to two rats of average size. Thus, a pair and their chicks (usually between 4 and 5 chicks) require 3000 rats each year. Therefore, they are very efficient in controlling the rat's infestation (The Star, 2023). Barn owl is protected in Malaysia under the Wildlife Conservation Act 2010, Act 716 Section 68(1).

3.2.2 Long-Tailed Macaque

Species: *Macaca fascicularis* (Plate 13.5).

Characteristics: Very easy to be distinguished from other species of macaque due to its long tailed. The matured adult size of the species is between 15 and 22 inches and usually, the male is larger compared to the female. The upper dorsal part was goldish gray. Meanwhile, the face and the frontal part is gray in color.

Symptoms: Usually, the damage is done when their group foraging for food. They will feed on the fruits, flowers and the trees leave. Their movement in group from one tree to another, will break the tree branches.

Control and Management: Control and management of macaques are rather difficult. They are protected under the act and any killing is prohibited. Therefore, most of the control are involving the repellent of the species such as using the frightening devices.

Plate 13.5 *Macaca fascicularis*. Photo by Pixabay

3.2.3 Wild Boar

Species: *Sus scrofa* (Plate 13.6).

Characteristics: According to Meijaard et al. (2011), the wild boar has long and pointed head, face with short snout, characteristic snout disks. Upper canines form tusks that curved up and outward. Male wild boar generally has larger canines. The wild boar back is usually rounded with relatively long legs. The body usually is brown to black in color. Their young are brown in color with longitudinal blackish stripes at torso (losing it between 2 and 6 month of age).

Symptoms: Wild boars eat anything on the forest floor such as a seed, an egg and a lizard. They are highly destructive in large numbers in part because they rip out thousands of trees and churn up the soil. According to Andrian et al. (2011), wild boars cause damage to all types of cultivated crops, both food crops, horticulture, and

Plate 13.6 *Sus scrofa*. Photo by Pixabay

plantations. This is due to the habit of wild boar destroying the soil when foraging (rooting) and making nests.

Control and Management: Wild boars caused severe damage to agricultural, so they are hunted by many people (Choquenot et al., 1996). In control and management of wild boar, many approaches were taken to reduce the number of the pests. One of the most effective ways to control wild boar population is the use of chemical pesticide such as zinc phosphide. According to Andrian et al. (2011), toxic baits and hooked baits, and the suitability of bait types for pest control of wild boars need to be ensured the effectiveness of pest control to wild boars, with toxic baits and traps. Trapping was also used especially to catch the boar alive. The snare traps used are traditional traps consisting of bamboo, ropes, rubber bands from used inner tubes, and wooden trigger snare.

3.2.4 Asian Elephant

Species: *Elephas maximus* (Plate 13.7).

Characteristics: In general, Asian elephant is smaller than African elephant. Skin color is usually gray and may be masked by soil because of dusting and wallowing. Their wrinkled skin is movable and contains many nerve centers. On average, a fully grown individual can size up to 2.75 m tall at the shoulder and 4.0 tons in weight, meanwhile their cows are smaller at about 2.40 mat the shoulder and 2.7 tons in weight. Asian elephant's trunk has only one fingerlike tip, that will be used to wrap

Plate 13.7 *Elephas maximus*. Photo by Pixabay

the food before squeezing into their mouth. One elephant adult consumes up to 150 kg of plant per days.

Symptoms: Degradation and fragmentation of elephant habitat, due to the opening of the lands for agriculture has led to the conflict between elephant and human. Human–elephant conflicts occur when elephants raid crops of plantation, which are scattered over a large area interspersed with forests.

Control and Management: Control and management of the elephant encroachment were carried out in many ways. In general, most of the approach are using the mechanisms of repelling the elephants. In some plantations, fence was established to secure the area from elephant's intrusion. Usually, the electrified fence was used to ensure the efficiency of repelling the elephant from the plantation area. Besides that, other technique to repel the elephant is the establishment of beehive at the border of plantation to repel the elephant (Badrul Amin et al., 2023). This technique is successful to reduce the elephant encroachment into the plantation, but to sustain the bee colonies is facing problems due to factors such as lack of foods source and pests infestation.

3.2.5 Malayan Porcupine

Species: *Hystrix brachyura.*

Characteristics: Porcupine is one of the largest rodent species in the world (Zainal Zahari, 2006). The size of a matured porcupine is measured between 51.5 and 85 cm from head to tail. The body of porcupines are covered in about 30,000 of sharp and rigid structures of quills. Those on their upper body parts are rough with black with white or yellow stripes. They weigh around 10–18 kg.

Symptoms: In agriculture sector, such as in a plantation, Malayan porcupine is considered as an important pest species because they feed on seedling in the plantation especially oil palm. This behavior has made them as unwanted and always hunted and killed.

Control and Management: In plantation, trapping and killing are the methods that always been used to control porcupine. However, some porcupines are killed using the similar rodenticides that are used to control rat's infestation.

3.3 *Reptiles*

3.3.1 Snake

Control and Management: Most of the management and control are situation base. Snakes usually will avoid human; except they are being threatened. However, many cases reported that snakes intruded the house and other human areas that required mechanisms to remove them. However, it is very important to know the species of snakes especially the venomous and non-venomous, so that the pre-caution action could be taken (Table 13.5).

Several approaches could be taken to avoid snake issues such as the establishment of snake-proof system during the designing the building structure, especially at the rural or near the forest areas. The type of building structure is to ensure that the house could not be encroach by the snakes. Bushes, shrubs, grass and debris around the house need to be regularly cleared to remove snake hiding places. It is highly recommended to not catching the snakes, but contacting the agencies such as Fire and Rescue Department or Civil Defense Force to help of removing the snakes from the premises. Most of the snakes captured by the agencies, will be released back into their wild environments, which is far from human activities.

4 Effect of Control and Management Toward Vertebrate Pests

According to Broom (1999), the methods that were used for the control and management of vertebrate pests might greatly contributed to consequences for the welfare of vertebrate pests species. Methods such as killing, trapping, the application of chemical and electricity to catch, before translocation, remove resource temporarily, physically exclusion, scare use repellent, changing physiologically such using vaccine or sterilizing agent might affect the welfare of the vertebrate which considered as pests. Therefore, it is important to highlight that the requirement to consider the welfare of the vertebrate pests during the management and control should not be neglected. During the management, good welfare for the animal must be assessed.

Table 13.5 Common snake species in Malaysia

	Species	Common name	Venomous / Non-venomous	Diet
1	*Ophiophagus hannah*	King Cobra Ular Tedung Selar	Venomous	Snakes and monitor lizards
2	***Malayopython reticulatus***	Reticulated Python Ular Sawa Batik	Non-venomous	Mammals and birds
3	*Naja naja*	Cobra Ular Tedung Senduk	Venomous	Small rodents, amphibians, birds, lizards and eggs
4	*Bungarus fasciatus*	Malayan Krait Ular Katang Tebu	Venomous	Small rodents, amphibians, birds, lizards and eggs
5	***Chrysopelea ornata***	Golden tree snake Ular Pokok Emas	Non-venomous	Bats, lizards, and small rodents
6	***Calloselasma rhodostoma***	Malayan Pit Viper	Venomous	Small rodents, amphibians, birds, lizards and eggs

References

Ahmed, K., & Ibrahim, M. Y. (2019). The silent expansion of rabies in Borneo. *Borneo Journal of Medical Sciences*. https://doi.org/10.51200/bjms.v0i0.1837

Andrian, J., Priyambodo, S., Dadang, (2011). Mechanical and chemical management of wild boar (*Sus scrofa* L.) pests in the Bireuen District of Aceh Province. *IOP Conference Series: Earth and Environmental Science*, 1–9. https://doi.org/10.1088/1755-1315/1208/1/012011

Ariff Ateed, M. N., Muhammad Syafiq, M. Z., Wan Mohd Hafezul, W. A. G., Abu Hassan, A., & Hasber, S. (2023). Field efficacy of anticoagulant rodenticides against rat infestation in oil palm plantation. *Journal iof Oil Palm Research, 35*(2), 365–375

Badrul Amin A. M., Norasmah B., Hisham H., Wan Nur Amina W. M., Faizol Shamrie M. S., Anis Nazirah, A., Wahizatul Afzan, A., & Mohamed Nor, Z. (2023). Sustaining beehive-fence to reduce elephant encroachment in oil palm plantation. *Journal of Oil Palm Research,* 35(2), 330–339.

Badrulhadza, A., Ku Muhammad Hasifi, K. Y., & Razali, B. (2022). Management of bird pests in rice field. *Buletin Teknologi MARDI, 34*, 29–44.

Broom, D. M. (1999). The welfare of vertebrate pests in relation to their management. In P. D. Cowan & C. J. Feare (Eds.), *Advances in vertebrate pest management* (pp. 309–329). Filander.

Choquenot, D., McIlroy, J., & Korn, T. (1996). *Managing vertebrate pests: Feral pigs*. Australian Government Publishing Service.

Conover, M. R. (2001). *Resolving human-wildlife conflicts: The science of wildlife damage management* (pp. 1–16). Lewis Publishers.

Giunchi, D., Albores-Bajas, Y. V., Valdacinni, N. E., Vanni, L., & Soldatini, C. (2012). Feral pigeons: Problems, dynamics and control methods. In Soloneski & Larramnedy (Eds.), *Integrated pest management and pest control: Current and future tactics*. InTech Publisher.

Lim B. L. (2016). The porcupines, the common bamboo rat, squirrels and the tree-shrew as secondary pests of agriculture in Malaysia. The secondary pests of agriculture are components of our wildlife heritage that feed on crop plants intermittently and rarely cause serious damage. *UTAR Agriculture Science Journal,* 1(2).

Marsh, R. E., & Salmon, T. P. (2010). *Vertebrate pest control chemicals and their use in urban and rural environments*. Elsevier.

Meijaard, E., d'Huart, J. P., & Oliver, W. L. R. (2011). Family suidae (pigs). In *Handbook of the mammals of the world 2, hoofed mammals. Lynx Edicions, Barcelona* (pp. 248–292)

Mowbray, S. (2022). Study assesses wildlife exposure to rat poison on oil palm plantations. *Mongabay Series: Global Palm Oil*. https://news.mongabay.com/2022/07/study-assesses-wildlife-exposure-to-rat-poison-on-oil-palm-plantations/.

Seprido, S., & Andriani, D. (2022). Rats community, attack, and its control efforts on smallholder oil palm in Kuantan Singingi. *Indonesian Journal of Oil Palm Research, 30*(3), 141–152.

Singleton, G. (2003). Impacts of rodents on rice production in Asia. In *IRRI Discussion Papers 287607*. International Rice Research Institute (IRRI). https://doi.org/10.22004/ag.econ.287607

Siti Aishah, J., Umarqayum, A. B., Farah Shafawati, M. T., & Jasmine Elanie, K. (2022). House crows (Corvus splendens): The carrier of pathogenic viruses or the misunderstood bird? *Journal of Applied Animal Research, 50*(1), 678–686

Sridhara, S. (2006). An overview of vertebrate pests in India. In R. M. Timm & J. M. O'brien (Eds.), Proceeding 22nd vertebrate pests conference (pp. 510–521). University of California, Davis.

The Star. (2023). *Barn owls for pest control*. https://www.thestar.com.my/metro/metro-news/2023/03/20/barn-owls-for-pest-control. Retrieved on January 18th, 2024.

Zainal Zahari, Z. (2006). Ternakan komersil hidupan liar Malaysia Landak Raya (*Hystrix brachyura*). In *Jabatan Perlindungan Hidupan Liar dan Taman Negara Semenanjung Malaysia (PERHILITAN)* (p. 54).

A Sustainable Rat Management Model in Rice Field by Natural Propagation of a Local Predator, the Barn Owl, *Tyto javanica*

Hafidzi Mohd Noor, Mohd Na'im, and Maisarah Burhanuddin

Abstract The barn owl (*Tyto javanica*) is a prevalent species found in Malaysia and has a worldwide distribution, with the exception of the Arctic area. Due to their ability to prey on rodents, they are considered very efficient biological control agents for managing rodent pests in agroecosystems. This chapter provides a concise analysis of the ecological function of *Tyto javanica* in agricultural areas in Malaysia and their propagation in rice fields. This text examines the possible use of these organisms as a part of integrated pest management (IPM) and ecologically based rodent management (EBRM). The efficacy of using this species may be evidenced by the assessment of rodent-induced agricultural damage, the computation of the rodent populace in the impacted area, and the surveillance of nest box occupancy. Other subjects for discussion include the study of barn owl ecology and the core principles of increasing the barn owl population in rice fields. It is advisable to combine the use of barn owls for rat control with other approaches, such as biorodenticides and cultural controls. Nevertheless, it is important to use prudence while employing chemical control methods in order to mitigate the potential hazard of secondary poisoning for barn owls or other nearby predators.

Keywords Biological control · Rice crop · Rodent · Nest boxes · Integrated pest management (IPM)

H. Mohd Noor (✉) · M. Na'im
Department of Plant Protection, Faculty of Agriculture, Universiti Putra Malaysia, Serdang, Malaysia
e-mail: hafidzi.mohdnoor@gmail.com

M. Burhanuddin
Sinar Mas Agro Resources Technology Research Institute, Pekanbaru, Indonesia
e-mail: maisarahmb@upm.edu.my

M. Wong (ed.), *Advances in Tropical Crop Protection*,
https://doi.org/10.1007/978-3-031-59268-3_14

1 Introduction

Rodent is an important pest in major granaries across Southeast Asia and can cause substantial losses (Diagne et al., 2023; John, 2014). The damage from rodent activities can typically range from 5%, as recorded in Malaysia, to 12% in Indonesia (Singleton et al., 2010) but can be considerably higher in a stochastic event. A complete loss of crops was experienced in the past, whereby rats appeared like a plague and destroyed crops overnight. The magnitude of the damage and its unpredictability have prompted an aggressive effort at control that relies on rodenticides when they are available (Castaño et al., 2023; Ravindran et al., 2022).

The advent of anticoagulants was a breakthrough in the 1950s as a means of bringing down rat infestations and populations to a manageable number (Bentley, 1972). However, it led to bait resistance as greater proportions of the local populations exhibited greater tolerance or resistance to the toxicity of the rodenticide (Noh et al., 2023). This led to the formulation of a class of anticoagulant Second Generation Anti-Coagulant Rodenticide (SGAR) of a higher toxic level (up to as much as 10 times). However, the rice field ecosystem is one that is rich in wildlife diversity, particularly the prey–predator interaction, a built-in sustainable system that maintains the carrying capacity of the rat population. Secondary poisoning of predators from rodenticide applications will occasionally lead to a higher carrying capacity of the prey from reduced predation pressure (Langford et al., 2013; Yan et al., 2016).

Beginning in the late 1980s, the idea of natural and facilitated propagation of a resident predator, i.e. the barn owl *Tyto javanica* that was promising in the oil palm was tried out in Selangor and Perak. The outcomes were fascinating, as in plots where pilot projects of installations of nest boxes showed a drop in crop damage from rat activities to a mere 1 to 2% within a year of its implementation (Hafidzi & Na'im, 2003a, 2003b). The need for rodenticide application decreased by 88% from what was conventionally practiced. Studies on the fecundity, feeding behaviour, foraging, and ranging activities of the barn owl proved its viability as a biological control agent in rice fields. (Hafidzi & Na'im, 2003a, 2003b). The role of the barn owl in the context and practice of IPM has proven that the barn owl, in combination with the First-Generation Anti-Coagulant Rodenticide (FGAR), can maintain a moderately low rat population that will sustain a viable population of the former and, more importantly, a sustainable economic return to the farmers.

2 Predation in the Context of Biological Control

Predation is an important component of pest management. As a process that forms an important food chain, predation acts to maintain the stability of ecosystems in a habitat. By understanding prey–predator relationships and interactions, the predation process can be taken into account in formulating integrated pest control efforts (Riddick, 2023; Saito et al., 2023). Biological control using predators (Cinel et al.,

2020; Khallaf et al., 2023; Taniwaki et al., 2022), parasites (Manochaya et al., 2022), and microbial agents (Abdul-Ghani et al., 2012) has been widely practiced in the management of insect pests. Ooi and Kelderman (1977), Ooi () have formulated several biological control efforts on selected insects and pests in Malaysia. However, biological control in efforts to control vertebrate pests received less attention until the success of the 'encourager' programme of *Tyto javanica* stamps in controlling rats in oil palm plantations (Duckett, 1991a, 1991b; Smal, 1990).

The first approach increases the number of predators, both directly and indirectly. The presence of predators can also be increased by maintaining the specific characteristics of the environment, which is the home habitat of a predator (Brown et al., 2023; Muñoz-Cárdenas et al., 2017). For example, the conservation of peripheral forest areas known to be the refuges or hiding spots of predators. Perhaps forest clusters can be deliberately created to obscure certain predators. The role of environmental habitats in accommodating predator populations has been discussed by de Thoisy et al. (2016) and Zhang et al. (2021). Likewise, the preservation of original features of habitats such as creeks and water sources. The presence of small 32 insect, fish, amphibian, and reptile communities is an attraction and provides alternative food to a wide variety of predators, such as snakes and monitor lizards. The presence of alternative foods has been proven in several studies as a factor contributing to the impact of established predation on target pests (Erlinge, 1972; Harwood & Obrycki, 2005; Samanta et al., 2014; Van Baalen et al., 2001).

In addition, the abundance of predators can be maintained by reducing mortality factors. Measures must be taken to reduce the mortality rate of predators due to chemical control efforts using poison baits. Half of the identified area accommodating a high number of predators should be free from chemical control. These include forest fringe areas and other places of predator passage. Poison bait application should also be controlled so that baiting activity is restricted to areas with serious levels of infestation. Poison bait types also have different effects on predators. *Tyto javanica*'s population is generally not affected by the use of first-generation rodenticides such as warfarin (Hafidzi, 2019) but shows serious secondary poisoning effects resulting from the use of second-generation rodenticides such as brodifacoum and flucoumafen (Madenhall & Pank, 1980; Noh et al., 2023; Salim et al., 2014). Other control efforts, such as using physical barriers, should be closely monitored to reduce injuries to non-target animals. The direct removal of predators by hunting must be stopped and any activity detected in violation of the Wildlife Act must be dealt with strictly in accordance with the provisions of the law. The habit of capturing or killing predators for any purpose should be justified through legal means and preventive efforts.

The second approach involves providing infrastructure in habitats that can improve the performance and effectiveness of predators. The preparation of nest boxes to increase *Tyto javanica* populations proved particularly effective in increasing predation rates on rodents (Charter & Rozman, 2022; Klein et al., 2007; Smal, 1990). In addition, providing a perch can also improve the ability of predatory birds to catch their prey (Meaney et al., 2021; Raine et al., 2019; Reinert, 1984). 'Clearing' habitat by controlling ground cover plants, managing as well as gathering organic materials such as fronds, leaves, and post-harvest remnants and destroying nest spots will also increase the effectiveness of predators as they expose pests indirectly.

3 *T. javanica* and Biological Control of Rats

One example of biological control of vertebrate pests in Malaysia is rat control in palm oil plantations by using *Tyto javanica* (see Fig. 14.1). *T. javanica* is believed to have migrated from Java (Lenton, 1980) and was first reported in Malaysia around 1900 (Gibson-Hill, 1949, 1950). It was very rare until the 1970s, when the oil palm cultivation sector experienced a rapid expansion in which the oil palm crop area in Peninsular Malaysia increased from about 54,000 hectares in 1960 to over 221,000 hectares in 1970 (Buckle et al., 1997; Nambiappan et al., 2018). This development also resulted in serious rat infestation problems on oil palm plantations (Chung, 2012). Rats and other small mammals are food for natural predators such as common cobras (*Naja* sp.), pythons (*Phyton* sp.), monitor lizards (*Varanus* sp.), civet cats (Family: Viverridae), and raptors such as owls (Order: Strigiformes), black-shouldered kites (*Elanus axillaris*), and eagles (family: Accipitridae). High and rich food resources have converted a well-established habitat, such as an oil palm plantation, into an ecosystem that holds a complex food chain.

Fig. 14.1 Adult barn owl, *Tyto javanica*

The first known breeding of *T. javanica* was observed in 1969 (Wells, 1972). In the early stages of breeding *T. javanica* in oil palm fields, all nests were built in the roof crevice of a resident worker's house (Duckett, 1982). This situation causes *T. javanica* to fall out of favour due to the nocturnal disturbance of the owlets' noise (Duckett, 1986). The absence of suitable and safe nesting places is a major limiting factor in the spread of this species in oil palm habitats. Lenton (1980) has studied the possibility of preparing nest boxes to encourage the breeding of *T. javanica*. Artificial nest boxes measuring 0.5 m × 0.5 m × 1.0 m were placed on a telegraph pole in an oil palm field (see Fig. 14.2). Studies conducted at three locations, one in Johor and two in Selangor, showed that *T. javanica* had a high propensity to use the artificial nest boxes (Lenton, 1980; Wendt & Johnson, 2017). For example, during the study period, 10 nest boxes housed in a plantation in Johor produced 35 young birds. An effort to move adult birds as well as their roof-nesting youngs of a house into a similar nestbox has also resulted in success (Duckett, 1982).

In 1985, Palm Oil Research Institute of Malaysia (PORIM) tested the suitability of this bird as a sustainable rat control agent in oil palm plantations. The study included the construction of 88 nest boxes on a farm in Kuala Selangor and another 370 nest boxes covering two farms in Bahau, Negeri Sembilan (Smal, 1990). This experiment has succeeded in increasing the high population of *T. javanica* while reducing the rat population, the need for baiting and maintaining a low level of rodent infestation in oil palm plantations (Smal, 1990). An analysis of the results of a study in Kuala

Fig. 14.2 Samples of nest boxes built for breeding barn owls, *T. javanica*

Selangor based on the 'prey-predator modelling' approach showed that 29 to 40% of the reduction in the need for baiting in plantations was a direct effect of the predation of *T. javanica.*

3.1 Ecology of *T. javanica* in Oil Palm Plantations

Based on field surveys (see Table 14.1), over 98% of *T. javanica* prey in oil palm plantations consisted of rats (Hazaidah, 2005; Lenton, 1980; Norsham, 1988; Smal, 1988, 1990). This estimate is based on an analysis of the regurgitation pellets obtained around the nest. Like all species of owls in the order Strigiformes, this bird also swallows its prey or parts of it. Once the prey's tissues have been digested, the bones and feathers are thrown up in the form of feeding castings or pellets (see Fig. 14.3). From these pellets, the food composition of *T. javanica* can be determined. The remaining food of *T. javanica* in oil palm plantations consists of a variety of other small mammals, various bird species, and amphibians. Lenton (1980) found that a small portion of the prey consisted of the tree shrew (Tupaia sp.; Order Scandentia), which is 1.08%. Hazaidah (2005) found the common tree shrew comprised of a mere 0.03% ($n = 316$) of the prey identified. Smal (1988) estimated that a *T. javanica* in an oil palm field needs 90 g of food a day, the equivalent of two small adult rats or a medium-sized adult rat. Norsham (1988) observed the size of the rats brought home in several nest boxes and found that the mean size of rats eaten in oil palm fields was 89.65 + 33.3 g. Lim et al. (1991) found that *T. javanica* showed a tendency to hunt large rats. They also found prey were biased towards male rats. This may be associated with male rats being more active as well as foraging over a wider area, making them more susceptible to predation. This assumption was also based on the large size of the remains of pellets because, generally, the size of male rats is larger than that of females.

In terms of the number of rats, Lenton estimated that *T. javanica* ate 0.80 ± 0.42 rats a day. Norsham (1988), on the other hand, found the range of the number of rats eaten to be quite large, as he observed from three nest boxes, each containing 0.9 ± 0.3, 0.72 ± 0.41, and 4.11 ± 0.95 rats per day, respectively. This indicates the number of rats preyed on by *T. javanica* was higher than its food needs (Duckett & Karuppiah, 1990). According to Duckett (1991a, 1991b), a pair of birds, along with

Table 14.1 Comparison of % of mice in *T. javanica* foods

Pallet count	Location	% Rat	Source
2389	Johor and Selangor	98.2	Lenton (1980)
14,173	Kuala Selangor, Selangor	99.4	Smal (1989)
22,496	Kuala Selangor, Selangor	99.4	Smal (1990)
N/A	Kuala Selangor, Selangor	100	Norsham (1988)
300	Keluang, Johor	98.1	Hazaidah (2005)

Fig. 14.3 Feeding casts or pellet spewed by *T. javanica*

the owlets, destroy around 2,000 rats a year. Browning et al. (2016) put forward a larger estimate of no less than 3000 rats a year. He estimated that a pair of owls with four youngs in the first breeding season and three youngs in the second breeding season could destroy around 4680 rodents per year.

In terms of rat species composition, Rattus tiomanicus represents the main species in *T. javanica* food on oil palm plantations (Hazaidah, 2005). Species based on the characteristics of the skull and mandibles of the prey. He found that 51%—162 of the 316 preys (a total of 300 pellets) collected in an oil palm plantation were R. tiomanicus. The remaining pellets could not be ascertained by their species due to an incomplete or broken skull. However, based on the rats captured in the field, all 192 rats trapped were *R. tiomanicus*, supporting the assumption that 100% of the rats consumed were *R. tiomanicus*. Duckett (1991a, 1991b) estimated that around 89.8% of rats consumed by *T. javanica* in oil palm plantations were *R. tiomanicus*. Aside from *R. tiomanicus, R. argentiventer* and shew species, *Suncus murinus* was also discovered as a barn owl diet in oil palm plantations (Puan et al., 2011). Current studies on owl pallets in Malaysia show that acclimatised owls in urban areas feed on commensal rodent pests, among other critters (Saufi et al., 2020) (see Table 14.2).

3.2 Ranging Behaviour of *T. javanica*

A home range is an area used by animals for daily activities such as foraging, nesting, breeding, and seizing or feeding prey (perching for birds as part of their hunting

Table 14.2 Comparison of rodent species in *T. javanica* food in oil palm plantations and urban areas

Area	*R. tiomanicus* (%)	Other *Rattus* sp. (%)	Shrew (%)	Source
Oil palmplantation	89.78[a]	9.22	–	Lenton (1980)
	51[b]	–	–	Hazaidah (2005)
	3.5[b]	94.8	5.2	Puan et al. (2011)
Urban area	–	51.39	35.42	Saufi et al. (2020)

[a] Species are determined by using multivariate isolation function
[b] Species are determined from the mandible characteristic which is the presence of anterolabial stylids on the lower and upper molars. The remainder of the other rat prey cannot be ascertained due to incomplete sculls and mandibles

strategy) (Makarieva et al., 2005). Lenton (1980) carried out the first study to determine the *T. javanica* residential/movement range in oil palm plantations using radio-telemetry. He found that the area of movement of non-breeding male and female birds ranged from 20 to 78 ha. There were 142 ha of breeding male birds in the area. He found the combined areas used by two pairs of birds were 173.3 ha and 250.4 ha, respectively. The daily movement ranged from 0.6 to 73.7 ha.

Norsham (1987) found that the daily ranging area was 0.42–14.87 ha. The total home range recorded was between 1.51 and 41.5 ha (see Table 14.3). Generally, these studies found that the area covered by male birds was larger than that of female birds. This is because during the breeding season, male birds perform hunting activities and bring food back to the nest. This also explains why the ranges covered by male birds in the breeding season are larger than they are outside the breeding season (Cain et al., 2023).

Naim et al. (2012) compared the residential ranges of male and female birds at the beginning of the breeding season in an oil palm plantation by applying three types of rodenticides: warfarin, brodifacoum, and biorodenticides. They found the ranges of female bird dwellings on the three plots and one rodenticide-free plot (as a control) were 69.39 ha, 52.50 ha, 49.85 ha, and 28.80 ha, respectively, corresponding to 60.51, 36.95, 15.22, and 18.19 ha for male birds. This suggests that early in the breeding season, female birds need to build up enough fat storage to lay eggs and roost. Taylor (2009) discovered that the weight of the female birds had an impact

Table 14.3 A comparison of the daily home ranges and total area used by *T. javanica* in oil palm plantations (Norsham, 1989)

Sex	Min area per night (ha)	Total home range
Male	14.874 + 14.76	41.471
Male	0.428 + 0.31	1.507
Male	Large	N/A
Female	0.740 + 0.36	2.165
Female	Small	N/A
Female	Small	N/A

on the size of the clutch. Durant et al., (2000) suggests female birds begin laying eggs once sufficient fat storage builds up. Adequate fat storage is very important to produce an appropriate clutch size and increase the reproductive rate (Blem, 1976; Lee et al., 2023). Female birds need to have adequate fat storage as a guarantee if faced with food shortages (Newton, 1993).

The substantial difference in male and female residential ranges is dependent on the reproductive phase. In the early phase of breeding, females are more actively foraging than males. On the contrary, males are more active and cover a wider range of area when the female is in the phase of egg laying, incubating, and egg hatching to provide adequate food for the parent and brood. Naim et al. (2011) found female residential ranges early on in the breeding season were larger (average 117 ha) than males (average 71 ha). In the incubating phase, male dwelling ranges were much larger (average 347 ha) than those of the females (average 36 ha). Salim et al. (2014) also found the dwelling range of males during brooding was casually larger, i.e. 44.1 $\pm$ 0.49 ha ($n = 6$) compared to females (26.2 $\pm$ 1.12 ha) ($n = 10$).

Another peculiar aspect is that *T. javanica* does not exhibit exclusive territoriality in oil palm habitats. Lenton (1980) found hunting ranges overlapped between pairs without any apparent territorial defence activity except around the nest boxes. *T. javanica* received the presence of other individuals without any antagonistic behaviour. Population increases as a direct consequence of nest box provisioning have produced an interesting effect on the social behaviour of *T. javanica.* They formed communal clusters containing up to 40 individuals at any one time, particularly after hunting (Lenton, 1980). Individuals reaching the age of breeding but having not yet secured a nest also have the tendency to form such communal clusters (Lenton, 1984). This carries implications for the fact that oil palm plantations have the potential to accommodate a high population of owls, which will further increase the predation rate.

T. javanica has a high rate of reproduction in oil palm plantations. The average clutch size in Malaysia is 6.6, compared to 5.5 in Europe (Lenton, 1984). From September to January, the nest boxes on two Malaysian palm oil plantations underwent a census that revealed a range of clutch sizes, from 3.12 to 5.31. (Hafidzi et al., 2003), by comparison, the censuses made in August and December 2000 in another study in oil palm plantations observed an average clutch size of 5.3 and 2.9, respectively (Hafidzi et al., 2003). The drier August weather, where rats are more concentrated in oil palm plantations that provide food than other habitats around farms, may have an impact on this significant difference. Lenton (1980) used to record up to 9 eggs from a single nest; 32% or 39 of the 122 enumerated nest boxes in an oil palm plantation, recorded a clutch size of 7–10 eggs. In the same study, he found the highest clutch size was 13, consisting of 3 owlets and 10 eggs.

T. javanica breeds two or three times a year in Malaysia, compared to only once in Europe (Lenton, 1983). According to Na'im et al. (2011), *T. javanica* in the oil palm breed occurs twice between July and October and from November to February, sometimes followed by the third breeding season, from February to June. The difference in breeding rates is the direct effect of the high food supply in oil palm plantations, which does not fluctuate due to the change of seasons as in temperate climates. The lack of nesting places, however, has limited the spread of *T. javanica* in oil palm plantations. They typically nest in the hollows of tree trunks that are hard to come by and are vulnerable to weather and predation (Duckett & Karuppiah, 1990).

Early on, *T. javanica* was not considered an effective predator on oil palm plantations. This assumption is based on observations in the Americas and Europe where *T. javanica* hunts in open fields (Bunn et al., 1982; Mikkola, 2013). Palm fronds forming a continuous canopy complicate the movement of the owls in search of prey as *T. javanica* conducts aerial sightings of prey. In the oil palm habitat, *T. javanica* observed their potential prey from a perch site or vantage point such as palm fronds, truncated fronds, and the stumps of dead wooden trees (Duckett & Karuppiah, 1990). Thus, 'T-stands' can be provided to facilitate *T. javanica* hunting in open areas such as young oil palm stands and replanting areas. Adaptation of prey hunting behaviours according to the nature of the habitat renders *T. javanica* an effective rodent biological control agent in oil palm plantations (Séchaud et al., 2021; Smal, 1989).

3.3 Some Ecological Principles in Rat Control with *T. javanica*

Erlinge (1972) have discussed two different predation effects on prey populations in particular small mammals. Based on prey–predator models pursued by Volterra (1926), prey–predator interactions resulted in an abundance of prey populations and predators that mutually formed oscillations at different phases (see Fig. 14.4).

Stable and predictable prey–predator population swings will result in a complex system where there is periodic immigration activity by both prey and predators (Nazmul et al., 2023; Saha et al., 2023). If there is a long-term follow-up for a predator to act on the increase in prey numbers, then a prey population boom will

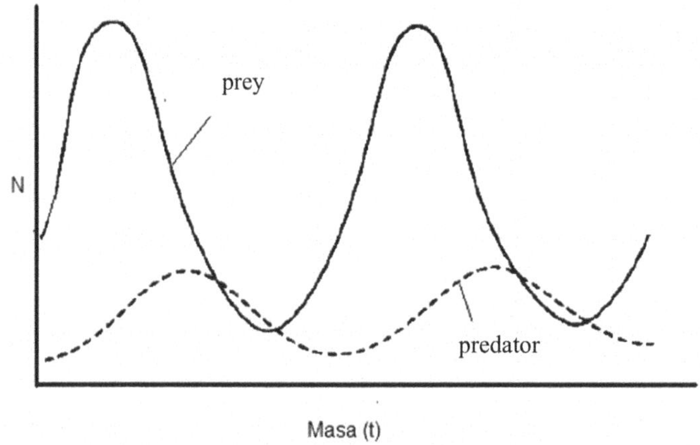

Fig. 14.4 Prey–predator relationships based on the Lotka-Voltera model

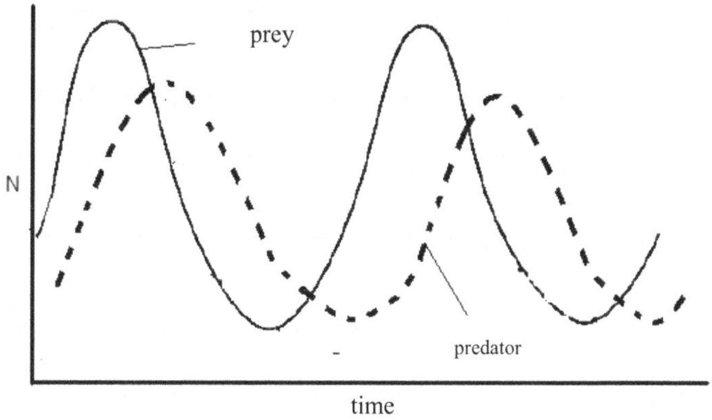

Fig. 14.5 A prey–predator relationship according to Lotka-Voltera in which the predator acts functionally (hunting the main prey when the number is abundant/easily obtained and switching to other prey when the main prey diminishes

occur until it reaches the infestation stage. Typically, this occurs when predators react numerically, which is through increasing numbers. Ricklefs (1973) shows that fluctuation in prey–predator systems such as those produced by the Lotka-Volterra model can be reduced when predators hunt prey when prey abundance is high and turn to other prey when primary prey diminishes.

A more stable system also results when the number of predators is limited by other factors, such as predation flexibility. Such predation forms a more established prey–predator interaction system in the context of pest management. In contrast to the first system, in which predators act by increasing the number of predators, which takes time and causes time lags in predation, in the second system, each individual predator acts functionally, increasing their respective predation capacity (Batabyal, 2023; Das et al., 2023). The result is that predation rates closely follow the abundance of prey, which provides continuous control (see Fig. 14.5).

How does the role of *T. javanica* in rat management fit the prey–predator system discussed above? There are actually several ecological principles that need to be understood in assessing the effectiveness of *T. javanica* in controlling rat infestation. Oil palm provides a super-optimal habitat to accommodate the extremely high rat population size. It is estimated that rat populations in oil palm plantations can reach between 300 and 400 per hectare (Wood, 1971). It is the high abundance that is actually reflected by the studies on prey composition of *T. javanica.* The high percentage represented by rats does not indicate that *T. javanica* has a specific choice or a high preference for hunting rats, but it is the food of choice due to its high numbers. In Europe the barn owl (*T. alba*) switches prey that is more readily available during a given season by virtue of the number and presence of that particular prey (Smal, 1990). Since *T. javanica* depends on an adequate supply of rodents, it is very important to maintain a sustainable population of prey and sustained level of predation. There are two methods to achieve this: providing alternative food from time to time and limiting the abundance of owls. In Texas, United States, *T. alba* which typically captures small animals also supplement its food requirement with larger alternative prey (Hamilton & Neill, 1981; Zuñiga et al., 2022). In oil palm plantations, the tendency to hunt for food other than rodents are very low. However, the types of recorded prey such as squirrels, tree shrews, frogs, small reptiles, and birds on small number of pellets indicate that oil palm may provide adequate alternative food when rats are difficult to obtain. The potential of oil palm providing alternative food for *T. javanica* population has not been assessed yet.

The second method is limiting the abundance of predator. Experience in oil palm plantations shows *T. javanica* does not exhibit flexibility in prey choice. This is not a direct effect of adequate food supply. Nomadic birds of prey such as *T. javanica* typically do not form excessive predation (Ims & Steen, 1990; Toleubayev et al., 2014; Weidensaul, 2015) and act absolutely positive towards rat numbers by increasing individual numbers through reproduction. This can cause over predation because the abundance of owls is not limited by the act of prey flexibility. This problem can however be overcome by taking into account the difficulty of the owls to acquire nesting places in habitats such as oil palm plantations and paddy fields. The nest boxes can act as a limiting factor where the abundance of owls in a field can be controlled. Smal (1990) proposed an abundance of one nest box or a pair of *T. javanica* per 10 hectares of palm oil to provide optimal rat control in an integrated pest management programme. Based on the model of Lotka-Volterra, *T. javanica*-rodent prey provides a unique predation system in oil palm habitat. A nomadic species, *T. alba* easily occupies a rich habitat of food such as oil palm plantation. *T. javanica* also shows a positive numerical response that increases the rate of reproduction when prey abundance is high. The combination of these features with communal tendencies can results in over predation cycles and explosion of rodent populations. However, the lack of nesting places acting as a limiting factor and the possible availability of alternative foods in palm oil plantations when rats are difficult to obtain produces an established predation system in the context of pest management. Experience in oil palm plantations in Malaysia shows that predation is manageable to produce optimal pest control (Zainal Abidin et al., 2021).

3.4 Rat Pest Management Using *T. javanica* in Paddy Fields

T. javanica is a common species found in rice paddies habitats in Malaysia. (Hafidzi & Na'im, 2003a, 2003b; Naim et al., 2010; Zainal Abidin et al., 2021). Buildings and premises such as Surau and community hall provide ideal nesting places for *T. javanica*. However, landscape and construction changes made it increasingly difficult for this owl to nest and breed. The modern surau and the concrete mosque have replaced the old wooden surau, which usually has a space between the ceiling and the roof. In addition, due to increased mechanisation in land preparation and rice harvesting, islands scattered in the middle of the field, either housing houses or trees, were removed to produce a continuous field. This also changed the landscape of the habitat for *T. javanica*, which reduced the availability of 'roosting' perches for hunting searches.

Like oil palm plantations, rice paddies provide suitable habitat and bountiful food to rats (Htwe et al., 2012; Lam, 1988; Phung et al., 2012). The ricefield rat *Rattus argentiventer* that inhabits soft-soaked fields is particularly suitable for living in rice paddies habitat. Since it nests in a burrow, a structure consisting of a network of borders consisting of raised bunds provides a natural nesting place and is safe to breed and protect its young. The high food supply since breeding season is closely following the phenology of the rice crop, which produces a boom in the rat population after the harvesting season and the cultivation of rice seeds in the following season. This condition causes serious damage to all levels of crops, especially at the level of booting and ripening. Reports reveal that the effects of rodent infestation under normal conditions can destroy around 10–15% of a season's rice yield. At the epidemic level (serious infestation), total destruction can occur.

So much like oil palm habitat, nesting is a limiting factor for *T. javanica* populations. In order to overcome this problem, the barn owl population-boosting project was also implemented in the rice field. The Selangor Barat-Laut Project in the Kuala Selangor paddy cultivation area was selected for the implementation of the pilot project in 1988. The rice crop area of Parit Buntar was also selected for the same purpose. Within a year of the implementation of the programme, the level of damage could be reduced from 12 to less than 2% (see Fig. 14.6 and 14.7).

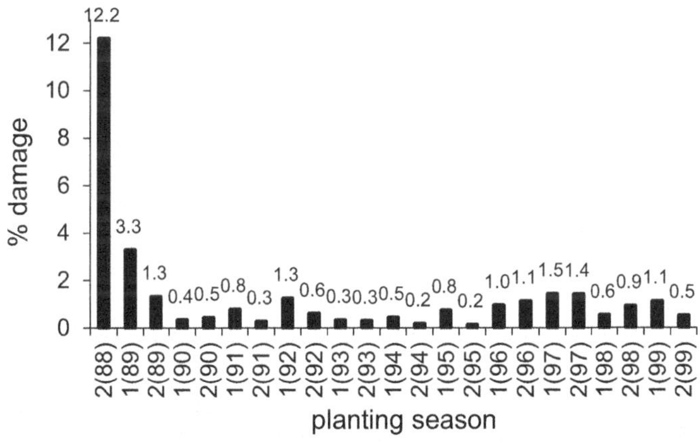

Fig. 14.6 Percent of paddy yield damage in the sawah sempadan, Tanjung Karang decreased from 12 to 1.5% or lower starting in the second season of the crop in 1989, a year after the implementation of the *T. javanica* natural propagation project through the provision of fabricated nest boxes (*Source* Department of Agriculture Malaysia)

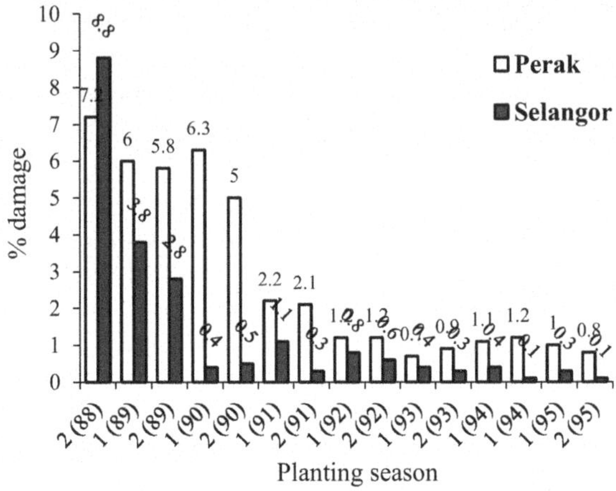

Fig. 14.7 Comparison of the average percent damage to rice crops in Perak and Selangor where the *T. javanica* propagation project was implemented (*Source* Department of Agriculture Malaysia)

3.5 Ecology T. javanica in Rice Fields

Farmers, who typically bait for rodents eight times a season, then had to bait only once per season of paddy planting. In addition to increasing harvests, control costs also decreased drastically, thus increasing the income of farmers. Rat populations influence the reproductive cycle of *T. javanica* in rice fields. The breeding cycle of rats is, in turn, influenced by the phenology of rice crops. The number of rats is low at the growth level of rice plants due to the inactive breeding status of male rats, although female rats are always in an active state of reproduction (Htwe et al., 2012; Murakami et al., 1990). The first litter was born during the paddy booting stage; the second appeared during the ripening stage; and the third was born immediately after harvesting. This condition causes the rat population to increase rapidly after harvest.

The breeding of owls usually occurs at the level of ripening, where the number of rodents begins to increase after the emergence of the second litter. The eggs of the owl hatch after a month, which is about a month or two after the rice is harvested. This is similar to the phase in which the number of rat populations peaked, which also occurred one or two months after harvesting (Murakami et al., 1990). Data from the nest box occupancy census in the rice area of Tanjung Karang conducted by the Department of Agriculture Malaysia showed nest occupancies, clutch, and brood sizes were high after harvest and before rice seeding (cultivation) in the following season (see Figs. 14.8, 14.9 and 14.10).

The high number of rodents after harvesting season also gives an advantage to owls for hunting unobstructed by rice crops. The end of the breeding season of the owl is in tandem with and closely follows the breeding season of the rodents. When the number of rats decreases, *T. javanica* spreads to other areas to meet its food requirements. The high bird numbers and declining rodent population are two factors that cause the re-distribution of birds to other areas (Altwegg et al., 2003; Sparks & Soper, 1989; Żmihorski et al., 2020).

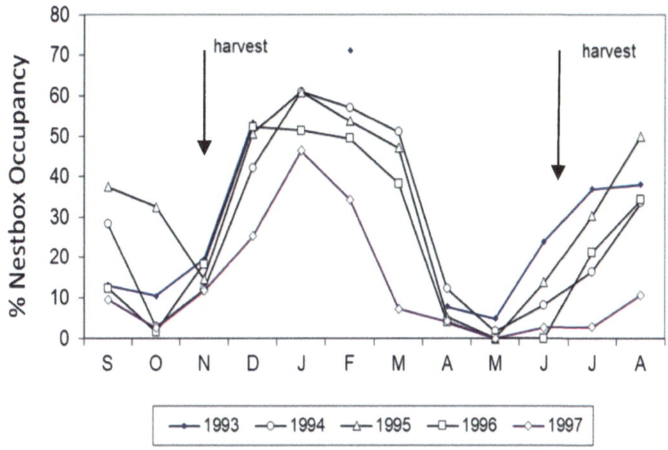

Fig. 14.8 Percentage of occupancy of the nest box by owls enumerated by the Department of Agriculture, Tanjung Karang from the period 1993–1997. Occupancy rates increase after harvest along with increasing numbers of rats in rice fields

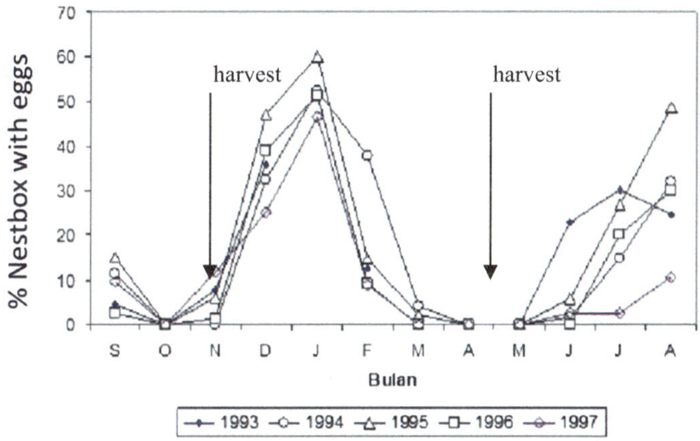

Fig. 14.9 A census of nest boxes with eggs in Tanjung Karang from 1993–1997 (*Source* Department of Agriculture Malaysia)

The nest box census in Parit Buntar, Perak, showed that nest boxes are constantly populated throughout the year, with an average occupancy of around 58% from September 2003 to July 2004 and around 89% from August 2004 to May 2005 (see Fig. 14.11). This suggests that occupancy rates vary among regions. This condition may be affected by rat population size, rainfall, and even non-concurrent and extended planting schedules. The presence of eggs and owlets peaked in February 2004, December 2004, and April 2005 (see Fig. 14.12).

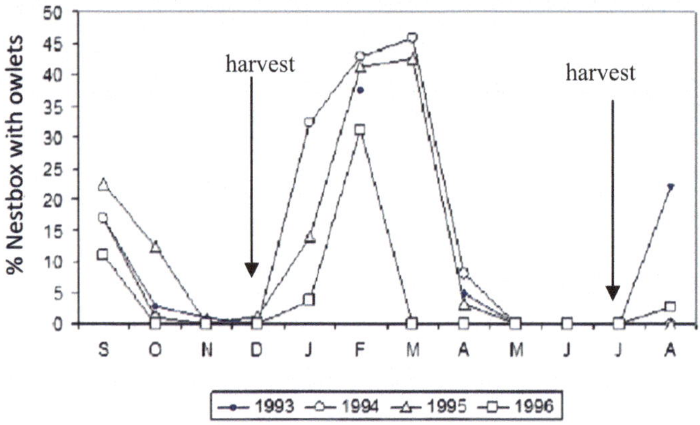

Fig. 14.10 A census of nest boxes with owlets in Tanjung Karang from 1993–1996 (*Source* Department of Agriculture Malaysia)

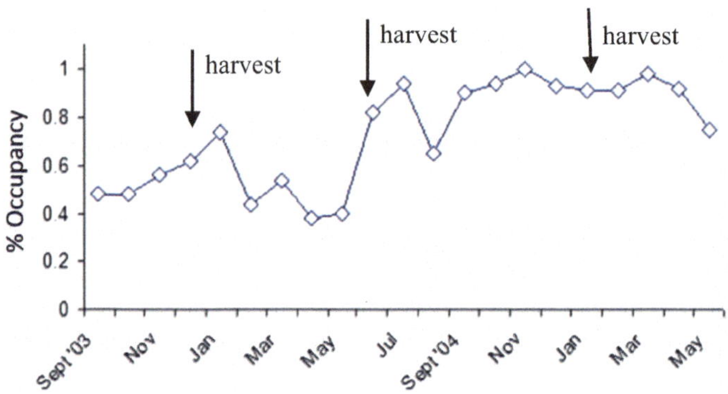

Fig. 14.11 Census of nest box occupancy in Parit Buntar, Perak from September 2003 to May 2005

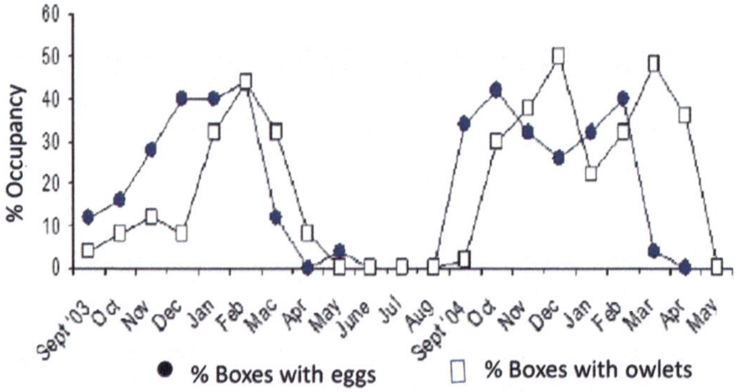

Fig. 14.12 Census of nest box occupancy with eggs and youngs in Parit Buntar, Perak from September 2003 to May 2005

3.6 *Effect of Nest Box Density on Home Range Size and Damage Levels*

Four nest box densities have been arranged in a study plot in a rice field in Tanjong Karang, Selangor: Plot A (5 ha/box), Plot B (10 ha/box), and Plot C (20 ha/box) (see Fig. 14.13). An additional plot (Plot D) with the existing approximate original density of 45 ha/box was included for comparison of crop damage assessment. A radio-telemetry study was conducted in Plots A, B, and C and the distance from each nestboxes was around 650 m (see Fig. 14.14). The range areas of barn owls were negatively influenced by nest box density (Table 14.4). Female owls hunted near nest boxes with range sizes of 5.79 ha in Plot A (5 ha/box), 14.98 ha in Plot B (10 ha/box) and an average home range size of 18.84 ha (17.05 ha and 20.63 ha) in Plot C (20 ha/box). From observations, females stay close to their nest box (167–212 m) and rarely fly into another female's territory. Males, on the other hand, ranged over a greater area, engaged in more exploratory flights (200–348 m from the nest box) and encompassed several nest boxes. Males in Plot A and Plot B had range sizes of 34.18 ha and 39.65 ha, respectively. These data suggest that males are less influenced by nest box density.

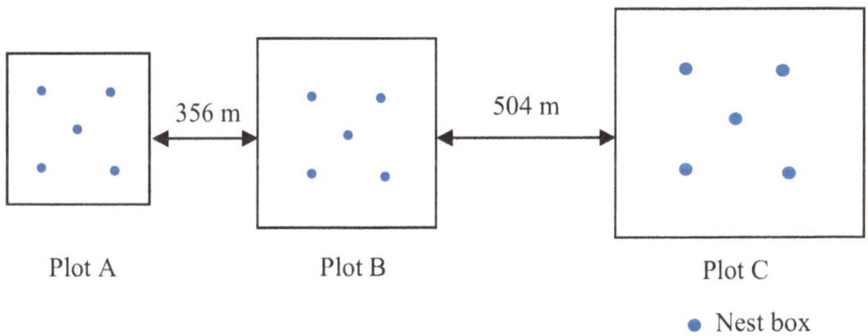

Fig. 14.13 Experimental plots representing three nest box densities. Plot A (5 ha/box), Plot B (10 ha/box), and Plot C (20 ha/box). Blue dots represents nest box

Fig. 14.14 Distance from each nestboxes was around 650 m

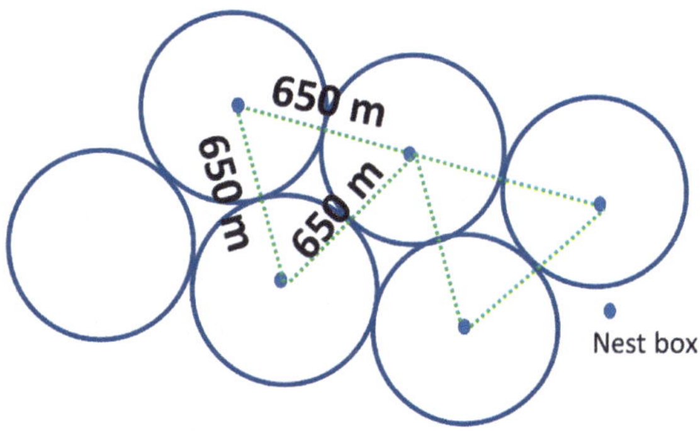

Table 14.4 The effects of nest box density on crop loss

Plot	A (5 ha/box)	B (10 ha/box)	C (20 ha/box)
Area (m^2)	50,000	100,000	200,000
Interbox distance	223 m (~ 500 m)	316 m (~ 650 m)	447 m (~ 900 m)
% Damage	0.62%	0.78%	1.57%

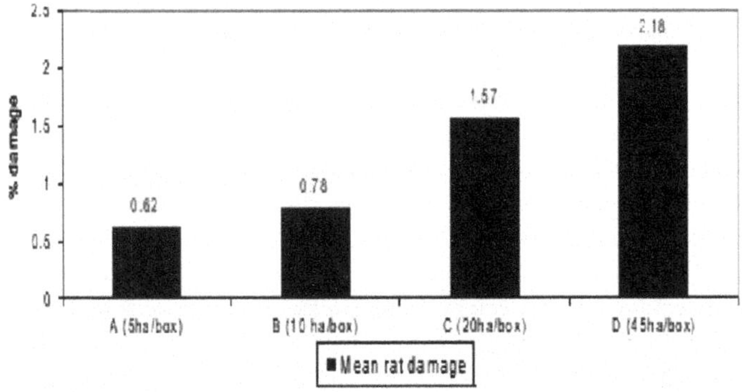

Fig. 14.15 Average % damage to rice crop in 4 different nest box densities

Figure 14.15 shows that the average damage levels for all plots in the three growth stages were 0.62, 0.79, 1.57, and 2.18% for blocks A, B, C, and D, respectively. This suggests that rat density, as reflected by the level of damage to rice crops, is more influenced by the density of female than male owls. Female owls are known to maintain an exclusive range around the nest box (Hafidzi et al., 2003), which is also suggestive of the small core area measuring < 1.5 ha in this study.

The two males, on the other hand, included at least six nest boxes and three nest boxes in the 5 ha/box area and 10 ha/box area, respectively, in their extensive home range, measuring > 30 ha. This study substantiated an earlier finding whereby males do not maintain exclusive home ranges and may encompass several nest boxes (Hafidzi et al., 2003). Therefore, females are the limiting factor and indirectly influence male hunting in rats.

4 Conclusions

Based on the ecological basis of *Tyto javanica* propagation in oil palm plantations, similar principles can be applied in the management of rats in rice fields using *T. javanica*. The effectiveness and sustainability of rat pest management through the natural propagation of *T. javanica* have been proven in rice fields and have been established in oil palm plantations. The optimum nestbox density in rice fields is lower, as the rat population can be higher than in oil palm. Aside from that, a meticulous programme should be planned when one decides to utilise barn owl in their plantation. The use of chemical control should be applied heedfully to reduce any cases of secondary poisonings towards barn owls and available predators in the area.

References

Abdul-Ghani, R., Al-Mekhlafi, A. M., & Alabsi, M. S. (2012). Microbial control of malaria: Biological warfare against the parasite and its vector. *Acta Tropica, 121*(2), 71–84. https://doi.org/10.1016/j.actatropica.2011.11.001

Altwegg, R., Roulin, A., Kestenholz, M., & Jenni, L. (2003). Variation and covariation in survival, dispersal, and population size in Barn Owls Tyto alba. *Journal of Animal Ecology, 72*(3), 391–399. https://www.jstor.org/stable/3505674

Batabyal, A. (2023). Predator–prey systems as models for integrative research in biology: the value of a non-consumptive effects framework. *Journal of Experimental Biology, 226*(19), jeb245851.

Bentley, E. W. (1972). A review of anticoagulant rodenticides in current use. *Bulletin of the World Health Organization, 47*, 275–280.

Blem, C. R. (1976). Patterns of lipid storage and utilization in birds. *American Zoologist, 16*(4), 671–684.

Brown, J. A., Lockwood, J. L., Piana, M. R., & Beardsley, C. (2023) Introduction of artificial light at night increases the abundance of predators, scavengers, and parasites in arthropod communities. *iScience, 26*, 106203. https://doi.org/10.1016/j.isci.2023.106203

Browning, M., Cleckler, J., Knott, K., & Johnson, M. (2016). Prey consumption by a large aggregation of barn owls in an agricultural setting. In *Proceedings of the Vertebrate Pest Conference* (Vol. 27, No. 27).

Buckle, A. P., Chia, T. H., Fenn, M. G. P., & Visvalingam, M. (1997). Ranging behavior and habitat utilisation of the Malayan wood rat (*Rattus tiomanicus*) in an oil palm plantation in Johore, Malaysia. *Crop Protection, 16*(5), 467–473, ISSN 0261-2194. https://doi.org/10.1016/S0261-2194(97)00010-0

Bunn, D. S., Warburton, A. B., & Wilson, R. D. (1982). *The barn owl*. A&C Black.

Cain, S., Solomon, T., Leshem, Y., Toledo, S., Arnon, E., Roulin, A., & Spiegel, O. (2023). Movement predictability of individual barn owls facilitates estimation of home range size and survival. *Movement Ecology, 11*(1), 10.

Charter, M., & Rozman, G. (2022). The importance of nest box placement for barn owls (*Tyto alba*). *Animals, 12*(20), 2815.

Castaño, P. A., Hanson, C. C., Campbell, K. J., Carrion, V., Fisher, P., Ruell, P., Will, D., & Siers, S. (2023). Invasive rodent eradication on islands: Assessment and mitigation of human exposure to rodenticides. *Biological Invasions, 25*, 653–671. https://doi.org/10.1007/s10530-022-02940-1

Chung, G. F. (2012). 6—Effect of pests and diseases on oil palm yield. O.-M. Lai, C.-P. & Tan, C.C. Akoh (Eds.). *Palm oil* (pp. 63–210), AOCS Press. ISBN 9780981893693. https://doi.org/10.1016/B978-0-9818936-9-3.50009-5

Cinel, S. D., Hahn, D. A., & Kawahara, A. Y. (2020). Predator-induced stress responses in insects: A review. *Journal of Insect Physiology, 122*, 104039, ISSN 0022-1910. https://doi.org/10.1016/j.jinsphys.2020.104039.

Das, B. K., Sahoo, D., Santra, N., & Samanta, G. (2023). Modeling predator–prey interaction: effects of perceived fear and toxicity on ecological communities. *International Journal of Dynamics and Control*, 1–33.

de Thoisy, B., Fayad, I., Clément, L., Barrioz, S., Poirier, E., & Gond, V. (2016). Predators, prey and habitat structure: Can key conservation areas and early signs of population collapse be detected in neotropical forests? *PLoS ONE, 11*(11): e0165362. https://doi.org/10.1371/journal.pone.0165362

Diagne, C., Ballesteros-Mejia, L., Cuthbert, R. N., Bodey, T. W., Fantle-Lepczyk, J., Angulo, E., Bang, A., Dobigny, G., & Courchamp, F. (2023). Economic costs of invasive rodents worldwide: The tip of the iceberg. *PeerJ, 11*, e14935. https://doi.org/10.7717/peerj.14935

Duckett, J. E. (1982). Barn owls (*Tyto alba*)—A proven natural predator of rats in oil palm. In E. Pushparajah & P. S. Chew (Eds.), *The oil palm in agriculture in the eighties* (pp. 461–472). Incorporated Society of Planters.

Duckett, J. E. (1986). The barn owl in Malaysia. *Planter, Kuala Lumpur, 62*, 531–535.

Duckett, J. E. (1991a). Management of the barn owl (*Tyto alba javanica*) as predator of rats in oil palm (*Elaeis quineensis*) plantations in Malaysia. *Birds of Prey Bulletin, 4*, 11–24.

Duckett, J. E. (1991b). The establishment and increase of barn owl (*Tyto alba javanica*) populations for rat control in Malaysian oil palm plantations. In *Paper presented at the International Conference on Biological Control in Tropical Agriculture*, August 27–30, 1991, Kuala Lumpur. Organizer: MAPPS.

Duckett, J. E., & Karuppiah, S. (1990). A guide to the planter in utilizing barn owls (*Tyto alba*) as an effective biological control in mature oil palm plantations. In *Proceedings of the 1989 International Palm Oil Development Conference-Agriculture* (pp. 357–372). Palm Oil Research Institute of Malaysia.

Durant, J. M., Massemin, S., Thouzeau, C., & Handrich, Y. (2000). Body reserves and nutritional needs during laying preparation in barn owls. Journal of Comparative Physiology B, 170, 253-260.

Erlinge, S. (1972). Interspecific relations between otter *Lutra* Decline of otters in *lutra* L. and mink *Mustela vison* Schreber in Sweden. *Oikos, 23*, 327–335.

Gibson-Hill, C. A. (1949). An annotated checklist of the birds of Malaya. *Bulletin of the Raffles plus, 20*, 3–299.

Gibson-Hill, C. A. (1950). *A checklist of birds of Singapore Island*. Raffles Museum.

Hafidzi, M. N., & Na'im, M. (2003a). The use of the barn owl, Tylo alba, to suppress rat damage in rice fields in Malaysia. *ACIAR Monograph Series, 96*, 274–276.

Hafidzi, M. N., & Na'im, M. (2003b). The use of the bam owl, *Tylo alba*, to suppress rat damage in rice fields in Malaysia. In G.R. Singleton, L.A. Hinds, C.J. Krebs, D.M. Spratt (Eds.), *Rats, Mice and People: Rodent Biology and Management* (pp. 274–276), ACIAR Monograph No. 96, 564 pp.

Hafidzi, M. N., Hamzah, M. H., & Jamaluddin, M. L. (2003). Ranging behaviour of Tyto alba in a ricefield from radio telemetry studies. *Journal of Malaysian Applied Biology, 32*(1), 47-51.

Hafidzi, M. N. (2019). Sustainable control of rats by rodenticide application and natural propagation of barn owls (*Tyto Javanica*). In *Owls*. IntechOpen.

Hamilton, K. L., & Neill, R. L. (1981). Food habits and bioenergetics of a pair of barn owls and owlets. *American Midland Naturalist,* 1–9.

Harwood, J. D., & Obrycki, J. J. (2005, September). The role of alternative prey in sustaining predator populations. In Proceedings of Second International Symposium on Biological Control of Arthropods (Vol. 2, pp. 453–462).

Hazaidah, A. H (2005) *Komposisi pemakanan burung pungguk jelapang (Tyto alba) di ladang kelapa sawit Sindora, Kluang Johor berdasarkan kepada analisis pellet* (Unpublished undergraduate thesis). Universiti Putra Malaysia

Htwe, N. M., Singleton, G. R., Hinds, L. A., Propper, C. R., & Sluydts, V. (2012). Breeding ecology of rice field rats, Rattus argentiventer and R. tanezumi in lowland irrigated rice systems in the Philippines. *Agriculture, Ecosystems & Environment, 161*, 39–45. ISSN 0167-8809. https://doi.org/10.1016/j.agee.2012.07.023.

Ims, R. A., & Steen, H. (1990). Geographical synchrony in microtine population cycles: a theoretical evaluation of the role of nomadic avian predators. *Oikos,* 381–387.

John, A. (2014). Rodent outbreaks and rice pre-harvest losses in Southeast Asia. *Food Security, 6*, 249–260.

Khallaf, M. A., Sadek, M. M., & Anderson, P. (2023). Predator efficacy and attraction to herbivore-induced volatiles determine insect pest selection of inferior host plant. *iScience, 26*, 106077. https://doi.org/10.1016/j.isci.2023.10607

Klein, Á., Nagy, T., Csörgő, T., & Mátics, R. (2007). Exterior nest-boxes may negatively affect Barn Owl Tyto alba survival: An ecological trap. *Bird Conservation International, 17*(3), 273–281. https://doi.org/10.1017/S0959270907000792

Lam, Y. M. (1988, March) Rice as a trap crop for the rice field rat in Malaysia. In *Proceedings of the Thirteenth Vertebrate Pest Conference* (p. 26).

Langford, K. H., Reid, M., & Thomas, K. V. (2013). The occurrence of second-generation anticoagulant rodenticides in non-target raptor species in Norway. *Science of the Total Environment, 450–451*, 205–208. ISSN 0048-9697. https://doi.org/10.1016/j.scitotenv.2013.01.100

Lee, M., Viegas, I., Norte, A. C., Ramos, J. A., & Araujo, P. M. (2023). Assessing the fatty acid profile of migratory birds with different fueling strategies. *Ibis, 165*(1), 297–304.

Lenton, G. M. (1980). *The ecology of barn owls (Tyto alba) in the Malay Peninsular with reference to their use in rodent control* (Ph.D. thesis) (253 p). University of Malaya.

Lenton, G. (1983). Wise owls flourish among the oil palms. *New Scientist, 97*(1345), 436-437.

Lenton, G. M. (1984). The feeding and breeding ecology of Barn Owls Tyto alba in peninsular Malaysia. *Ibis, 126*(4), 551–575.

Lim, J.L., M.G.P. Fenn, A.P. Buckle, M. Visvalingam and J.K. Ward. 1991. Prey selection by barn owls (Tyto alba) and its impact on rat control in oil palm plantations. In: (Y. Basiron, S. Jalani, K.C. Chang, S.C. Cheah, I.E. Henson, N. Kamarudin, K. Paranjothy, N. Rajanaidu, D. Tayeb and D. Ariffin, eds.) Proceedings of the 1991 PORIM International Palm Oil Conference – Progress, Prospects & Challenges Towards the 21st Century, Module I Agriculture. 9–14 September 1991. Palm Oil Research Institute of Malaysia, Kuala Lumpur. pp. 458–465.

Makarieva, A. M., Gorshkov, V. G., & Li, B. L. (2005). Why do population density and inverse home range scale differently with body size? *Implications for Ecosystem Stability. Ecological Complexity, 2*(3), 259–271.

Meaney, K. M., Peacock, D. E., Taggart, D., & Smith, J. (2021). Rapid colonization, breeding and successful recruitment of eastern barn owls (*Tyto alba delicatula*) using a customised wooden nest box in remnant mallee cropping areas of southern Yorke Peninsula, South Australia. *Wildlife Research, 48*(4), 334–344.

Mendenhall, V. M., & Pank, L. F. (1980). Secondary poisoning of owls by anticoagulant rodenticides. *Wildlife Society Bulletin*, 311–315.

Manochaya, S., Udikeri, S., Srinath, B. S., Sairam, M., Bandlamori, S. V., & Ramakrishna, K. (2022). In vivo culturing of entomopathogenic nematodes for biological control of insect pests: a review. *Journal of Natural Pesticide Research, 1*, 100005. ISSN 2773-0786. https://doi.org/10.1016/j.napere.2022.100005

Mikkola, H. (2013). *Owls of the world: A photographic guide.* A&C Black.

Muñoz-Cárdenas, K., Ersin, F., Pijnakker, J., van Houten, Y., Hoogerbrugge, H., Leman, A., Pappas, M. L., Duarte, M. V. A., Messelink, G. J., Sabelis, M. W., & Janssen, A. (2017). Supplying high-quality alternative prey in the litter increases control of an above-ground plant pest by a generalist predator. *Biological Control, 105*, 19–26. https://doi.org/10.1016/j.biocontrol.2016.11.004

Murakami, O., Priyono, J., & Triastiani, H. (1990). Population management of the ricefield rat in Indonesia. In G.R. Quick (Ed.), *Rodents and rice* (pp. 49–60). International Rice Research Institute.

Naim, M., Hafidzi, M. N., Azhar, K., & Jalila, A. (2010). Growth performance of nestling barn owls, Tyto alba javanica in rat baiting area in Malaysia. *ARPN Journal of Agricultural and Biological Science, 5*(6), 1–13.

Naim, M., Hafidzi, M. N., Azhar, K., & Jalila, A. (2011). Comparison of the breeding performance of the barn owl Tyto alba javanica under chemical and bio-based rodenticide baiting in immature oil palms in Malaysia. *Dynamic Biochemistry, Process Biotechnology and Molecular Biology, 5*(2), 5–11.

Naim, M., Umar, J., & Hafidzi, M. N. (2012). The ranging behaviour of Tyto alba in oil palm under baiting with anticoagulant rodenticides, warfarin and brodifacoum and a biorodenticides *sarcocystis singaporensis* (Zaman & Colley, 1975). *Pertanika Journal of Tropical Agricultural Science, 35*(2), 209–221.

Nambiappan, A., Ismail, A., Hashim, N., Ismail, N., Shahari, D. N., Idris, N. A. N, Omar, N., Salleh, K. M., Hassan, K. A. M., & Kushairi, A. (2018). Malaysia: 100 years of resilient palm oil economic performance. *Journal of Oil Palm Research, 30*(1), 13–25. https://doi.org/10.21894/jopr.2018.0014

Nazmul, S. K., Mondal, B., Thirthar, A. A., Alqudah, M. A., & Abdeljawad, T. (2023). Bistability and tristability in a deterministic prey–predator model: Transitions and emergent patterns in its stochastic counterpart. *Chaos, Solitons & Fractals, 176*, 114073.

Newton, S. F. (1993). Body condition of a small passerine bird: Ultrasonic assessment and significance in overwinter survival. *Journal of Zoology, 229*(4), 561–580.

Noh, A. A. M., Ahmad, A. H., & Salim, H. (2023). Efficacy of cholecalciferol rodenticide to control wood rat, Rattus tiomanicus and its secondary poisoning impact towards barn owl *Tyto Javanica Javanica. Scientific Reports, 13*(1), 2854.

Norsham, M.R. (1987). Lakuan dan pergerakan burung hantu Tyto alba dan tikus Rattus sp. Latihan ilmiah. Universiti Kebangsaan Malaysia. (Unpublished.)

Norsham, R. (1988). Kajian pergerakkan burung pungguk dan tikus dengan menggunakan radio telemetri. Universiti Kebangsaan Malaysia. Master of Science thesis. Unpublished. 80 pp

Ooi, P. A. C., & Kelderman, W. (1977). A parasite of the diamond-back moth in Cameron Highlands, Malaysia. *Malaysian Agricultural Journal, 51*, 187–190.

Ooi, P. A. C. (1979). The natural enemies of *Plutella xylostella* (L.) in Cameron Highlands, Malaysia. *Malaysian Agricultural Journal, 52*, 77–84.

Ooi, P. A. C. (1981). Microbial control of the diamond-back moth in Cameron Highlands. *Malaysia. Malaysian Applied Biology., 10*, 49–56.

Phung, N. T. M., Brown, P. R., & Leung, L. K. P. (2012). Changes in population abundance, reproduction and habitat use of the rice-field rat, Rattus argentiventer, in relation to rice-crop growth stage in a lowland rice agroecosystem in Vietnam. *Wildlife Research, 39*(3), 250-257.

Puan, C. L., Goldizen, A. W., Zakaria, M., Hafidzi, M. N., & Baxter, G. S. (2011). Absence of differential predation on rats by Malaysian barn owls in oil palm plantations. *Journal of Raptor Research, 45*(1), 71–78.

Raine, A. F., Vynne, M., & Driskill, S. (2019). The impact of an introduced avian predator, the Barn Owl Tyto alba, on Hawaiian seabirds. *Marine Ornithology, 47*, 33–38.

Reinert, S. E. (1984). Use of introduced perches by raptors: Experimental results and management implications. *Raptor Research, 18*(1), 25–29.

Ravindran, S., Mohd Noor, H. & Salim, H. (2022) Anticoagulant rodenticide use in oil palm plantations in Southeast Asia and hazard assessment to non-target animals. *Ecotoxicology*. 10. 1007/s10646-022-02559-x

Ricklefs, R. E. (1973). Patterns of growth in birds. II. Growth rate and mode of development. *Ibis, 115*(2), 177–201.

Riddick, E. W. (2023). Chapter 2—Production of coleopteran predators. J.A. Morales-Ramos, M. Guadalupe Rojas, & D.I. Shapiro-Ilan (Eds.), *Mass production of beneficial organisms* (2nd ed., pp. 13–36). Academic Press. ISBN 9780128221068. https://doi.org/10.1016/B978-0-12-822106-8.00013-0

Saha, S., Sahoo, D., & Samanta, G. (2023). Role of predation efficiency in prey–predator dynamics incorporating switching effect. *Mathematics and Computers in Simulation, 209*, 299–323.

Salim, H., Noor, H. M., Hamid, N. H., Omar, D., Kasim, A., & Abidin, C. M. R. Z. (2014). Secondary poisoning of captive barn owls, Tyto alba javanica, through feeding with rats poisoned with chlorophacinone and bromadiolone. *J Oil Palm Res, 26*(1), 62–72.

Saufi, S., Ravindran, S., Hamid, N. H., C. M., Rizuan, Z. A., Ahmad, A. H., & Salim, H. (2020). Establishment of Barn Owls (Tyto alba javanica) in an Urban Area on Penang Island, Malaysia. *Journal Raptor Research, 54*(3), 265–272. https://doi.org/10.3356/0892-1016-54.3.265

Singleton, G. R., Belmain, S., Brown, P. R., Aplin, K., & Htwe, N. M. (2010). Impacts of rodent outbreaks on food security in Asia. *Wildlife Research, 37*, 355–359.

Sparks, J., & Soper, T. (1989). *Owis their natural and unnatural history*. David & Charles.

Samanta, S., Mandal, A. K., Kundu, K., & Chattopadhyay, J. (2014). Control of disease in prey population by supplying alternative food to predator. *Journal of Biological Systems, 22*(04), 677–690.

Smal, C. M. (1988). Barn owl (Tyto alba) for the control of rats in agricultural crops in the tropics. In Symposium on Biological Control of Pests in Tropical Agricultural Ecosystem (21 p.), 1-3 June. Bogor. Seameo- Biotrop.

Smal, C. M. (1989). Research on the use of barn owls Tyto alba for biological control of rats in oil palm plantation: 1986–1989. In *PORIM International Conference*. Agriculture.

Smal, C. M. (1990). Predictive modelling of rat populations in relation to use of rodenticides or predators for control. In *PORIM Occasional Paper No. 25*. Kuala Lumpur: Palm Oil Research Institute Malaysia.

Saito, T., Buitenhuis, R., & Brownbridge, M. (2023) Use of the generalist predator *Anystis baccarum* in greenhouse IPM: Interactions with other biological control agents, a laboratory study. *Biological Control, 177*, 105127, ISSN 1049-9644. https://doi.org/10.1016/j.biocontrol.2022.105127

Séchaud, R., Schalcher, K., Machado, A. P., Almasi, B., Massa, C., Safi, K., & Roulin, A. (2021). Behaviour-specific habitat selection patterns of breeding barn owls. *Movement Ecology, 9*(1), 1–11.

Toleubayev, A., Zhumatayev, R., & Baimuhamedova, D. (2014). Image of an eagle in the art of the early nomads. *Procedia-Social and Behavioral Sciences, 122*, 240–244.

Taniwaki, T., Watanabe, K., Komine, H., Tochigi, K., Yamane, M., & Koike, S. (2022). Response of specialist and generalist predators to nonprogressive annual fluctuations in herbivorous insect populations. *Biological Control, 165*, 104810. ISSN 1049-9644. https://doi.org/10.1016/j.biocontrol.2021.104810

Taylor, I. R. (2009). How owls select their prey: A study of Barn Owls Tyto alba and their small mammal prey. *Ardea, 97*(4), 635–644.

Van Baalen, M., Křivan, V., van Rijn, P. C., & Sabelis, M. W. (2001). Alternative food, switching predators, and the persistence of predator-prey systems. *The American Naturalist, 157*(5), 512–524.

Volterra, V. (1926). Fluctuations in the abundance of a species considered mathematically. *Nature, 118*, 558–560.

Weidensaul, S. (2015). *Owls of North America and the Caribbean*. Houghton Mifflin Harcourt.

Wells, D.R. (1972). Bird Report: 1969. *Malayan Nature Journal, 25*, 43-61.

Wendt, C. A., & Johnson, M. D. (2017). Multi-scale analysis of barn owl nest box selection on Napa Valley vineyards. *Agriculture, Ecosystems & Environment, 247*, 75–83. ISSN 0167-8809. https://doi.org/10.1016/j.agee.2017.06.023

Wood, B. J. (1971). Investigations of rats in rice fields demonstrating an effective control method giving substantial yield increase. *PANS Pest Articles & News Summaries, 17*(2), 180–193.

Yan, H., Zhu, L., Zhuo, X., Shen, M., & Xiang, P. (2016). Anticoagulant rodenticide intoxication in east China: a three-year analysis. *Forensic Research, 1*(1), 22–27. 7. https://doi.org/10.1080/20961790.2016.1242042

Zainal Abidin, C. M. R., Mohd Noor, H., Hamid, N. H., Ravindran, S., Puan, C. L., Kasim, A., & Salim, H. (2021). Comparison of effectiveness of introduced barn owls, *Tyto javanica*, and rodenticide treatments on rat control in oil palm plantations. *Journal of Pest Science*, 1–14.

Żmihorski, M., Kowalski, M., Cichocki, J., Rubacha, S., Kotowska, D., Krupiński, D., Rosin, Z. M., Šálek, M., & Pärt, T. (2020). The use of socio-economy in species distribution modelling: Features of rural societies improve predictions of barn owl occurrence. *Science of the Total Environment, 741*, 140407. ISSN 0048-9697. https://doi.org/10.1016/j.scitotenv.2020.140407

Zhang, R., Liu, H., Zhang, Q., Zhang, H., & Zhao, J. (2021). Trophic interactions of reef-associated predatory fishes (*Hexagrammos otakii* and *Sebastes schlegelii*) in natural and artificial reefs along the coast of North Yellow Sea, China. *Science of The Total Environment, 791*, 148250. ISSN 0048-9697. https://doi.org/10.1016/j.scitotenv.2021.148250

Zuñiga, A. H., Rau, J. R., Fuenzalida, V., & Sandoval, R. (2022). Diet composition of the Austral Pygmy Owl in a peri-urban protected area in south-central Chile. *Zoodiversity,* 56(5)

Invasive Apple Snails (Ampullariidae): Threats and Management

Anis Syahirah Mokhtar, Arfan Ahmed Gilal, and Rita Muhamad

Abstract Alien invasive *Pomacea canaliculata* Lamarck and *Pomacea maculata* Perry were introduced from their native habitats of South America in various regions of the world for variable purposes. However, they were discarded in wild either because of mismanagement or failed to achieve desired goals and became well established in the introduced countries because of their invasive characteristics. Once established, they pose great threats to natural wetlands, macrophytes, agricultural crops (particularly rice and taro), other vertebrates such as worms, insects, crustaceans and snails, water quality of ponds and streams, as well as vectors important human and animal diseases. Billions of dollars have been lost either because of direct losses of *Pomacea* spp. or cost incurred on their management. Therefore, various control methods such as cultural, mechanical, and biological have been used to restrict the spread and population of these snails, but none of them get the desired results. Hence, chemical control is considered as the most reliable and quick method to manage their population. However, use of molluscicides such as fentin acetate, a banned but widely used pesticide, has created alarming threats to humans and environment. But, recently, botanical molluscicides and their derivatives has shown promising effectiveness against *Pomacea* spp. Therefore, a comprehensive overview of potential threats of invasive *Pomacea* spp. and their management options is given here that could help growers and policy makers to restrict the spread and losses of these snails.

Keywords Apple snails · Botanicals · Invasive species · Management · Molluscicides · Pomacea

A. S. Mokhtar (✉) · R. Muhamad
Department of Plant Protection, Faculty of Agriculture, Universiti Putra Malaysia, 43400 UPM Serdang, Selangor, Malaysia
e-mail: anissyahirah@upm.edu.my

A. A. Gilal
Department of Entomology, Faculty of Crop Protection, Sindh Agriculture University, Tandojam, Sindh, Pakistan
e-mail: aagilal@sau.edu.pk

M. Wong (ed.), *Advances in Tropical Crop Protection*,
https://doi.org/10.1007/978-3-031-59268-3_15

1 Introduction

The biodiversity in the world is constantly threatened by invasive alien species (IAS) (Bellard et al., 2016; Duenas et al., 2021; Howard et al., 2022; Simberloff, 2013). The IAS also causes billions of dollars of economic losses in various countries (Bradshaw et al., 2016; Howard et al., 2022; Pimentel et al., 2005; Sepulveda et al., 2020). The spread of IAS in the world is facilitated by ever-increasing human travel, rapid economic growth, and the changing paradigm of climate (Blumenthal, 2006; Jourdan et al., 2018; Rahel & Olden, 2008). Moreover, the physical features of the invaded areas (Davis et al., 2000) and biological features of IAS (Kolar & Lodge, 2001) have helped in their introduction, spread, and establishment.

Invasive apple snails (*Pomacea* spp.), natives of South America, have become successful invaders in various world countries during the late twentieth century (Cowie, 2002; Hayes et al., 2009, 2015). Particularly, two apple snail species, i.e., *Pomacea canaliculata* and *P. maculata*, have successfully invaded various cultivated (especially rice) and aquatic macrophytes along with natural wetlands after their introduction in new locations (Cowie et al., 2017). Generally, the term "golden apple snails (GAS)" was primarily used to describe these invasive species in the invaded regions of Southeast Asia either because of their bright orange-yellow, apple-like shells (Joshi & Sebastian, 2006; Lai et al., 2005) or the expected financial gains with their introduction in these countries (Cowie et al., 2006, 2017; Hayes et al., 2008).

Various invasive features, such as high fecundity, fast growth, omnivorous feeding ability, and strong resistance to various environmental stresses, have contributed to the invasiveness of these two species in their introduced locations (Cowie, 2002; Estebenet & Martin, 2002). Moreover, stronger reproductive potential and faster growth have been reported among the introduced population of *P. canaliculata* compared to their native populations. This difference may be attributed to evolutionary changes in the invaded areas concerning their feeding habits and more favorable climatic conditions (Zhang et al., 2023). Thus, these two *Pomacea* spp. have posed significant environmental, economic, and health threats not only to wetlands and aquatic crops but also to humans and other vertebrates (Cowie, 2002; Nghiem et al., 2013). Notably, *P. canaliculata* is a generalist omnivorous feeder with great flexibility that lacks natural enemies in the invaded areas and has a high reproductive potential and the capability to survive under diversified environmental conditions (Cowie, 2002; Lach et al., 2000). Moreover, both species also serve as vectors for parasitic species of nematodes that can cause severe human health issues (Cowie et al., 2006; Hollingsworth & Cowie, 2006; Lv et al., 2011). Considering such substantial negative impacts, one species, i.e., *P. canaliculata*, has been included among the world's 100 worst invasive species (GISD, 2023).

2 Origin and Distribution

Freshwater apple snails (Family: Ampullariidae: Order: Architaenioglossa, Class: Gastropoda) mostly have a Gondwanan origin in Africa, but they are widely distributed in humid tropics and subtropics of South and Central America, Africa, and Asia (Anis et al., 2019; Hayes et al., 2009, 2015; Mokhtar, 2016). However, members of the genus Pomacea are considered native to freshwater habitats in South and Central America (Cowie et al., 2017; Hayes et al., 2008, 2009). Among the introduced invasive species, *P. canaliculata* mostly has its natural range, which includes the Lower Paraná, Uruguay, and La Plata basins in South America. However, it may also occur in the lower regions of the Upper Paraná and parts of southern Brazil (Cowie et al., 2017; Hayes et al., 2012; Martín et al., 2017).

In comparison to *P. canaliculata, P. maculata* has a relatively sizable native range that extends from the lower Paraná River basin in Argentina and Uruguay to the north of Amazonia, passing through Paraguay and northwards in Brazil. This range overlaps with most of the native range of *P. canaliculata* (Cowie et al., 2017; Hayes et al., 2008, 2009, 2012; Thiengo et al., 2013).

During the late 1970's and early 1980's, repeated and intentional introduction of both invasive species, i.e., *P. canaliculata* and *P. maculata*, in various East and Southeast Asian countries for variable purposes has proved disastrous to cropping systems, wetlands, and human populations of these countries (Hayes et al., 2008; Lv et al., 2013; Naylor, 1996). Among the reasons for their introduction, novel human food sources for the local people and their export potential were the most important reasons for the spread and establishment of invasive *Pomacea* spp., followed by their aquarium trade potential (Cowie et al., 2017). Although it was a well-known pest of rice in coastal areas of Surinam during the 1950's (Acosta & Pullin, 1991), it was still introduced into rice-growing countries like Taiwan, Philippines, and Japan during the early 1980's because of the promising economic returns. Subsequently, it was introduced to Cambodia, China, Indonesia, Malaysia, South Korea, and Thailand (Lee & Oh, 2006; Marwoto et al., 2018; Preap et al., 2006; Suharto et al., 2006; Yahaya et al., 2017), and later in the 1990s into Laos (Anderson, 1993; EFSA et al., 2020). However, multiple factors, such as the unpalatability of the snails, import restrictions due to health issues from foreign countries, and a decline in the local market, resulted in the negligence and release or escape of both *Pomacea* spp. in the irrigation systems and general water pathways (Naylor, 1996). Afterward, a rapid spread was witnessed in the rice-growing areas of the mentioned countries. Within five years, the infested rice areas by *Pomacea* spp. in Taiwan and the Philippines reached up to 15 and 25%, respectively (Ketelaar, 1993; Mochida, 1991). Meanwhile, they invaded three-quarters of rice-growing areas in Japan and Vietnam (Naylor, 1996; Yusa & Wada, 2002). The spread of *Pomacea* spp. may have reached northern Japan, as the cooler temperature has not limited their movement (Ito, 2002). However, their behavioral adaptations and acclimation may facilitate their potential spread in the north (Matsukura et al., 2008; Wada & Matsukura, 2007).

Among other reasons for introducing *Pomacea* spp. are their potential in the aquarium trade and their ability to manage local weeds or snails that vector schistosomes (Cowie & Hayes, 2012). The introduction of *P. canaliculata* in the Pacific Islands, particularly Hawaii, was either through the aquarium trade around 1989 (Cowie et al., 2007; Levin et al., 2006) or the importation of infested parts of taro (Levin et al., 2006) and boats used in the transportation (EFSA, 2012). In North America, *P. canaliculata* is now present in California, Arizona, Florida, Georgia, Texas (Howells et al., 2006), Alabama (Martin et al., 2012), and South Carolina (Benson, 2018).

The introduction of *Pomacea* spp. in the Asian subcontinent countries, i.e., Pakistan (Baloch, 2017), Myanmar (Myint & Ye, 2017), and Sri Lanka (Epa, 2006) was also because of their aquarium trade potential. However, their potential escape and spread into the neighboring countries (India and Bangladesh) may have severe consequences for the crops and the environment (Djeddour et al., 2021).

The voracious feeding of ampullariidae on aquatic plants also facilitates the introduction of invasive *Pomacea* spp., particularly *P. canaliculata* into Southeast Asian countries (Joshi & Sebastian, 2006). Another reason for their possible introduction was to control snail vectors of *Schistosoma* spp. through competition and predation (Cowie & Hayes, 2012).

Although invasive *Pomacea* spp. were introduced for various purposes, they failed to capitalize on the market and were discarded in the wild to become severe pests of aquatic plants and wetlands (Cowie et al., 2017; Horgan et al., 2014a). Now, both invasive *Pomacea* spp. are reported from different continents of the world, i.e., Asia (Cambodia, China, Egypt, Indonesia, Israel, Japan, Kenya, Laos, Malaysia, Myanmar, Pakistan, Philippines, Singapore, South Korea, Taiwan, Thailand, Vietnam), Africa (South Africa, Kenya, Mauritius), Australia (Australia, New Zealand, Papa New Guinea), Europe (Spain, Switzerland, France, and Russia), South America (Chile, Ecuador), and North America (USA, Dominican Republic, Haiti, Puerto Rico) along with Pacific islands especially Hawaiian islands and Guam (Baloch, 2017; Buddie et al., 2021; Burks et al., 2011; Collier et al., 2011; Cowie et al., 2017; Djeddour et al., 2021; EFSA, 2020; Ip et al., 2014; Khin & Khin, 2017; López et al., 2010; Ng et al., 2017; Rodriguez et al., 2017; Rosario & Moquete, 2006; Shivambu et al., 2020; Yahaya et al., 2017; Yang et al., 2018).

Despite their widespread invasion into many countries around the world, successful eradication efforts have been made in the Ebro Delta, Spain (Joshi & Parera, 2017), northern Spain (EFSA et al., 2020), and some municipalities in France and Switzerland (EFSA et al., 2020).

3 Feeding Habits, Damage, and Threats of Invasive *Pomacea* spp.

The invasive *Pomacea* spp. exhibit great flexibility in their feeding behaviors, capable of shredding (using their jaws to get submerged pieces of materials), scraping (utilizing their radula to collect materials adhering to the submerged surface), and collecting (using their foot to gather materials floating at the water surface) (Cummins & Klug, 1979; Djeddour et al., 2021). Moreover, they can also use all three mechanisms in quick succession to get vegetal, animal, microbial, or detrital materials (Cowie, 2002; Morrison & Hay, 2011). Hence, such an uncommon ability of invasive *Pomacea* spp. help them feed on various foods rapidly with effective processing to amplify mechanical digestion because of the presence of powerful enzymes (Martín et al., 2019). Nevertheless, during their initial development after hatching from eggs, they mostly feed on submerged algal or soft-bodied detritus materials. However, as they grow, they become capable of feeding on higher plants, including rice (Arfan et al., 2015).

Considering such diversified feeding potential, damage and threats of both *Pomacea* spp. can be described below:

3.1 *Damage to Rice and Other Macrophytes*

Both invasive *Pomacea* spp. mostly inhibit slow-moving or lentic water sources in tropical to warm temperate regions. Hence, aquatic macrophytes are identified as their primary feeding source (Djeddour et al., 2021). Among the macrophytes fed upon by both *Pomacea* spp. are crops of varying importance, such as rice, taro, morning glory, mat rush, lotus, Chinese mat grass, Japanese parsley, watercress, water chestnuts, azolla, wild rice, and water lilies (Burks et al., 2011; Burlakova et al., 2009; Horgan, 2018; Yusa, 2006). Generally, most of these plants are quite unpalatable to invasive *Pomacea* spp., but these are the only trophic options for them in monocultures (Djeddour et al., 2021).

Among the macrophytes mentioned, rice is the most essential crop damaged by *Pomacea* spp. Significant yield losses have been reported in many Southeast Asian countries, i.e., Thailand, Philippines, Malaysia, Indonesia, and Myanmar, along with China, Japan, Taiwan, and Vietnam (Anis et al., 2019; Arfan et al., 2014; Burlakova et al., 2009; Mokhtar, 2016; Suharto et al., 2006). Missing rice hills (see Fig. 15.1) and floating rice fragments are the common symptoms of damage of Pomacea spp. in the field (Gilal et al., 2016; Sanico et al., 2002).

The density and size of snails profoundly impact the damage to rice (Arfan et al., 2017; EFSA, 2012; Gilal et al., 2016;). A snail density of 8 snails per m^2 can cause 90% damage to rice production (Lee et al., 2010; Naylor, 1996), and a single snail can feed up to 24 young rice seedlings daily (Cowie, 2002). Moreover, snails with more than 5 cm shell height are three times more damaging than snails below 2 cm,

Fig. 15.1 Apple snail feeds on rice tillers and causes empty patches

whereas juveniles generally feed on soft algal and detritus materials in the fields (Arfan et al., 2015; Naylor, 1996). The extent of damage to rice by *Pomacea* spp. also depends on the rice sowing method (direct or transplanted) and its growth stage. They can completely destroy both direct-seeded and 14-day transplanted seedlings (Arfan et al., 2017; Lee et al., 2010; Sanico et al., 2002). However, after 21 and 28 days, the transplanted rice seedlings become tolerant to *Pomacea* spp. damage, particularly after the 5th week of transplanting (Arfan et al., 2017; Horgan et al., 2014b; Sanico et al., 2002).

The water level in rice fields also plays a significant role in determining the level of *Pomacea* spp. damage to rice as water level above their shell height can facilitate movement and damage of the snails (Gilal et al., 2016; Liang et al., 2013; Sanico et al., 2002; Sin, 2003). Therefore, it has been suggested to drain the irrigation water up to saturation level after sowing direct-seeded or 21-day-old seedlings to minimize the damage caused by *Pomacea* spp. (Arfan et al., 2017; Sin, 2003). Among the factors mentioned earlier, the water level in rice fields has been identified as the critical factor in determining the impact of *Pomacea* spp. level on rice crops, followed by the snail density (Sin, 2003).

Both the invasive *Pomacea* spp. are capable of using pedal surface collecting, i.e., capturing surface material to continue their survival when their favorable and palatable aquatic plants have been depleted (Saveanu & Martín, 2015). Moreover, Qiu et al. (2011) also reported the ability of *P. canaliculata* to feed on both fresh and decaying leaves of aquatic plants to continue their survival throughout the year.

3.2 Threat to Natural Ecosystems

Invasive *Pomacea* spp. has not only threatened rice food security in different countries, but their indiscriminate feeding has depleted the palatable aquatic vegetation to favor the dominance of unpalatable species. This phenomenon has caused a disturbance to the normal functionality of aquatic ecosystems (Horgan et al., 2014a; Nghiem et al., 2013). It has been reported that voracious feeding of *P. canaliculata* has resulted in the complete loss of higher aquatic plants, resulting in enhanced phytoplankton densities and turbidities of the water that leads to eutrophication (Carlsson, 2017; Martín et al., 2019). Accordingly, clear and wetland-rich waters have turned into turbid waters, greatly disturbing the normal functionality of aquatic ecosystems that provide food and habitats for many aquatic organisms (Carlsson, 2017). The voracious feeding of *P. maculata* in North America has facilitated the fast growth and dispersion of Hydrilla verticillate, an invasive weed (Meza-Lopez & Siemann, 2015; Monette et al., 2017). *Pomacea* spp. are also capable of bioaccumulating pesticides (Coat et al., 2011), toxins (Berry & Lind, 2010), and heavy metals (Dummee et al., 2012; Peña, 2017; Vega et al., 2012) in aquatic habitats.

A recent study by Liu et al. (2023) highlighted the potential threat posed by *P. canaliculata* to mangrove species such as *Acanthus ilicifolius* and *Sonneratia apetala* in Guangzhou, China, by exploiting the debris on the surrounding ground areas. They also reported the positive effect of phosphorus contents and ash of decaying leaf litter of these mangrove species on the feeding and growth of *P. canaliculata.* In contrast, lignin, tannin, and flavonoid contents were found to exhibit negative impacts. Wei et al. (2018) and Liu et al. (2022) also confirmed the invasion of mangrove areas of Hainan Island and Guangzhou, China.

3.3 Threats to Human Health

Many hazardous human diseases that are caused by zoonotic nematode parasites, such as *Angiostrongylus cantonensis*, *A. costaricensis*, *Echinostoma revolutum*, *E. ilocanum*, and *Gnathostoma spinigerum*, are vectored by invasive *Pomacea* spp. (Komalamisra et al., 2009; Yang et al., 2013). *Pomacea* spp. also vectored *Angiostrongylus cantonensis* (cause of eosinophilic meningitis and meningoencephalitis) and *A. costaricensis* (cause of angiostrongyliasis) in humans and other vertebrates. These diseases are transferred mainly by feeding un-cooked or raw snails (Kim et al., 2014). The outbreaks of eosinophilic meningitis in China (Yang et al., 2013) and other regions of the world are particularly due to the consumption of *Pomacea* spp. (Kim et al., 2014; Thiengo et al., 2013). Schistosomiasis, an infection of the urinary tract or intestine of humans, is also vectored by *Pomacea* spp. when they carry the causal organisms, i.e., schistosomiasis worms (Cantanhede et al., 2014). Invasive *Pomacea* spp. and *Pila ampullacea* also vector gnathostomiasis that is caused by *Gnathostoma spinigerum* and characterized by intermittent subcutaneous swellings and inflammatory reactions associated with eosinophils (Komalamisra et al., 2009).

In addition to the above diseases, the empty shells of deceased snails, particularly after pesticide application in rice and other crops, pose severe health hazards. This is because they can potentially cut the feet of individuals working in such fields for various purposes (Cowie, 2002; Douangboupha & Khamphoukeo, 2006). Burkett-Cadena and Unnasch (2013) have reported that in Florida, empty shells of snails can be used by the Asian tiger mosquito, *Aedes albopictus*, for larval development and further spread. Moreover, overuse and indiscriminate use of pesticides to control invasive *Pomacea* spp. may also cause severe human health and environmental safety issues (Cowie, 2002). The capability of *Pomacea* spp. to bioaccumulate toxins and other metal pollutants may also potentially threaten humans and their surrounding ecosystems (Berry & Lind, 2010).

3.4 Threat to Other Invertebrates

The potential of invasive *Pomacea* spp., particularly *P. canaliculata*, to exploit other trophic levels (carrion feeding) enables them to persist, colonize, and elicit impacts in the invaded areas (Manara et al., 2016). Accordingly, colonization of invasive species in the invaded areas has severely affected benthic populations such as snails, worms, and crustaceans. The situation worsens for such creatures when macrophytes, algal, or detritus materials are lacking in aquatic habitats (Horgan et al., 2014a; Kwong et al., 2009). Furthermore, high trophic flexibility, large size, voracious feeding, and higher growth of both invasive *Pomacea* spp. (Hayes et al., 2015; Saveanu et al., 2016) have shown severe adverse effects on the growth and development of native *Pila scutata* in Thailand (Chaichana & Sumpan, 2014) and *P. palusoda* in the USA (Posch et al., 2013). Moreover, *P. canaliculata*, capable of predating on the egg masses of other snail species, has also been reported because of its strong radula and jaws (Turner et al., 2007). The invasive *Pomacea* spp. can also consume fresh or partially rotten vertebrates and invertebrates' particles (Saveanu et al., 2016). Hence, they exhibit significant competition with other macro-invertebrate detritivores, especially insects, due to their specialized feeding behavior in contrast to the more generalized feeding of *Pomacea* spp. (Fenoglio et al., 2014). The ingestion of poisonous eggs of *P. canaliculata* has also shown detrimental effects on opportunist amphibian predators (Brola et al., 2021). Moreover, the indiscriminate and heavy usage of pesticides in rice and other aquatic habitats has resulted in a sharp decline in the populations of native snails (Chaichana & Sumpan, 2014; Cowie et al., 2017).

3.5 Quantitative Losses Due to Invasive Pomacea spp.

The information on the quantitative losses caused by the invasive *Pomacea* spp. in the invaded areas is lacking (Arfan & Muhamad, 2020) and, if available, not adequate because they did not include the costs related to human health and environmental consequences (Naylor, 1996; Nghiem et al., 2013). However, rough estimates of the monetary losses of *P. canaliculata* in the Philippines during its first decade (early 1980s to mid-1990s) of invasion were around US$1.2 billion (Naylor, 1996; Sin, 2003). Anderson (1993) reported that farmers in the Philippines spend US$10 million only on pesticides against *Pomacea* spp. However, even then, snails remained a significant problem in rice, as in 2003, 1.4 million out of the 3 million rice fields were infested with them (Adalla & Magsino, 2006). In Taiwan, the losses to rice showed a sharp increase from US$ 2.7 million in 1982 to US$ 30.9 million in 1986. The annual losses of US$ 74.8 million due to invasive *Pomacea* spp. in rice have been reported in Thailand and Vietnam based on the gross production of rice in these countries (Nghiem et al., 2013). Moreover, annual losses of US$ 200 and 175.6 million have been reported by Cheng and Kao (2006) and Yang et al. (2006), respectively, for the costs of damage caused by *Pomacea* spp. to the agriculture and ecological environment.

4 Management of Invasive *Pomacea* spp.

In general, there is no method to eradicate the entire pest population. It can only be controlled by reducing the population (Arfan & Muhamad, 2020). Various management methods, such as cultural, mechanical, biological, and chemical control, have been practiced all over the world to minimize the problems caused by apple snails in the fields. Among all, using chemical molluscicide is said to be more effective in terms of application and timing (Mokhtar, 2016). As recommended in the Integrated Pest Management (IPM) program, chemical control should be the last option because it is more harmful despite being quick and effective. However, the best strategy is to combine some of these methods. Nevertheless, as of now, no combination has proven to be quite effective, and their impact on crop yield is still unknown (Byers et al., 2002).

The preventive and corrective strategy manage the population of *Pomacea* spp. are discussed below.

4.1 Cultural and Mechanical Control

Sowing Method: The problem of pomacean snails becomes much more severe in direct-sown rice fields than in transplanted fields (Wada, 2004) as they feed vigorously on sprouts and young seedlings. Thus, the selection of rice planting practices does impact the severity of this pest infestation. Transplanting older rice is more effective, as demonstrated by Wada (1997) and Yusa and Wada (1999). Based on their findings, excellent management of apple snails has generally yielded positive results in transplanted rice fields in Japan because rice gradually becomes tolerant to the snails as it grows. In Indonesia, most farmers practice transplanting cultivation systems using 30 to 40-day-old seedlings since they are less susceptible to snail attack than direct seeding. Adjustment in time of sowing has also proved to be effective in the reduction of snail damage to rice as more damage is recorded in direct-seeded and transplanting 2-week-old seedlings than 4- to 6-week-old seedlings in Malaysia (Gilal et al., 2016; Sanico et al., 2002; Teo, 2003). Compared to complete loss caused by direct-seeded and 14-day-old transplanted rice, no more loss is reported in 21 and 28-day-old seedlings after week five of transplanting (Gilal et al., 2016). Moreover, increasing the number of seedlings per hill and hand-picking has effectively reduced snail damage (Sanico et al., 2002; Yanes et al., 2014).

Unfortunately, not all countries can opt for a transplanting system as it is costly. Different direct seeding methods, such as dry and wet seeding, can also help reduce the snail population in rice fields. Snail damage does not occur in dry-seeded fields, where rice seeds are sown, and young plants are kept dry. Furthermore, dry direct seeding effectively reduced seedling loss at the beginning of rice growth (Liang et al., 2014; Wada, 2004). Even though this method is quite promising, rice farmers often do not adopt dry seeding mainly because of weed problems and water leaking after irrigation. The most commonly implemented method in Asian countries is wet seeding, in which rice seeds are sown in wet puddled fields. For wet seeding, water management is very crucial to prevent apple snail infestation.

Water Management: Proper management of irrigation, including water level in the rice field, is proven to be effective in limiting the movement of apple snails and minimizing damage (Gilal et al., 2016; Takashi, 2004; Teo, 2003). Apple snails have difficulty moving in water less than 2 cm deep, so maintaining the water level below 2 cm during the vulnerable stages of the rice plant is recommended. According to Ngoc (2002) and Wada (2004), maintaining water level at a shallow depth (2–3 cm) and water management after sowing and seeding rate is highly influential in snail management. This result is also supported by Gilal et al. (2016), whereby cultivating rice using 28 days old transplanting seedlings at a 2 cm water level is recommended to reduce the damage by *Pomacea* spp. and get a good yield. Moreover, constructing slightly deeper ditches or small canals, for example, 15–25 cm wide and 5 cm deep in or around the paddy field, can facilitate the collection and destruction of snails by hand or by treating them with a pesticide. This is because the snails tend to assemble around ditches when paddy fields are thoroughly drained, making manual collection or extermination easier (Cowie, 2002).

Use of Attractant and Hand-Picking: Hand-pick snails and crush egg masses. This is best done in the early morning and late evening when the snails are most active. Other than that, place bamboo or wooden stakes to provide sites for egg laying that allow easy collection of snail eggs for destruction. Although time-consuming and labor-intensive, this method is the most effective non-chemical method to reduce snail's population in the field (Kunimoto & Nishikawa, 2008; Suharto et al., 2006). Herbage attractants or plants that attract snails, such as papaya and cassava leaves, can be used as bait to make hand-picking easier. Teo (1999) reported that leaves of sweet potato, papaya, and tapioca were effective as snail attractants. Studies in Malaysia showed the potential of jackfruit skin and damaged pomelo (Syamsul et al., 2016) along with papaya (Badrulhadza & Hafizi, 2014) to attract higher numbers of both adult and juvenile snails, hence help in their collection and destruction. Baiting using fermented fruits can also attract snails in the field, as demonstrated by Chan et al. (2014) where *Candida ethanolica*, a local yeast strain isolated from fermenting pineapple, is efficient as bio attraction of golden apple snails.

Prevent Field Entry: Snails can invade fields from canals, rivers, and reservoirs; thus, it is important to be aware of water flow. Placement of wire-mesh grills, woven bamboo, or mesh bags at the main rice inlets and outlet can act as a barrier to snails' movement from one place to another. They can easily be collected and destroyed at these barriers (Arfan & Muhamad, 2020; Teo, 2003). However, this method is inefficient in capturing the young stage (hatchlings) due to its relatively small size.

4.2 Biological Control

Snails can be harvested, cooked and eaten, or sold as animal feed. However, it is important to take extra care and cook the snails thoroughly, as they are known carriers of Angiostrongyliasis, the rat lungworm that could lead to fatality. In the biological control of apple snails, ducks such as William and Siam Peking have been introduced into the rice field with the recommended density of five to 10 ducks per hectare. According to Teo (2001), a suitable duck density may reduce the snail population to less than one snail per m^2 after a month. Several wild bird species can also be adapted to feed on snails. They can be placed into fields during final land preparation or after crop establishment when plants are around 30–35 days after transplant (DAT) (Cagauan & Joshi, 2002). However, keeping the ducks requires a high cost and becomes impractical when the plant matures due to grain production. Nile Tilapia (*Oreochromis niloticus*), black carp (*Mylopharyngodon piceus*), and common carp (*Cyprinus carpio*) are also very effective in controlling apple snails. Carp species with pharyngeal teeth have a high potential for preying on snails and exhibit effective predation behavior, as demonstrated in field experiments (Halwart, 1994; Ichinose et al., 2002). Common carp at a density of 2041 fish per hectare could reduce the *P. canaliculata* population significantly, but constructing a separate pond was suggested to increase the survivability and efficacy of the fish (Arfan & Muhamad, 2020; Teo, 2006). However, using fish may not be practical since fish culture requires maintaining deep water in fields, which is often incompatible with modern farming methods. Red ants feed on the snail eggs, while ducks (and sometimes rats) will eat young snails. Fire ants, *Solenopsis geminata*, firefly larvae, *Luciola lateralis* Motschulsky were also introduced as good egg predators (Ng & Tan, 2011; Yusa, 2001). *Solenopsis geminata* and *Pheidologeton* spp. can cause up to 50% predation of *P. canaliculata*, as recorded in the Philippines (Yusa, 2001, 2006;). Among micro-organisms, two isolates, i.e., CKP-012 and CKP-032 of the entomopathogenic fungi, *Paecilomyces lilacinus*, have shown promising results in controlling the snail eggs, whereas only CKP-012 showed mortality among newly hatched hatchlings. However, the effectiveness of CKP012 against juveniles decreased with age (Arfan & Muhamad, 2020). Natural enemies in the paddy field are not promising to minimize snail populations due to their poor performance.

4.3 Chemical Control

Synthetic Pesticides: The use of synthetic pesticides began in the United States of America (USA) in the 1930s. Its popularity soared after World War II, and its use became widespread all over the USA and different nations. Of all the control measures that have been introduced, synthetic pesticides are the most preferred by farmers (Anis et al., 2019; Cowie, 2002; Mokhtar, 2016). This is because it is highly effective in controlling snails in a short period. Chemicals such as niclosamide, metaldehyde, endosulfan, copper sulphate, calcium cyanamide, sodium pentachlorophenolate, niclosamide; various organo-tin compounds, cartap hydrochloride, isazophos, and fentin acetate (Anis et al., 2019) have been used and evaluated to reduce ampullariidaes population in the field (Joshi et al., 2001; Wada, 2004). Many of these synthetic compounds have severe human and environmental hazards and are banned in many countries of the world (Halwart, 1994; Schnorbach et al., 2006). Among those compounds, fentin acetate has been chosen for apple snail extermination because it is very effective, albeit toxic to humans and aquatic species. This is the main problem of over-using synthetic chemicals for pests, as it negatively impacts paddy field components more than provides benefits. Thus, it is not recommended, as it will give rise to more problems, such as resistance and pesticide residues in plants and the environment. Pesticide application will also result in a quick recovery of the snail population, according to Halwart (1994). In addition, according to Pingshih et al. (2006), the worst scenario that can happen if chemical molluscicides are continuously used is that the farmers will risk developing skin lesions, nail losses, sight abnormalities and death. Joshi et al. (2004) reported that niclosamide 250 EC also has a detrimental effect on germinated seeds in direct-seeded rice cultures, whereby a reduction in rice shoot and root length was detected. Niclosamide and metaldehyde are presently more appropriate molluscicides as the bait-type application of metaldehyde could reduce the snail population in the field even during heavy rains due to its slow release of active chemicals (Schnorbach et al., 2006; Takashi, 2004). Above all, the application of any synthetic chemicals following the recommended rate and calendar spraying and using complete personal protection equipment (PPE) during application may reduce snail resistance problems in the field, protect the user, and minimize all the hazards caused by synthetic chemicals.

Botanical Pesticides: Botanical pesticides have long been discovered as effective alternatives to synthetic chemicals due to their low toxicity to non-target organisms, slow biodegradability, and development of resistance against phytochemical mixtures (Anis et al., 2019; El-Wakeil, 2013; Mokhtar, 2016; Prakash et al., 2008). Moreover, botanical pesticides are locally available and more associated with indigenous self-sufficient pest snail control strategies than their imported synthetic pesticides. Several plant compounds have been discovered as an alternative to synthetic molluscicides. Few compounds, such as saponins, quinoa saponins, azadirachtin, and other phenolic compounds like tannins and terpenes, are widely studied plant chemicals against apple snails. Various extracts of neem¸ *Azadirachta indica* showed molluscicidal activities against *P. canaliculata*, comparable to niclosamide (Noor & Mohd, 2017). A recent study also evaluated the efficacy of methanol crude extracts of *Andrographis paniculata*, *Entada spiralis*, *Ficus deltoidea*, *Furcraea selloa*, and *Ipomoea batatas* along with niclosamide against *P. maculata*. These studies reported that *F. selloa* and *E. spiralis* showed similar feeding deterrent potentials to niclosamide. Moreover, *F. selloa* crude extracts also exhibited maximum (80%) mortality of the targeted *P. maculata* at 15 ppm (Anis et al., 2019). Besides having insecticidal and molluscicidal properties, saponin bioactive compound is also known for its anti-feedant activity and saponin containing plant extract with concentration of 0.05% by weight or higher may serve as an anti-feedant, thus preventing terrestrial molluscs from feeding on living plants (Anis et al., 2019). In addition, more recently, extracts of several plants have been exploited against various stages of *Pomacea* spp. with varying degrees of success. *Azadirachta Indica*, *Cymbopogon citratus*, *Carica papaya*, *Artocarpus integer*, *Allium sativum*, *Piper bitle* and *Curcuma longa* were reported to cause mortality of different stages of apple snails using at different effective concentrations (Garin et al., 2019; Harun et al., 2016; Ibrahim et al., 2017; Latip et al., 2018; Mansur et al., 2018).

5 Perspectives

The invasive ampullariidaes species, *P. canaliculata*, and *P. maculata* have caused tremendous losses to humans after their introduction to countless countries of the world from their native in South America. Multiple factors contributed to their successful invasion. Thus, basic information about their biology, feeding habits, damage, threats, and possible management measures discussed here could provide a comprehensive understanding of their invasiveness. Therefore, it is crucial for societies including governments, farmers, and agrochemical industries to implement the efforts to restrict further spread of *Pomacea* spp. and adopt suitable cultural and mechanical measures to reduce apple snail populations below threshold levels.

Acknowledgements The authors would like to express gratitude to the Ministry of Higher Education Malaysia for the financial support provided under Fundamental Research Grant Scheme (FRGS) Grant, VOT number: 5540218.

References

Acosta, B. O., & Pullin, R. S. (1991). *Environmental impact of the golden snail (Pomacea sp.) on rice farming systems in the Philippines: Summary report of the workshop held at the Freshwater Aquaculture Center, Central Luzon State University, Nueva Ecija, Philippines, 9–10 November 1989* (Vol. 28).

Adalla, C. B., & Magsino E. A. (2006). Understanding the golden apple snail (*Pomacea canaliculata*): Biology and early initiatives to control the pest in the Philippines. In R. C. Joshi, & L. S. Sebastian (Eds.), *Global advances in ecology and management of golden apple snails* (pp. 199–213). Philippine Rice Research Institute.

Anderson, B. (1993). The Philippine snail disaster. *The Ecologist, 23*, 70–72.

Anis, S. M., Muhamad, R., Omar, D., & Asib, N. (2019). Molluscicidal and feeding deterrent activity of crude plant extracts on *Pomacea maculata* Perry. *Current Research Journal of Biological Sciences, 11*(2), 6–12.

Arfan, A. G., & Muhamad, R. (2020). A review on basic biology and ecology of invasive apple snails (Pomacea spp.) for effective management. *Journal of ISSAAS, 26*(2), 66–85.

Arfan, A. G., Muhamad, R., Omar, D., Azwady, A. N., & Manjeri, G. (2014). Distribution of two *Pomacea* spp. in rice fields of Peninsular Malaysia. *Annual Research and Review in Biology, 4*, 4123–4136.

Arfan, A. G., Muhamad, R., Omar, D., Nor Azwady, A. A., & Manjeri, G. (2015). Comparative life cycle studies of *Pomacea maculata* and *Pomacea canaliculata* on rice (*Oryza sativa*). *Pakistan Journal of Agricultural Sciences, 52*(4), 1075–1079.

Arfan, A. G., Muhamad, R., Omar, D., Azwady, A. N., & Manjeri, G. (2017). Sowing methods and water levels influence apple snail damage to rice and its yield in peninsular Malaysia. *Experimental Agriculture, 53*(4), 528–538.

Badrulhadza, A., & Hafizi, Y. (2014). Evaluation of several plant-based attractants for apple snail management. *Acta Biologica Malaysiana, 3*(2), 49–57.

Baloch, W. A. (2017). Country report (Pakistan) on non-native apple snails. In R.C. Joshi, R. H. Cowie, & L. S. Sebastian (Eds.), *Biology and management of invasive apple snails.* (pp. 213–219). Philippine Rice Research Institute

Bellard, C., Cassey, P., & Blackburn, T. M. (2016). Alien species as a driver of recent extinctions. *Biology Letters, 12*(2), 20150623.

Benson, A. J. (2018). Non-indigenous aquatic species: *Pomacea maculata*. USGS Non-indigenous Aquatic Species Database, Gainesville, Florida. Available at: https://nas.er.usgs.gov/queries/FactSheet.aspx?speciesID=2633. Accessed June 24, 2023.

Berry, J. P., & Lind, O. (2010). First evidence of "paralytic shellfish toxins" and cylindrospermopsin in a Mexican freshwater system, Lago Catemaco, and apparent bioaccumulation of the toxins in "tegogolo" snails (*Pomacea patula* catemacensis). *Toxicon, 55*(5), 930–938.

Blumenthal, D. M. (2006). Interactions between resource availability and enemy release in plant invasion. *Ecology Letters, 9*(7), 887–895.

Bradshaw, C. J., Leroy, B., Bellard, C., Roiz, D., Albert, C., Fournier, A., & Courchamp, F. (2016). Massive yet grossly underestimated global costs of invasive insects. *Nature Communications, 7*(1), 12986.

Brola, T. R., Dreon, M. S., Fernández, P. E., Portiansky, E. L., & Heras, H. (2021). Ingestion of poisonous eggs of the invasive apple snail *Pomacea canaliculata* adversely affects bullfrog *Lithobathes catesbeianus* intestine morphophysiology. *Malacologia, 63*(2), 171–182.

Buddie, A. G., Rwomushana, I., Offord, L. C., Kibet, S., Makale, F., Djeddour, D., & Day, R. K. (2021). First report of the invasive snail *Pomacea canaliculata* in Kenya. *CABI Agriculture and Bioscience, 2*(1), 1–10.

Burkett-Cadena, N. D., & Unnasch, T. R. (2013). Apple snails and tiger mosquitoes: a curious association between two invasive species in Florida, USA. In *Paper presented at Entomology 2013, 61st Annual Meeting of the Entomological Society of America, 10–13 November 2013, Austin Convention Center, Austin, Texas.*

Burks, R. L., Hensley, S. A., & Kyle, C. H. (2011). Quite the appetite: Juvenile island apple snails (*Pomacea insularum*) survive consuming only exotic invasive plants. *Journal of Molluscan Studies, 77*(4), 423–428.

Burlakova, L. E., Karatayev, A. Y., Padilla, D. K., Cartwright, L. D., & Hollas, D. N. (2009). Wetland restoration and invasive species: Apple snail (*Pomacea insularum*) feeding on native and invasive aquatic plants. *Restoration Ecology, 17*(3), 433–440.

Byers, J. E., Reichard, S., Randall, J. M., Parker, I. M., Smith, C. S., Lonsdale, W. M., Atkinson, I. A. E., Seastedt, T. R., Williamson, M., Chornesky, E., & Hayes, D. (2002). Directing research to reduce the impacts of nonindigenous species. *Conservation Biology, 16*, 630–640.

Cagauan, A. G., & Joshi, R. C. (2002). Golden apple snail *Pomacea* spp. in Philippines. In T. Wada, Y. Yusa, & R. C. Joshi (Eds.), *Proceedings of the Special Working Group on the Golden Apple Anail (Pomacea spp.) at the 7th ICMAM* (pp. 1–38) KONARC, Kumamoto, Japan.

Cantanhede, S. P. D., Fernandez, M. A., Mattos, A. C. D., Montresor, L. C., Silva-Souza, N., & Thiengo, S. C. (2014). Freshwater gastropods of the Baixada Maranhense Microregion, an endemic area for schistosomiasis in the State of Maranhão, Brazil: I-qualitative study. *Revista Da Sociedade Brasileira De Medicina Tropical, 47*, 79–85.

Carlsson, N. O. (2017). Invasive apple snails are threatening natural ecosystems in Southeast Asia, Europe and North America. In: R. C. Joshi, R. H. Cowie, & L. S. Sebastian (Eds.), *Biology and management of invasive apple snails* (pp. 45–61) Philippine Rice Research Institute.

Chaichana, R., & Sumpan, T. (2014). The potential ecological impact of the exotic snail *Pomacea canaliculata* on the Thai native snail *Pila scutata. Science Asia, 40*(1), 11–15.

Chan, G. F., Othman, F., Zulkiffli, M. H., Yousif, R. H., Yusof, A. M., & Rashid, N. A. A. (2014). Yeasts as the novel attractant of *Pomacea canaliculata. International Journal of Sciences: Basic and Applied Research, 18*, 51–60.

Cheng, E. Y., & Kao, C. H. (2006). Control of golden apple snail *Pomacea canaliculata* (Lamarck), in Taiwan. In R. C. Joshi, & Sebastian, L. S. (Eds.), *Global advances in ecology and management of golden apple snails* (pp. 155–167). Philippine Rice Research Institute.

Coat, S., Monti, D., Legendre, P., Bouchon, C., Massat, F., & Lepoint, G. (2011). Organochlorine pollution in tropical rivers (Guadeloupe): Role of ecological factors in food web bioaccumulation. *Environmental Pollution, 159*(6), 1692–1701.

Collier, K. J., Demetras, N. J., Duggan, I. C., & Johnston, T. M. (2011). Wild record of an apple snail in the Waikato River, Hamilton, New Zealand and their incidence in freshwater aquaria. *New Zealand Natural Sciences, 36*, 1–9.

Cowie, R. H. (2002). Apple snails (Ampullariidae) as agricultural pests: Their biology, impacts and management. Molluscs as crop pests. In G. M. Barker (Ed.), *Molluscs as crop pests*. CABI.

Cowie R. H., & Hayes K. A. (2012). Apple snails. In R. A. Francis (Ed.), *A handbook of global freshwater invasive species* (pp. 207–217). Earthscan.

Cowie, R. H., Hayes, K. A., & Thiengo, S. C. (2006). What are apple snails? Confused taxonomy and some preliminary resolution. In R. C. Joshi, & L. S. Sebastian (Eds.), *Global advances in ecology and management of golden apple snails* (pp. 3–23). Philippine Rice Research Institutes.

Cowie, R. H., Hayes, K. A., Tran, C. T., & Levi, P. (2007). Distribution of the invasive apple snail *Pomacea canaliculata* (Lamarck) in the Hawaiian Islands (Gastropoda: Ampullariidae). *Bishop Museum Occasional Papers, 96*, 48–51.

Cowie, R. H., Hayes, K. A., Strong, E. E., & Thiengo, S. C. (2017). Non-native apple snails: Systematics, distribution, invasion history and reasons for introduction. In R. C. Joshi, & L. S.

Sebastian (Eds.), *Global advances in ecology and management of golden apple snails* (pp. 3–32) Philippine Rice Research Institute.
Cummins, K. W., & Klug, M. J. (1979). Feeding ecology of stream invertebrates. *Annual Review of Ecology and Systematics, 10*(1), 147–172.
Davis, M. A., Grime, J. P., & Thompson, K. (2000). Fluctuating resources in plant communities: A general theory of invasibility. *Journal of Ecology, 88*(3), 528–534.
Djeddour, D., Pratt, C., Makale, F., Rwomushana, I., & Day, R. (2021). The apple snail, *Pomacea canaliculata*: An evidence notes on invasiveness and potential economic impacts for East Africa. In *CABI Working Paper 21* (77 pp.). https://doi.org/10.1079/CABICOMM-62-8149
Douangboupha, B., & Khamphoukeo, K. (2006). Golden apple snail damage and management practices in rice farmers' fields in the Lao PDR. In R. C. Joshi, & L. S. Sebastia (Eds.), *Global advances in ecology and management of golden apple snails* (pp. 275–280). Philippine Rice Research Institute.
Duenas, M. A., Hemming, D. J., Roberts, A., & Diaz-Soltero, H. (2021). The threat of invasive species to IUCN-listed critically endangered species: A systematic review. *Global Ecology and Conservation, 26*, e01476. https://doi.org/10.1016/j.gecco.2021.e01476
Dummee, V., Kruatrachue, M., Trinachartvanit, W., Tanhan, P., Pokethitiyook, P., & Damrongphol, P. (2012). Bioaccumulation of heavy metals in water, sediments, aquatic plant and histopathological effects on the golden apple snail in Beung Boraphet reservoir, Thailand. *Ecotoxicology and Environmental Safety, 86*, 204–212.
EFSA (European Food Safety Authority) Panel on Plant Health. (2012). Scientific opinion on the evaluation of the pest risk analysis on *Pomacea insularum*, the island apple snail, prepared by the Spanish Ministry of Environment and Rural and Marine Affairs. *EFSA Journal, 10*(1), 2552.
EFSA (European Food Safety Authority) Plant Health Newsletter. (2020). Media Monitoring, No. 43, October 2020 EFSA-Q-2020-00069. https://doi.org/10.2903/sp.efsa.2020.EN-1943
EFSA (European Food Safety Authority), Schrader, G., Delbianco, A., & Vos, S. (2020). Pest survey card on Pomacea spp. *EFSA Supporting Publications, 17*(6), 1877E. https://doi.org/10.2903/sp.efsa.2020.EN-1877
El-Wakeil, N. E. (2013). Botanical pesticides and their mode of action. *Gesunde Pflanz, 65*, 125–149.
Epa, U. P. K. (2006). Alien invasive snail, *Pomacea* sp., in Sri Lanka. In R. C. Joshi, & L. S. Sebastian (Eds.), *Global advances in ecology and management of golden apple snails* (pp. 319–324). Philippine Rice Research Institute.
Estebenet, A. L., & Martín, P. R. (2002). Minireview: *Pomacea canaliculata* (Gastropoda: Ampullariidae): Life-history traits and their plasticity. *Biocell, 26*, 83–89.
Fenoglio, S., Merritt, R. W., & Cummins, K. W. (2014). Why do no specialized necrophagous species exist among aquatic insects? *Freshwater Science, 33*(3), 711–715.
Garin, J. R. G., Alcantara, J. M. A. N., Biagtan, C. F. G., & Mangapot, J. L. L. (2019). Molluscicidal activity of garlic (*Allium sativum*) bulb aqueous extract against golden apple snail (*Pomacea canaliculata*). *Philosophical Journal Natural Social and Science, 1*(1.2), 26.
Gilal, A. A., Muhamad, R., Omar, D., Aziz, N. A. A., & Gnanasegaram, M. (2016). Foes can be friends: Laboratory trials on invasive apple snails, *Pomacea* spp. preference to invasive weed, *Limnocharis flava* (L.) Buchenau compared to rice, *Oryza sativa* L. *Pakistan Journal of Zoology, 48*(3), 673–679.
GISD (Global Invasive Species Database). (2023). Species Profile: *Pomacea canaliculata*. Global Invasive Species Database. Available at: http://www.iucngisd.org/gisd/species.php?sc=135. Accessed June 24, 2023.
Halwart, M. (1994). The golden apple snail *Pomacea canaliculata* in Asian rice farming systems: Present impact and future threat. *International Journal of Pest Management, 40*, 199–206.
Harun, N. B., Redhuan, M., Eliyana, N., Khaing, A. T., & Mohamed, W. Z. (2016). Molluscicidal properties of *Azadirachta indica* on *Pomacea canaliculata*. *International Medicine of Journal, 23*(6), 702–705.
Hayes, K. A., Joshi, R. C., Thiengo, S. C., & Cowie, R. H. (2008). Out of South America: Multiple origins of non-native apple snails in Asia. *Diversity and Distributions, 14*(4), 701–712.

Hayes, K. A., Cowie, R. H., & Thiengo, S. C. (2009). A global phylogeny of apple snails: Gondwanan origin, generic relationships, and the influence of outgroup choice (Caenogastropoda: Ampullariidae). *Biological Journal of the Linnean Society, 98*(1), 61–76.
Hayes, K. A., Cowie, R. H., Thiengo, S. C., & Strong, E. E. (2012). Comparing apples with apples: Clarifying the identities of two highly invasive Neotropical Ampullariidae (Caenogastropoda). *Zoological Journal of the Linnean Society, 166*(4), 723–753.
Hayes, K. A., Burks, R. L., Castro-Vazquez, A., Darby, P. C., Heras, H., Martín, P. R., & Cowie, R. H. (2015). Insights from an integrated view of the biology of apple snails (Caenogastropoda: Ampullariidae). *Malacologia, 58*(1–2), 245–302.
Hollingsworth, R .G., & Cowie, R. H. (2006). Apple snails as disease vectors. In R. C. Joshi, & L. S. Sebastian (Eds.), *Global advances in ecology and management of golden apple snails* (pp. 121–132). Philippine Rice Research Institute.
Horgan, F. G. (2018). The ecophysiology of apple snails in rice: Implications for crop management and policy. *Annals of Applied Biology, 172*(3), 245–267.
Horgan, F. G., Stuart, A. M., & Kudavidanage, E. P. (2014a). Impact of invasive apple snails on the functioning and services of natural and managed wetlands. *Acta Oecologica, 54*, 90–100.
Horgan, F. G., Figueroa, J. Y., & Almazan, M. L. P. (2014b). Seedling broadcasting as a potential method to reduce apple snail damage to rice. *Crop Protection, 64*, 168–176.
Howard, L., van Rees, C. B., Dahlquist, Z., Luikart, G., & Hand, B. K. (2022). A review of invasive species reporting apps for citizen science and opportunities for innovation. *NeoBiota, 71*, 165–188.
Howells, R. G., Burlakova, L. E., Karatayev, A. Y., Marfurt, R. K. & Burks, R. L. (2006). Native and introduced Ampullariidae in North America: History, status, and ecology. In R. C. Joshi, & L. S. Sebastian (Eds.), *Global advances in ecology and management of golden apple snails* (pp. 73–112). Philippine Rice Research Institute.
Ibrahim, R., Haiyee, Z. A., & Latip, S. N. H. M. (2017). The antifeedant activity of essential oil from *Cymbopogon citratus* and *Piper bitle* for controlling golden apple snail, *Pomacea Canaliculata. Journal of Funding Application Science, 9*(6S), 39–49.
Ichinose, K., Tochihara, M., Wada, T., Suguiura, N., & Yusa, Y. (2002). Influence of common carp on apple snail in a rice field evaluated by a predator—prey logistic model. *International Journal of Pest Management, 48*(2), 133–138. https://doi.org/10.1080/09670870110100703
Ip, K. K., Liang, Y., Lin, L., Wu, H., Xue, J., & Qiu, J. W. (2014). Biological control of invasive apple snails by two species of carp: Effects on non-target species matter. *Biological Control, 71*, 16–22.
Ito, K. (2002). Environmental factors influencing overwintering success of the golden apple snail, *Pomacea canaliculata* (Gastropoda: Ampullariidae), in the northernmost population of Japan. *Applied Entomology and Zoology, 37*(4), 655–661.
Joshi, R. C., & Parera, X. V. (2017). The rice apple snail in Spain: A review. *International Pest Control, 59*(2), 106–108.
Joshi, R. C., & Sebastian, L. S. (2006). *Global advances in ecology and management of golden apple snails* (600 pp.). Philippine Rice Research Institute.
Joshi, R. C., Dela Cruz, M. S., Martin, A. R., Cabbigat, J. C., Bahatan, R. F., Bahatan, A. D., Choy-Awo, J., Chilagan, N. P., & Cayong, A. B. (2001). Current status of golden apple snail in the Ifugao Rice Terraces, Philippines. *Journal of Sustainable Agriculture, USA*.
Joshi, R. C., Desamito, M. S., Martin, A. R., & Sebastian, L. S., Coupland, J. B. (2004). Detrimental effects of niclosamide 250EC at preseeding in direct-seeded rice culture. *International Rice Research Notes, 29*, 36–37.
Jourdan, J., O'Hara, R. B., Bottarin, R., Huttunen, K. L., Kuemmerlen, M., Monteith, D., & Haase, P. (2018). Effects of changing climate on European stream invertebrate communities: A long-term data analysis. *Science of the Total Environment, 621*, 588–599.
Ketelaar, J. W. H. (1993). *Strategies for solving the Philippine snail problem: A system perspective* (MSc thesis) (100 p.). Wageningen Agricultural University, Wageningen, The Netherlands.

Khin, K. M. M., & Khin, N. Y. (2017). Apple snails (*Pomacea* spp.) in Myanmar: Current status and integrated management activities. In R. C. Joshi, R. H. Cowie, & L. S. Sebastian (Eds.), *Biology and management of invasive apple snails* (pp. 209–211). Philippine Rice Research Institute.

Kim, J. R., Hayes, K. A., Yeung, N. W., & Cowie, R. H. (2014). Diverse gastropod hosts of *Angiostrongylus cantonensis*, the rat lungworm, globally and with a focus on the Hawaiian Islands. *PLoS ONE, 9*(5), e94969.

Kolar, C. S., & Lodge, D. M. (2001). Progress in invasion biology: Predicting invaders. *Trends in Ecology and Evolution, 16*(4), 199–204.

Komalamisra, C., Nuamtanong, S., & Dekumyoy, P. (2009). *Pila ampullacea* and *Pomacea canaliculata*, as new paratenic hosts of *Gnathostoma spinigerum*. *Southeast Asian Journal of Tropical Medicine and Public Health, 40*(2), 243–246.

Kunimoto, Y., & Nishikawa, M. (2008). Improvement of the catch efficiency of the apple snail, *Pomacea canaliculata* (Lamarck) (Gastropoda: Ampullariidae) by the trap crop. *Japanese Journal of Farm Work Resources, 43*, 75–82.

Kwong, K. L., Chan, R. K., & Qiu, J. W. (2009). The potential of the invasive snail *Pomacea canaliculata* as a predator of various life-stages of five species of freshwater snails. *Malacologia, 51*(2), 343–356.

Lach, L., Britton, D. K., Rundell, R. J., & Cowie, R. H. (2000). Food preference and reproductive plasticity in an invasive freshwater snail. *Biological Invasions, 2*(4), 279–288.

Lai, P. Y., Chang, Y. F., & Cowie R. H. (Eds). (2005). *Proceedings—APEC Symposium on the Management of the Golden Apple Snail, 6–11 September 2004* (153 pp.). National Pingtung University of Science and Technology, Pingtung, Taiwan.

Latip, S. N. H. M., Nawi, F. W. M., Shari, E. S., & Mansur, S. H. P. (2018). Potential of *Carica papaya* and *Artocarpus integer* extracts as botanical pesticides for controlling, golden apple snail, *Pomacea canaliculata*. In *Proceedings of Regional Conference on Science, Technology and Social Sciences (RCSTSS 2016)* (pp. 963–973). Singapore: Springer.

Lee, T. G., & Oh, K. C. (2006). Golden apple snails in Korea. In R. C. Joshi, & L. S. Sebastian (Eds.), *Global advances in ecology and management of golden apple snails* (pp. 516–525). Philippine Rice Research Institute.

Lee, G., Paik, C., Noh, T., Seo, H., & Choi, M. (2010). Analysis of damages and rice consumption by golden apple snails (*Pomacea canaliculata*: Ampullariidae) at growth stages of rice. *Korean Journal of Applied Entomology, 49*(4), 343–349.

Levin, P., Cowie, R. H., Taylor, J. M., Hayes, K. A., Burnett, K. M., & Ferguson, C. A. (2006). Apple snail invasions and the slow road to control: ecological, economic, agricultural and cultural perspectives in Hawaii. In R. C. Joshi, & L. S. Sebastian (Eds.), *Global advances in ecology and management of golden apple snails* (pp. 325–335). Philippine Rice Research Institute.

Liang, K., Zhang, J., Song, C., Luo, M., Zhao, B., Quan, G., & An, M. (2013). Integrated management to control golden apple snails (*Pomacea canaliculata*) in direct seeding rice fields: An approach combining water management and rice-duck farming. *Agroecology and Sustainable Food Systems, 38*(3), 264–282.

Liang, K., Zhang, J., Song, C., Luo, M., Zhao, B., Quan, G., & An, M. (2014). Integrated management to control golden apple snails (*Pomacea canaliculata*) in direct seeding rice fields: An approach combining water management and rice-duck farming. *Agroecology and Sustainable Food Systems, 38*(3), 264–282. https://doi.org/10.1080/21683565.2013.809562

Liu, J., Zhao, B., Li, Y., Deng, X., Qiao, Y., Xu, J., & Xu, S. (2022). Palatability of mangrove leaves to invasive apple snails: The relation between feeding electivity and multiple plant characteristics. *Aquatic Invasions, 17*(2), 277–299.

Liu, J., Chen, Z., Li, Y., Chen, D., He, Y., Zhao, B., & Guo, J. (2023). Potential detritivorous diet of the invasive apple snail (*Pomacea canaliculata* Lamarck, 1822) in Mangroves: The relationship between feeding indicators and chemical characteristics of decaying leaf litter. *Journal of Marine Science and Engineering, 11*(3), 672. https://doi.org/10.3390/jmse11030672

López, M. A., Altaba, C. R., Andree, K. B., & López, V. (2010). First invasion of the apple snail *Pomacea insularum* in Europe. *Tentacle, 18*, 26–28.

Lv, S., Zhang, Y. I., Steinmann, P., Yang, G. J., Yang, K. U. N., Zhou, X. N., & Utzinger, J. (2011). The emergence of angiostrongyliasis in the People's Republic of China: The interplay between invasive snails, climate change and transmission dynamics. *Freshwater Biology, 56*(4), 717–734.

Lv, S., Zhang, Y., Liu, H. X., Hu, L., Liu, Q., Wei, F. R., Guo, Y. H., Steinmann, P., Hu, W., Zhou, X. N., & Utzinger, J. (2013). Phylogenetic evidence for multiple and secondary introductions of invasive snails: Pomacea species in the People's Republic of China. *Diversity and Distributions, 19*(2), 147–156.

Manara, E., Saveanu, L., & Martín, P. R. (2016). Effects of ingestion of lithic particles on growth of the apple snail *Pomacea canaliculata* (Caenogastropoda, Ampullariidae). *Malacologia, 59*(2), 211–222.

Mansur, S. H. P., Latip, S. N. H. M., & Nawi, F. W. M. (2018). Effectiveness of *Curcuma longa* and *Cymbopogon citratus* leaf extracts as botanical pesticides for controlling golden apple snail, *Pomacea canaliculata.* In *Proceedings of Regional Conference on Science, Technology and Social Sciences (RCSTSS 2016)* (pp. 975–982). Singapore: Springer.

Martin, C. W., Bahya, K. M., & Valentine, J. F. (2012). Establishment of the invasive island apple snail *Pomacea insularum* (Gastropoda: Ampullaridae) and eradication efforts in Mobile, Alabama, USA. *Gulf of Mexico Science, 30*(1), 5. https://doi.org/10.18785/goms.3001.05

Martín, P. R., Seuffert, M. E., Tamburi, N. E., Burela, S., & Saveanu, L. (2017). Behaviour and ecology of *Pomacea canaliculata* from Southern Pampas (Argentina). In R. C. Joshi, R. H. Cowi, & L. S. Sebastian (Eds.), *Biology and management of invasive apple snails* (pp. 241–256). Philippine Rice Research Institute.

Martín, P. R., Burela, S., Seuffert, M. E., Tamburi, N. E., & Saveanu, L. (2019). Invasive Pomacea snails: Actual and potential environmental impacts and their underlying mechanisms. *CAB Reviews, 14*(42), 11. https://doi.org/10.1079/PAVSNNR201914042

Marwoto, R .M., Isnaningsih, N. R., & Joshi, R. C. (2018). The invasive apple snail (*Pomacea* spp.) in Indonesia. *Agriculture for Development, 35*, 41–48. https://www.cabi.org/isc/FullTextPDF/2019/20193044933.pdf

Matsukura, K., Okuda, M., Kubota, K., & Wada, T. (2008). Genetic divergence of the genus Pomacea (Gastropoda: Ampullariidae) distributed in Japan, and a simple molecular method to distinguish *P. canaliculata* and *P. insularum*. *Applied Entomology and Zoology, 43*(4), 535–540.

Meza-Lopez, M. M., & Siemann, E. (2015). Experimental test of the invasional meltdown hypothesis: An exotic herbivore facilitates an exotic plant, but the plant does not reciprocally facilitate the herbivore. *Freshwater Biology, 60*(7), 1475–1482.

Mochida, O. (1991). Spread of freshwater Pomacea snails (Pilidae, Mollusca) from Argentina to Asia. *Micronesica,* (3, Supplement), 51–62.

Mokhtar, A. S. (2016). Nano emulsion formulation of saponin to control black apple snails, *Pomacea maculata* Perry (Doctoral dissertation, PhD Thesis, Universiti Putra Malaysia, Serdang, Selangor).

Monette, D., Ewe, S., Dinkins, J. M., & Markwith, S. H. (2017). Interactions of exotic and native Pomacea with wetland vegetation structure in the Greater Everglades, Florida, USA. *Fundamental and Applied Limnology, 189*(4), 291–299.

Morrison, W. E., & Hay, M. E. (2011). Herbivore preference for native vs. exotic plants: Generalist herbivores from multiple continents prefer exotic plants that are evolutionarily naïve. *PLoS ONE, 6*, e17227. https://doi.org/10.1371/journal.pone.0017227

Myint, K. K. M., & Ye, K. N. (2017). Apple snails (*Pomacea* spp.) in Myanmar: Current status and integrated management activities. In R. C. Joshi, R. H. Cowie, & L. S. Sebastian (Eds.), *Biology and management of invasive apple snails* (pp. 209–211). Philippine Rice Research Institute.

Naylor, R. (1996). Invasions in agriculture: Assessing the cost of the golden apple snail in Asia. *Ambio, 25*, 443–448.

Ng, T. H., & Tan, S. K. (2011). Observations of land snails feeding on the eggs of *Pomacea canaliculata* (Lamarck, 1822) (Mollusca: Gastropoda). *Nature in Singapore, 4*, 79–83.

Ng, T. H., Tan, S. K., & Yeo, D. C. J. (2017). South American apple snails, *Pomacea* spp. (Ampullariidae), in Singapore. In R. C. Joshi, R. H. Cowie, & L. S. Sebastian (Eds.), *Biology and management of invasive apple snails* (pp. 221–239). Philippine Rice Research Institute.

Nghiem, L. T., Soliman, T., Yeo, D. C., Tan, H. T., Evans, T. A., Mumford, J. D., & Carrasco, L. R. (2013). Economic and environmental impacts of harmful non-indigenous species in Southeast Asia. *PLoS ONE, 8*(8), e71255.

Ngoc, H. K. (2002). Golden apple snail in Vietnam. In *Proceedings 7th International Congress on Medical and Applied Malacology*. SEAMEO Regional Center for Graduate Study and Research in Agriculture.

Noor, S. H. M. L., & Mohd, F. K. (2017). Neem, *Azadirachta indica*, as a potential biopesticide for controlling the apple snail, *Pomacea canaliculata*. In R. C. Joshi, R. H. Cowie, & L. S. Sebastian (Eds.), *Biology and management of invasive apple snails* (pp. 343–350). .

Peña, S. C. (2017). Accumulation of copper and other elements by the apple snail *Pomacea canaliculata*. In R. C. Joshi, R. H. Cowie, & L. S. Sebastian (Eds.), *Biology and management of invasive apple snails* (pp. 133–144). Philippine Rice Research Institute.

Pimentel, D., Zuniga, R., & Morrison, D. (2005). Update on the environmental and economic costs associated with alien-invasive species in the United States. *Ecological Economics, 52*(3), 273–288.

PingShih, Y., YuHui, C., WuChung, L., & YaHui, C. (2006). Golden apple snail management and prevention in Taiwan. In R. C. Joshi, R. H. Cowie, & L. S. Sebastian (Eds.), *Biology and management of invasive apple snails* (pp. 169–179). Philippine Rice Research Institute.

Posch, H., Garr, A. L., & Reynolds, E. (2013). The presence of an exotic snail, *Pomacea maculata*, inhibits growth of juvenile Florida apple snails, *Pomacea paludosa*. *Journal of Molluscan Studies, 79*(4), 383–385.

Prakash, A., Rao, J., & Nandagopal, V. (2008). Managing resistance to insecticides is a key for the future of crop protection. *Journal of Biopesticides, 2008*(1), 154–169.

Preap, V., Chanty, P., Bunnarith, K., & Hin, K. (2006). The golden apple snail situation in Cambodia. In R. C. Joshi, & L. S. Sebastian (Eds.), *Global advances in ecology and management of golden apple snails* (pp. 267–274). Philippine Rice Research Institute.

Qiu, J. W., Chan, M. T., Kwong, K. L., & Sun, J. (2011). Consumption, survival and growth in the invasive freshwater snail *Pomacea canaliculata*: Does food freshness matter? *Journal of Molluscan Studies, 77*(2), 189–195.

Rahel, F. J., & Olden, J. D. (2008). Assessing the effects of climate change on aquatic invasive species. *Conservation Biology, 22*(3), 521–533.

Rodriguez, C., Espinosa, E., & Coello, R. (2017). *Pomacea canaliculata* in Ecuador: A recent pest with multiple implications. In R. C. Joshi, R. H. Cowi, & L. S. Sebastian (Eds.), *Biology and management of invasive apple snails* (pp. 257–291). Philippine Rice Research Institute.

Rosario J., & Moquete C. (2006). The aquatic snail *Ampullaria canaliculata* L.—plague of irrigated lowland rice in the Dominican Republic. In R. C. Josh, & L. S. Sebastian (Eds.), *Global advances in ecology and management of golden apple snails* (pp. 514–515). Philippine Rice Research Institute.

Sanico, A. L., Peng, S., Laza, R. C., & Visperas, R. M. (2002). Effect of seedling age and seedling number per hill on snail damage in irrigated rice. *Crop Protection, 21*(2), 137–143.

Saveanu, L., & Martín, P. R. (2015). Neuston: A relevant trophic resource for apple snails? *Limnologica, 52*, 75–82.

Saveanu, L., Manara, E., & Martín, P. R. (2016). Carrion consumption and its importance in a freshwater trophic generalist: The invasive apple snail *Pomacea canaliculata*. *Marine and Freshwater Research, 68*(4), 752–759.

Schnorbach, H. J., Rauen, H. W., & Bieri, M. (2006). Chemical control of the golden apple snail, *Pomacea canaliculata*. In R. C. Joshi, & L. S. Sebastian (Eds.), *Global advances in ecology and management of golden apple snails* (pp. 419–438). Philippine Rice Research Institute.

Sepulveda, A. J., Nelson, N. M., Jerde, C. L., & Luikart, G. (2020). Are environmental DNA methods ready for aquatic invasive species management? *Trends in Ecology and Evolution, 35*(8), 668–678.

Shivambu, T. C., Shivambu, N., & Downs, C. T. (2020). Exotic gastropods for sale: An assessment of land and aquatic snails in the South African pet trade. *Management of Biological Invasions, 11*(3), 512–524.

Simberloff, D. (2013). *Invasive species: What everyone needs to know*. Oxford University Press. https://doi.org/10.1093/wentk/9780199922017.001.0001

Sin, T. S. (2003). Damage potential of the golden apple snail *Pomacea canaliculata* (Lamarck) in irrigated rice and its control by cultural approaches. *International Journal of Pest Management, 49*(1), 49–55.

Suharto, H., Marwoto, R. M., Heryanto, M., & Siwi S. S. (2006). The golden apple snail, *Pomacea* spp., in Indonesia. In R. C. Joshi, & L. S. Sebastian (Eds.), *Global advances in ecology and management of golden apple snails* (pp. 231–242). Philippine Rice Research Institute.

Syamsul, R. B., Muhamad, R., Arfan, A. G., & Manjeri, G. (2016). Effectiveness of various plant-based attractant traps against apple snail, *Pomacea maculata* (Gastropoda: Ampullariidae) in a rice field. *Pertanika Journal of Tropical Agriculture Science, 39*(2), 137–143.

Takashi, W. (2004). Strategies for controlling the apple snail *Pomacea canaliculata* (Lamarck) (Gastropoda: Ampullariidae) in Japanese direct-sown paddy fields. *Japanese Agricultural Research Quarterly, 38*(2), 75–80.

Teo, S. S. (1999). Control of the golden apple snail (*Pomacea* spp.) by handpicking using herbage as attractants. In *Proceedings of the MCB-MAPPS Plant Protection Conference'99 on "Sustainable Crop Protection Practices in the Next Millennium", Kota Kinabalu, Sabah, Malaysia.*

Teo, S. S. (2001). Evaluation of different duck varieties for the control of golden apple snail (*Pomacea canaliculata*) in transplanted and direct seeded rice. *Crop Protection, 20*, 599–604.

Teo, S. S. (2003). Golden apple snail (*Pomacea canaliculata* Lamarck, 1819) in Sabah, Malaysia Current situation and management strategy. In *Proceedings of Seventh International Congress on Medical and Applied Malacology. Los Baños, Laguna, Philippines*

Teo, S. S. (2006). Evaluation of different species of fish for biological control of golden apple snail *Pomacea canaliculata* (Lamarck) in rice. *Crop Protection, 25*(9), 1004–1012.

Thiengo, S. C., de Oliveira Simões, R., Fernandez, M. A., & Júnior, A. M. (2013). *Angiostrongylus cantonensis* and rat lungworm disease in Brazil. *Hawai'i Journal of Medicine and Public Health, 72*(6 Suppl 2), 18–22.

Turner, A. M., Turner, R. R., & Ray, S. R. (2007). Competition and intraguild egg predation among freshwater snails: Re-examining the mechanism of interspecific interactions. *Oikos, 116*(11), 1895–1903.

Vega, I. A., Arribére, M. A., Almonacid, A. V., Ribeiro, G. S., & Castro-Vazquez, A. (2012). Apple snails and their endosymbionts bioconcentrate heavy metals and uranium from contaminated drinking water. *Environmental Science and Pollution Research, 19*, 3307–3316.

Wada, T. (1997). Introduction of the apple snail *Pomacea canaliculata* and its impact on rice agriculture. In *Proceedings of International Workshop on Biological Invasions of Ecosystems by Pests and Beneficial Organisms* (pp. 170–180). National Institute of Agro-Environmental Sciences.

Wada, T. (2004). Strategies for controlling apple snail *Pomacea canaliculata* (Lamarck) (Gastropoda:Ampullariidae) in Japanese direct-seeded paddy fields. *Japan Agricultural Research Quarterly, 28*, 75–80.

Wada, T., & Matsukura, K. (2007). Seasonal changes in cold hardiness of the invasive freshwater apple snail, *Pomacea canaliculata* (Lamarck)(Gastropoda: Ampullariidae). *Malacologia, 49*(2), 383–392.

Wei, M., Mao, W., Wenqing, W., Yi, L., Liuqing, L., & Chaoyi, T. (2018). Biodiversity of mangrove mollusks in the west coast of Hainan Island, China. *Biodiversity Science, 26*(7), 707–716.

Yahaya, H., Badrulhadza, A., Sivapragasam, A., Nordin, M., Hisham, M. M., & Misrudin, H. (2017). Invasive apple snails in Malaysia. In R. C. Joshi, R. H. Cowie, & L. S. Sebastian (Eds.), *Biology and management of invasive apple snails* (pp. 169–195). Philippine Rice Research Institute.

Yanes, F. J., Almazan, M. L. P., & Horgan, F. G. (2014). Reducing seed-densities in rice seedbeds improves the cultural control of apple snail damage. *Crop Protection, 62*, 23–31.

Yang, P. S., Chen, Y. H., Lee, W. C., & Chen, Y. H. (2006). Golden apple snail management and prevention in Taiwan. In R. C. Joshi, & L. S. Sebastian (Eds.), *Global advances in ecology and management of golden apple snails* (pp. 169–179). Philippine Rice Research Institute.

Yang, T. B., Wu, Z. D., & Lun, Z. R. (2013). The apple snail *Pomacea canaliculata*, a novel vector of the rat lungworm, *Angiostrongylus cantonensis*: Its introduction, spread, and control in China. *Hawai'i Journal of Medicine and Public Health, 72*(6 Suppl 2), 23–25.

Yang, Q. Q., Liu, S. W., He, C., & Yu, X. P. (2018). Distribution and the origin of invasive apple snails, *Pomacea canaliculata* and *P. maculata* (Gastropoda: Ampullariidae) in China. *Scientific Reports, 8*(1), 1–8.

Yusa, Y. (2001). Predation on eggs of the apple snail *Pomacea canaliculata* (Gastropoda: Ampullariidae) by the fire ant *Solenopsis geminata*. *Journal of Molluscan Studies, 67*(3), 275–279.

Yusa, Y. (2006). Predators of the introduced apple snail, *Pomacea canaliculata* (Gastropoda: Ampullariidae): their effectiveness and utilization in biological control. In R. C. Joshi, & L. S. Sebastian (Eds.), *Global advances in ecology and management of golden apple snails* (pp. 345–361). Philippine Rice Research Institute.

Yusa, Y., & Wada, T. (1999). Impact of the introduction of apple snail and their control in Japan. *Naga, ICLARM Q, 22*(3), 9–13.

Yusa, Y., & Wada, T. (2002). Current status of the *Pomacea canaliculata* in Japan. In T. Wada, Y. Yusa, & R. C. Joshi (Eds.), *Proceedings of the Meeting of the Special Working Group on the Golden Apple Snail (Pomacea spp.) at the Seventh International Congress on Medical and Applied Malacology* (pp. 74–79). Los Baños, Laguna, Philippines: SEAMEO Regional Center for Graduate Study and Research in Agriculture.

Zhang, C., Gu, J., Saveanu, L., Martín, P. R., Shi, Z., & Zhang, J. (2023). Invasiveness of *Pomacea canaliculata*: The differences in life history traits of snail populations from invaded and native areas. *Agronomy, 13*(5), 1259. https://doi.org/10.3390/agronomy13051259

Bioherbicides: An Eco-Friendly Weed Management Strategy Towards Sustainable Agriculture

Mahmudul Hasan, Sarker Mohammad Rezaul Karim, Mst. Motmainna, and Muhammad Saiful Ahmad-Hamdani

Abstract Sustainable and eco-friendly weed management presents a significant challenge in the twenty-first-century agricultural sector. A substantial burden has been forced on agriculture, especially by the injudicious use of chemical herbicides, resulting in extensive environmental contamination. It is essential to have environmentally safe and cost-effective solutions with reduced use of chemical herbicides to improve productivity and sustainability in agriculture. Bioherbicides ought to possess active ingredients derived from living organisms, e.g., fungi, bacteria, and viruses, including the natural byproducts during their growth, or plant-derived products with phytotoxic properties along with ancillary compounds (e.g., surfactants, adjuvants, preserving agents, water retaining additives, inert fillers, and more) applied for eco-friendly weed control in cropping and non-cropping zones. Therefore, the importance of bioherbicide usage in sustainable agriculture cannot be overemphasized. This chapter provides insight into the concept and categories of bioherbicides, the history of the development of bioherbicides, methods of bioherbicide production, formulation of bioherbicides and their application, mode of actions of bioherbicides, factors affecting their efficacy, and the limitations of bioherbicides development and application for sustainable weed management.

1 Concept of Bioherbicides

Bioherbicides are products originating from living organisms, e.g., fungi, bacteria, and viruses or their natural metabolites, or products developed from allelopathic plants, which are applied to control weeds without corrupting the environment (Bailey, 2014). These are a competitive sub-class of herbicides derived from naturally occurring microorganisms or compounds produced from their byproducts or allelochemicals produced from allelopathic plants that suppress the growth and development of weeds through different action modes (Fenibo et al., 2021). The microbes

M. Hasan · S. M. R. Karim · Mst. Motmainna · M. S. Ahmad-Hamdani (✉)
Department of Crop Science, Faculty of Agriculture, Universiti Putra Malaysia, UPM 43400 Serdang, Selangor, Malaysia
e-mail: s_ahmad@upm.edu.my

M. Wong (ed.), *Advances in Tropical Crop Protection*,
https://doi.org/10.1007/978-3-031-59268-3_16

in bioherbicides possess invasive genes that defeat the defense genes of weeds and kill them. Bioherbicides exert their inhibitory effects through multiple mechanisms of action, such as affecting growth promotors, inducing the metabolism of toxic substances, and inhibiting the non-specific multiple sites (Dar et al., 2021). These multiple mechanisms of action against target weeds exterminate the development of resistance, which is common in chemical herbicides.

With the growing interest in organic foods and beverages and the rising consumption demand for organic foods, farmers are everchanging from using harmful chemical herbicides to more sustainable and greener farm practices, including the use of bioherbicides in this era of climate change. The high cost involved in developing and registering chemical herbicides, the development of herbicide resistance in many weed species, and the increasing awareness concerning chemical pesticides are also why researchers emphasize the development of additional weed control tools, including biological weed control using bioherbicides.

Bioherbicides have a significant role in controlling weeds in restricted areas, such as restored native ecosystems, national parks, wetlands, wildlife refuges, and riparian borders, where chemical herbicides are prohibited based on rules and regulations in these specific areas (Hasan et al., 2021a). As a component of the overall crop management program, bioherbicides can be successfully integrated into sustainable weed management since it is effective at different stages of weed growth: seed germination in soil, the early growth stage of weed population, the mid and competitive growth stage, and the seed production stage of the weeds.

2 Merits and Demerits of Bioherbicides

Bioherbicides have many merits in comparison to synthetic herbicides, which include:–

(i) A high degree of weed specificity,
(ii) Non-target plants or humans are unaffected,
(iii) No residue build-ups in the soil, water, and food environments,
(iv) Useful for controlling herbicide-resistant weed populations.

The benefits of bioherbicides include but are not exclusive to shifting the course of weed resistance, harmonizing input to chemical herbicides, environmentally friendly, weed specific, no negative impact on beneficial plants, crops, or humans, quick biodegradability, no hassle of post-harvest adulteration, and stability against abiotic pressures (Hasan et al., 2023), and compatibility in integrated weed management system (Motmainna et al., 2021a). In many cases, bioherbicides are usually not detrimental to humans due to the absence of the risk of mutagenic, genotoxic, and cytotoxic effects (Hossen et al., 2021). Some fungal pathogens produce toxins during the interaction with weeds which are undesirable for mammalian and avian.

Bioherbicides are beneficial under the following conditions:

(i) Weed resistance has evolved to a broad-spectrum herbicide,
(ii) Very high rates of herbicides are required to control a particular weed,
(iii) Control of parasitic weeds cannot be executed properly due to the dearth of selective herbicides. The selection of rhizosphere bacteria or fungi that produce amino acids has been suggested to control broomrape germination and infection in horticultural crops,
(iv) Synthetic herbicides are not preferred due to high costs and environmental restrictions, and
(v) Invasive weeds are well-established and recognized, and fewer herbicide options exist.

3 Category of Bioherbicides

Based on the source of herbicidal units/chemicals, the bioherbicides are categorized as plant-based bioherbicides, microorganism-based bioherbicides, and byproduct-based bioherbicides.

3.1 Plant-Based Bioherbicides

Although the development of bioherbicides started with microorganisms as active ingredients, more recent research initiatives have delved into producing bioherbicides from plant-based allelochemicals (Motmainna et al., 2021b). This development involves several steps: (i) searching for phytotoxic compounds from plants, (ii) testing their efficiencies to control different weeds in laboratory, greenhouse, and fields, (iii) developing effective and useful formulations, (iv) scaling up the technology, and (v) marketing the products. The importance of developing bioherbicides from the screened allelochemicals is based on the following facts: (i) since many allelochemicals naturally degrade in nature before it reaches the target weed, direct use of allelochemicals as herbicides is not successful, (ii) isolation of phytochemicals is a complex process, and the phytotoxic compounds exist in low quantities, making them too costly to use as herbicides, (iii) although direct application of allelopathic plant materials can control weeds effectively, it necessitates high amounts of plant material, and therefore, the technique does not match the current trend in agricultural production. Thus, synthesizing a simple compound form with strong phytotoxic activities is essential to develop novel bioherbicides.

Bashar et al. (2022) isolated seven effective allelochemicals from the *Parthenium hysterophorus* L., among which caffeic acid was found as the most phytotoxic to important rice weeds, e.g., *Digitaria sanguinalis* (L.) Scop. and *Eleusine indica* (L.) Gaertn. They demonstrated that the allelochemicals could be well formulated as an effective and beneficial bioherbicide to control rice weeds. A new bioherbicide referred to as WeedLock has recently been developed in Malaysia from a plant extract, which is a non-selective contact herbicide, and controls weed species, including *Ageratum conyzoides* L., *Euphorbia hirta* L., *E. indica, Axonopus compressus* (Sw.) P.Beauv., *Cyperus iria* L., and *Fimbristylis miliacea* (L.) Vahl (Hasan et al., 2021b).

Other plant-based bioherbicides available globally include BioWeed, derived from *Pinus radiata* D.Don and developed by Barmac, Lidcombe, Australia, and Beloukha, derived from *Brassica napus* L. and marketed by Grochem, Port Melbourne, Australia (Muñoz et al., 2020; Travlos et al., 2020). Verdeguer et al. (2020) reported that six commercial bioherbicides, such as Matratec, GreenMatch, Green-MatchEX, WeedZap, Weed Slayer, and Avenger Weed Killer, have also been registered and are available in the USA. Bioherbicides such as BioWeed, Avenger Weed Killer (Avenger Products, LLC, Buford, Georgia), and Weed Slayer (Agresearch International, LLC, McKinney, United States) successfully controlled *Ochna serrulate* Walp., *D. sanguinalis.*, and *Echinochloa crus-galli* (L.) P.Beauv., respectively (Travlos et al., 2020; Verdeguer et al., 2020).

3.2 Microorganism-Based Bioherbicides

As previously mentioned, most of the research and achievements focused on the development of microorganism-based bioherbicides. Therefore, a detailed discussion on the history of development, modes of actions, formulations, factors affecting efficacy, and more, are presented later under different headings.

3.3 Bioherbicides from Natural Byproducts

The third source of bioherbicides is byproducts of different industries. Most byproducts have herbicidal properties, which could be used as a source for bioherbicide molecules. The Distillers' Dried Grains with Solubles (DDGS) is a byproduct of the ethanol production industry that successfully controls the *Stellaria media* (L.) Vill. and *Poa annua* L. weeds and inhibits the seed germination of *Oxalis corniculata* L. (Hintz, 2007). Corn gluten meal (CGM), a byproduct from the wet-milling corn process, causes physiological changes leading to abnormalities in several weeds such as *Agrostis palustris* Huds., *Chenopodium album* L., *Portulaca oleracea* L., *Solanum nigrum* L., and *Rumex crispus* L. (Liu & Christians, 1997). Meanwhile, mustard seed meal (MSM), a byproduct of the mustard oil pressing industry, contains glucosinolates and can inhibit the growth of many weeds (Boydston et al., 2011). Snyder et al. (2009) used Brassicaceae seed meals (BSMs) in carrot fields to control different weeds and were found to affect weed growth.

4 Modes of Action of Bioherbicides

Modes of action are defined as specific actions owned by bioherbicides in reducing weed growth or termination. Some plants and microbes secrete metabolites depending on different environmental conditions, which inhibit weed seed germination and weed growth. The overall damage to the weeds due to the actions of bioherbicides is discussed in the subsequent headings below.

4.1 Effect on Cell Anatomy and Cell Division

Bioherbicides change the cell anatomy, increase the number of lipid globules, decrease the number of mitochondria, and destroy the mitochondrial and nucleic membranes (Motmainna et al., 2023). The fatty acids in bioherbicides (e.g., Pelargonic acid) disturb the cell membranes, which causes loss of cellular functions (Lederer et al., 2004). Since the root surface allows the transport of more phytotoxic compounds compared to other plant surfaces, the allelochemicals of bioherbicides alter the cell cycle and cell division of the receptor weeds. The effects of bioherbicides enhance the production of superoxide, hydrogen peroxide, and hydroxyl radicals, resulting in DNA and cellular membrane damage (Tigre et al., 2012). In addition, the allelochemicals reduce the cell size of metaxylems in the roots through changes in auxins, thereby inhibiting root elongation (Anese et al., 2015). The phytotoxins, α, β-dehydrocurbularin, inhibit mitosis in root tip cells and prevent root growth (Jiang et al., 2008).

4.2 Effect on Biochemical Processes

The essential oils in bioherbicides cause damage to the DNA, biochemical processes, mitosis, and meristematic cells, inhibiting seedling growth of weeds. The phenolic compounds in bioherbicides reduce the amylase activity of weeds, delaying seed germination as the reduced amylase activity slows the starch hydrolysis process (Šćepanović et al., 2022). Blocking of hydrolysis creates an imbalance in the osmotic pressure of cells that affects imbibition rates leading to inhibition in seed germination. *Enterobacteria sp.* 1–3, when in contact with weeds, secrete high concentrations of Indole Acetic Acid (IAA), which enhances the synthesis of IAA synthase and aminocyclopropane-1-carboxylate synthase that stimulate the biosynthesis of ethylene in plant bodies, which in turn retards weed growth (Park et al., 2015). Bioherbicides with *Cercospora* sp. spores produce cercosporin and beticolin to inhibit weed growth (Tessmann et al., 2008). Mycoherbicides containing *Ascochyta caulina* induce the production of phytotoxins, including N^2-β-D-glucopyranoside, trans-4-amino-ρ-proline, and aglycone of ascaulitoxins, which kill the weeds when appropriately applied (Netland et al., 2001).

4.3 *Effect on Nutrient Uptake*

Since bioherbicides change the structures and functions of cellular membranes, they affect the absorption of nutrients, especially Ca, K, Mg, and Fe, by weeds. The reduction in Ca inhibits root growth and induces leaf senescence (Weng et al., 2022). A decrease in K, Mg, and Fe content in plants causes chlorosis. The application of bioherbicides leads to a higher accumulation of Na, which disrupts nutrient and water uptake (Radhakrishnan & Lee, 2014).

4.4 *Effect on Seed Germination*

Microbial bioherbicides are used to control weeds through root infection. For example, *Phoma macrostoma* colonizes and enters the root hairs of weeds and grows in the intercellular spaces infecting the vascular trachea to prevent food supply to the weeds (Bailey et al., 2011). The extracellular lipases of the fungi utilize the lipids and proteins in the endosperm of weed seeds for their growth and consequently prevent weed seed germination. Many other bioherbicidal compounds produced by different bacteria and fungi include α, β-dehydocurvularin, 2,4-K Da protein, trichothecene, β-1, 4-exoglucanase, glucosidase, xylanase, and proteinase which inhibit seed germination of weeds and further growth after germination. The methanolic extracts of three weeds, *P. hysterophorus, Cleome rutidosperma* DC., and *Borreria alata* (Aubl.) DC., inhibited the germination of *A. conyzoides, E. hirta,* Weedy rice, *Echinochloa colona L., Oryza sativa* L., *Zea mays* L., *Abelmoschus esculentus* (L.) Moench, and *Amaranthus gangeticus* L. (Motmainna et al., 2021c).

4.5 *Effect on Weed Metabolism*

Many physiological processes, such as ion transport and stomatal opening, among others, are regulated by different amino acids. However, *Enterobacteria sp.* 1–3 based bioherbicides inhibit the synthesis of amino acids responsible for the processes in the host weeds. Macrocidin produced by *Phoma* sp. in bioherbicide formulation affects the ion synthesis of carotenoids, which enhance the production of phytons, eventually reducing the ratio of β-carotene to lutein. This mode of action inhibits chlorophyll synthesis and reduces the rate of photosynthesis (Hubbard et al., 2015).

4.6 Effect on Photosynthesis and Stomatal Conductance

Bioherbicides reduce the chlorophyll a/b-binding proteins causing interruption of chlorophyll synthesis and eventually affecting photosynthesis. For example, the reduction in oxygen-evolving enhancer protein 1 (OEEP 1) affects the release of oxygen by splitting H_2O, increasing thioredoxin activity, and protecting the tetra-manganese cluster and ionic contents (Kim et al., 2015). The reduction in OEEP 1 also lowers the concentration of Mg, thereby reducing chlorophyll synthesis since Mg is the central element of chlorophyll. Bioherbicides affect the production of pigments and carotenoids, which can increase photooxidation. Meanwhile, carotenoids help scavenge for reactive oxygen species (ROS) to protect plants from photooxidation (Hasan et al., 2022; Motmainna et al., 2021d). They also reduce the photochemical quenching coefficient, electron transport, and carbohydrate synthesis (Sousa et al., 2014).

Toxic compounds derived from plant species affect the stomatal conductance of receptor plants or weeds. Rhizome production of the zinger plant (*Zingiber officinale* Roscoe) was significantly reduced to a rate of 512 g/plan when treated with tamarind (*Tamarindus indica* L.) leaves. This reduction was associated with stomatal conductance, decreased leaf production and root length, and eventually, plant growth retardation (Abraham et al., 2016).

4.7 Effect on Plant Hormones

Bioherbicide application inhibits the gibberellin pathway and stimulates the accumulation of abscisic acid (ABA), jasmonic acid (JA), and salicylic acid (SA) (Radhakrishnan et al., 2018; Lee et al., 2015). Higher accumulation of ABA causes stomatal closure, low photosynthetic rate, and increased ROS production. As gibberellins are signaling molecules that regulate shoot growth, a reduction in gibberellins leads to reduced weed growth (Radhakrishnan et al., 2013). Increased SA production causes a reduction in electron transport and stomatal closure, inhibiting photosynthesis.

4.8 Effect on Functions of Genes

Phytochemicals in bioherbicides can affect the functions of genes responsible for different activities in radical tissues and the endosperm. Several genes associated with gibberellin and phytoalexin production, as well as Mitogen-activated protein kinase signaling, were downregulated by caffeic acid, resulting in growth inhibition of *Setaria viridis* (L.) P.Beauv (Chen et al., 2021).

5 Production Methods

To develop an effective bioherbicide, especially from pathogenic microbes, we need to comprehensively understand the pathogen(s), weed biology and population dynamics, the requirement for effective disease expansion, and the interactions between the host and the pathogen. The major phases in the development of bioherbicides are outlined below.

5.1 Discovery

The discovery phase includes (i) collection of diseased plant materials, especially from the group of plants for which the control weed is to be implemented, (ii) cultivation of and isolation of the pathogen, (iii) demonstration of Koch's Postulates, i.e., development of a useful benchmark for judging the cause-and-effect relationship between the microbes and the disease, (iv) identification of the pathogen, (v) culture of the pathogen in a suitable artificial medium, and (vi) maintenance of pathogen cultures in short-term and long-term storages.

5.2 Development

The development phase of herbicide production involves (i) the determination of the optimum conditions for spore/conidia production, (ii) the determination of optimum conditions for infection and disease development, (iii) the determination of host range, and (iv) elucidation of the mechanism of actions of the pathogen.

5.3 Deployment

Deployment of the herbicide production involves (i) the formulation of the bioherbicide using the most effective substances (e.g., surfactants, wetting agents, adjuvants), (ii) scaling up the technology to continue its development and to expand its market, and (iii) marketing of the product(s).

The procedures for large-scale production of a granular bioherbicide from a mycotoxin are outlined in the following schematic diagram (see Fig. 16.1), along with the description of different steps and activities. The process starts with laboratory culture cultivation, then mixing of components of medium, fermentation, extraction with a suitable solvent, purification of culture filtrate, fractionation of the effective chemical, pasta mixing with isolated chemicals to make the dough, extrusion of the dough, grinding of the dough, fluid bed drying, and lastly product packaging.

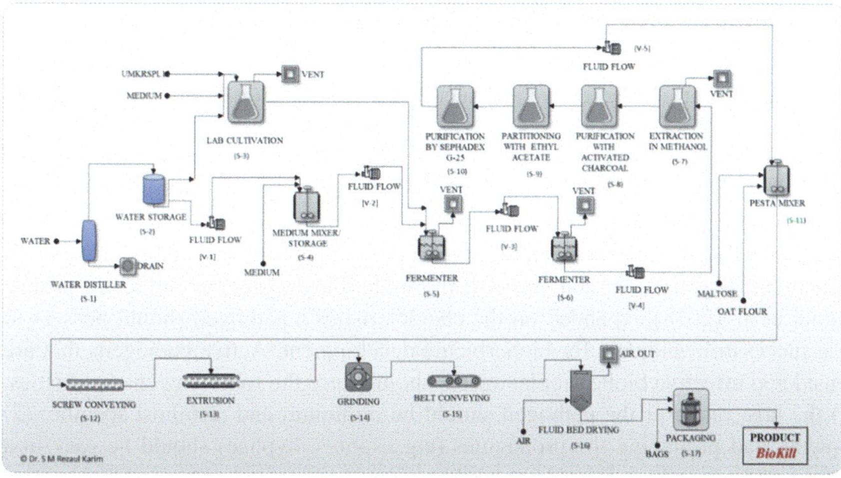

Fig. 16.1 Technological model for the production of bioherbicide from mycotoxins (modified from Mupondwa et al., 2015). S = Stages in production chain, V = Fluid flow control valve. BioKill is a proposed bioherbicide from the fungal isolate encoded as UMKRSPL1

(i) The fungus culture is isolated in a PDA medium in the laboratory (S-3), supported by a water supply from water storage (S-1, S-2).
(ii) Media broth is transferred to the fermenters (S-5), which receive Richard media (S-4) and water for the fermentation process.
(iii) The culture filtrate is transferred from the fermenters to the extraction and purification chambers (S-7 to S-10).
(iv) Extraction in methanol (S-7) is conducted followed by purification by activated charcoal (S-8).
(v) The partially purified filtrate undergoes partitioning with ethyl acetate (S-9).
(vi) Fractioning of the filtrate by Sephadex G-25 column (S-10) in the herbicide production chain.
(vii) This purified mycotoxin is the actual bioherbicide molecule in liquid formulation and can be used directly on the weed after being formulated properly. Specified doses can also be decided at this stage.
(viii) Granular formulation can be done by mixing the mycotoxin with oat flour and maltose (ratio of 3.3:1) in a pasta mixer (S-11).
(ix) The pasta dough is fed into a twin-screw extruder and granule cutter (S-12, S-13, S-14, S-15), where it is cut into granules and then dried properly in a fluid bed drying system (S-16).
(x) The end product is packed in airtight packets (S-17) for marketing and distribution.

The methods of bioherbicide production from different sources are described as follows.

5.4 Bioherbicides from Fungal Spores or Conidia

The development of an effective and successful bioherbicide from microbes should consider the following as suggested by Bedi et al. (2002): (i) if the propagules of the pathogen can be generated on low-cost substrates, (ii) if the inoculum requires extensive processing before application, (iii) if the product is deemed viable after production for a long time, (iv) if the product needs to compete with a well-established chemical herbicide, (v) whether the market size is enough to recover the cost for development and registration, and (vi) will the bioherbicide generate sufficient profits. Kumar et al. (2018) expanded on the characteristics a pathogen should possess to be a successful candidate for bioherbicide development. Active pathogens that are considered effective bioherbicidal agents should have the following characteristics: (1) the host range of the pathogen should be minimum and also host specific; (2) growing and producing the propagules (e.g., spores, hyphae) should be easy in a simple media to reduce cost; (3) it should be more virulent to have successful control of targeted host/weed; (4) it must have a strong destructive power on the host weed; (5) it should have less efficient dispersion mechanisms; (6) the product must be operative under varied conditions; (7) the inoculum can be produced abundantly under traditional liquid or solid state fermentation; (8) the effect of the product should be long-lasting; (9) it should have a longer shelf-life; (10) the application of formulation should be well-matched with the current agricultural system; (11) the product must be genetically stable; and (12) the pathogen must ensure perfect and graceful weed control.

The success of a bioherbicide development depends mainly on the technique of mass production of causing units, i.e., spores or conidia of the fungus. In their studies, Moraes et al. (2011) compared two spore production techniques of the fungus, *Bipolaris europhorbiae,* in solid and liquid media cultures to control the *Euphorbia heterophylla* L. weed in soybean fields. The victory of the most successful commercial bioherbicide with *Colletotrichum gloesporioides f. sp. aeschynomene* was because of low production cost, relatively simple formulation, and quick and secondary spread in the field.

5.5 Bioherbicide from Mycotoxins

The second category of bioherbicide production is to produce, isolate and formulate mycotoxins. This technique produces the mycotoxins through aerobic or anaerobic fermentation of a selected fungus in an appropriate growth medium before placing them in a bioreactor. The fermentation broths are collected, purified, and fractioned with appropriate solvents. Then, the toxins are identified using High-Performance Liquid Chromatography (HPLC) and Gas Chromatography-Mass Spectrometry (GC–MS). The isolated mycotoxins are then tested in the laboratory, greenhouse, and in fields before being formulated in preparation for commercial use. An example of this kind of bioherbicide preparation is mentioned in the study by Brun et al. (2016) involving *Phoma sp.* The formulated bioherbicide was observed to inhibit the germination of the test weed species completely.

5.6 Bioherbicide from Viruses

Viruses are also a good source for developing potential bioherbicides. A successful plant virus-based bioherbicidal unit is the tobacco mild green mosaic virus (TMGMV), which was used to control the tropical soda apple (*Solanum viarum* Dunal) weed (Charudattan, 2023). It is a problematic weed found in pastures and non-cropped areas in the southeastern and mid-southern areas of the USA. For the formulation of the TMGMV nanoparticles, an organosilicon adjuvant is needed, and a carborundum is added to assist as an abrasive and help the virus to enter the plant (Ferrell et al., 2008).

5.7 Bioherbicide from Phytocompounds

The fourth category of bioherbicides is plant-based bioherbicides, usually developed by extracting the allelochemicals present in selected allelopathic plants, weeds, or crops using the most efficient solvent (Motmainna et al., 2021e). The solution is filtrated using an appropriate filter paper or cheese cloth layer. The filtrate is then centrifuged in a centrifuge machine for 10 to 15 min and again filtrated. The filtrate is cleaned by passing through a layer of activated charcoals or other cleaning media. The filtrate materials are evaporated in a vacuum evaporator at 40 °C until the material is converted to dry biomass. The dry biomass is then diluted with distilled water to different concentrations and tested against different weed species. The most phytotoxic compound is then isolated and identified following appropriate chromatographic techniques (e.g., HPLC, GC–MS). After identifying the effective chemical, the chemical is purchased from the market and appropriately formulated for terminating weeds.

The metagenomic technique is also used to isolate herbicidal compounds. First, the DNA fragments are collected from the soil, and the genes are cloned in the appropriate vectors, which can produce phytotoxic compounds in the host plants (Kato-Noguchi & Peters, 2013).

6 Formulation of Bioherbicides

Bioherbicide formulations are purposefully developed materials containing the spores or conidia of the pathogen, viruses, or phytochemicals extracted from a plant as an active ingredient, and other ancillary components such as surfactants, adjuvants, preserving agents, wetting agents, and inert fillers, which are used for ease in handling, and application, storage, and transportation, increasing shelf-life and efficacy. The commonly used formulations are suspension concentrate (SC), wettable powder (WP), dry flowable, water-dispersible or wettable granules (WG), and non-disintegrable granules (G). The primary functions of formulation (especially for microbial bioherbicide) are to subject the weeds to infection by the pathogen present in the bioherbicide and to protect the infecting pathogen from environmental restraints.

6.1 Production of Solid Granular and Powder Formulation

A form of production procedure for granular formulated bioherbicides is described below.

The isolated fungal mycelia, conidia, or spores (culture filtrate) are mixed with a sodium alginate solution and other fillers, such as Kaolin clay. Then the solution is dripped into a solution of 0.25 M calcium chloride. When the $CaCl_2$ ions (Ca^{+2}) react with the sodium alginate, it forms gel beads. The beads are allowed to harden for several minutes in the calcium chloride solution (see Fig. 16.2). The hardened beads are collected, rinsed with water, and then air-dried. Different researchers used the alginate to prepare the bioherbicide in the form of biodegradable pellets in which fungi are entrapped. A good number of fungi namely *Alternaria cassia*, *Alternaria macrospora*, *Fusarium lateritium*, *Colletotrichum malvarum*, *Phyllosticta sp.*, *Fusarium solani,* and *Cercospora kikuchii,* have been used to formulate pellets (Chehri et al., 2015; de Almeida et al., 2020; Rovetto et al., 2023).

Another method of granular formulation is the mixing of semolina wheat flour with Kaolin clay and the pathogen in a liquid growth medium. The mixture is rubbed, forming a dough, then rolled with a pasta press to make thin sheets, and subsequently air-dried for 48 h. The sheets are then milled and sieved to produce uniform-sized granules (Boyette et al., 2007).

Solid formulation with a spray dryer is also a suitable technology for powder formulation of bioherbicides. In this technology, the bioherbicide molecules or pathogenic units (spore or conidia) are mixed with distilled water, adjuvants, and silicon, an inert ingredient. The component of the mixture is homogenized by stirring at 300 rpm for 15 min using a mechanical stirrer. The liquid form of the bioherbicide is then placed in the feeding system of a spray dryer (see Fig. 16.3). The spray drier is provided with a spray assembly made of stainless steel with two fluid nozzles. The function of the inner nozzle is to make a liquid jet, while the outer nozzle

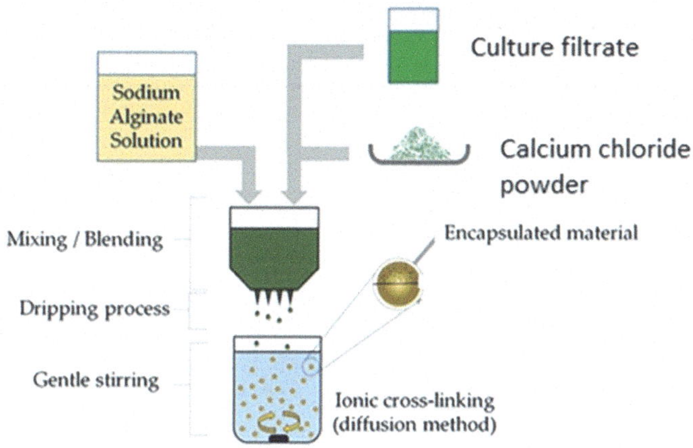

Fig. 16.2 Schematic diagram of mycoherbicide encapsulation (modified from da Silva et al., 2017)

develops the compressed air outlet. The spray assembly is also accompanied by a de-blocking device that controls the jet nozzle's blocking and is triggered by an integral compressor. The assembly also includes a feeding system, air heating system, drying chamber, inter fluid beds, cooling system, and powder sifter. The compressed air atomizes the liquid as it develops from the jet and finally forms the spray. The spray is carried out in a tower-shaped drying chamber with internal and external fluid beds where all the spays are deposited. Then, the hot air from the air heating system dries the sprays to form a powder. The atomizer is run under a pressure of 0.3 MPa and the air temperature at 125 °C. After drying the spray volumes properly, the yields of the dried solid are determined in comparison to the initial solid mass used.

6.2 Production of Liquid Formulation

The commonly used liquid formulations are water suspensions of spores containing a wetting agent (0.1% v/v), such as Tween 20, and silicone polyester, such as Silwet L-77. Wetting agents facilitate the spread of bioherbicides on foliar surfaces, while the copolymers aid in the penetration of microbial spores through stomata and lenticels. However, a simple aqueous suspension can also be used under ideal conditions for fungal infection in the fields (El-Baky & Amara, 2021). Some other materials are also used to decrease the loss of water, such as humectants (e.g., glycerol and sorbitol) (Poluektova et al., 2018). Complex polysaccharides and polymers like alginates, gums, and polyacrylamides are stickers in the formulations. formulation of bioherbicide (Source: Freund-Vector Corporation, USA; https://www.freund-vector.com/about/).

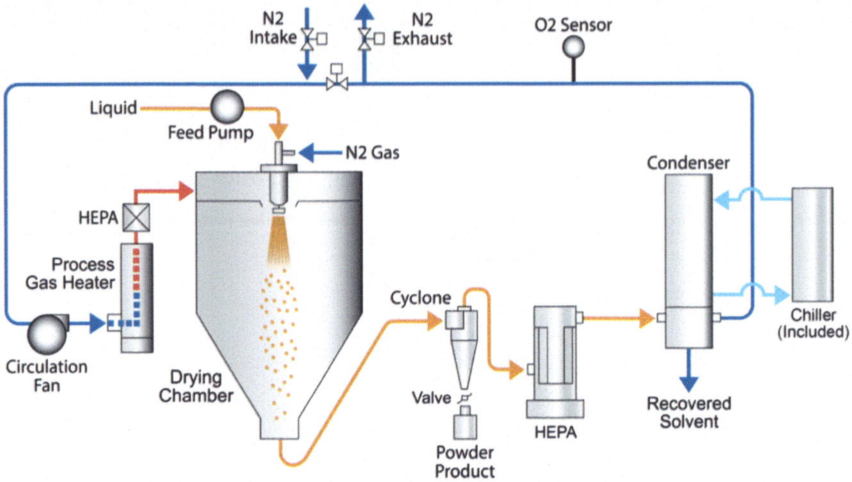

Fig. 16.3 Schematic diagram of Spray Drying System for producing powder

Another formulation is a vegetable oil emulsion containing 10% oil and 1% emulsifying agent. This type of emulsion is also modified by adding dextrose [1% (w/v)] and carboxymethyl cellulose [0.2% (w/v)] to the oil phase, which helps in reducing dew dependence. This formulation was used in controlling Bathurst burr (*Xanthium spinosum* L.) using *Colletotrichum orbiculare* (Auld, 1993). Zhang and Watson (1997) also used this modified emulsion with *Exserohilum monoceras* to control barnyard grasses (*Echinochloa* spp.) and achieved more than 90% mortality under simulated field conditions.

The third emulsion type is invert emulsion, with a continuous oil phase and water droplets. It helps reduce evaporation loss in bioherbicides, decrease dew requirements, and lessen the essential spore concentrations (Batta, 2016). However, it is expensive for its higher percentage of oil (>30%) and is very viscous. To solve this problem, special spraying equipment (e.g., air-assist nozzles) is used for effective application.

Some other ancillary materials are found helpful in the formulations for fungi-based bioherbicides, such as germination stimulants, e.g., iron chelators or nutrients. They help to reduce the period of susceptibility to low humidity (Zidack & Quimby, 1999). For better performance, some phytotoxins, enzymes, and enzyme inhibitors may need to be incorporated into the formulations (de Almeida et al., 2020).

Two well-known bioherbicides, DeVine and Collego, were successfully produced under submerged culture fermentation using bioreactors. The DeVine has been marketed as a liquid formulation with a short shelf-life, but in the case of Collego, they harvested the spores and dried them before packaging. Before application in the fields, the dried mass is rehydrated and sprayed.

6.3 *Production of Hot Foam Formulation*

Recently, a thermal-based weed control technology for hot foam application has been developed, where plant-based oil and sugar (e.g., alkyl polyglucoside surfactant) are used to generate a biodegradable foam that is heated in a boiler and applied to the weeds using a Foamstream MW Series machine (Klein, 2022). Hot foam formulation is a new concept of weed management by thermal weed control technology. In this technique, the foam insulates the weeds so the surrounding air cannot enter the weed body, transferring energy to the wanted plants. It is especially suitable in urban areas, e.g., to control weeds on pavements, roadside weeds, and more. It can also be used to control weeds in railways or weeds in wide-spaced cotton fields, olive groves, citrus groves, and orchards (Martelloni et al., 2019). The technique is effective, especially for controlling a number of broadleaf weeds, such as *Festuca arundinacea* Schreb., *Taraxacum officinale* (L.) Weber ex F.H.Wigg., and *Plantago lanceolata* L., which are difficult to control by other methods but can easily be managed by this technique.

7 Limitations/Constraints in Using Bioherbicides

The limitations of bioherbicide usage can be categorized into three constraints as outlined below:

7.1 *Biological Constraints*

Host inconsistency, host range resistance mechanism, and interactions with other microbes are the significant constraints that affect bioherbicide efficacy. As a result, most bioherbicides may deliver limited weed control, even under model environmental conditions. This discrimination in the host range is associated with the pathogen's bi-physiology and host variability and resistance. In order to address the issue, fungal pathogens having different strains with pathogenicity against several weeds should be developed through selective screening, genetic recombination, or hybridization (Kremer, 2005).

7.2 *Environmental Constraints*

The epidemiology of the causal organisms in bioherbicides and the performance of bioherbicide formulation depend on environmental conditions, particularly temperature and humidity, as the inoculum production depends on the sporulation of the microbes. A dew period of more than 12 h is required for better infection by a pathogen of several potential bioherbicides (Harding & Raizada, 2015). Similarly, the effectiveness of *C. orbiculare* for controlling *X. spinosum* is limited by high temperature after inoculating the weed (Chakraborty & Ray, 2021), limiting the application.

7.3 Technological Constraints

To ensure a successful venture of bioherbicides, the market size should be sufficiently bigger, which is not available in some areas. Patent protection, maintaining secrecy and developing regulations and acts regarding an effective bioherbicide require lots of effort and financial involvement (Harsha et al., 2018). Another critical constraint is the reduction of shelf-life, which may occur during the selection of pathogen strains, formulation methods, and the interaction between these parameters (Kostina-Bednarz et al., 2023). Testing compatibility between the formulation components (e.g., sunscreens, humectants, and starches) can take more time with increased costs (Geoffrey et al., 2019).

8 Future Directions

Despite recent advances in the study of bioherbicides, there are still many opportunities for researchers to investigate new solutions and improve the effectiveness of the existing and marketed bioherbicides. More emphasis should be made on isolating and identifying new pathogens or allelopathic substances. The bioactivity of the identified chemicals or pathogens should be thoroughly examined under laboratory and field conditions. If satisfied, those pathogens or compounds can be used as a tool for new natural herbicide development. More importance to bioherbicide formulation should be given in future research, which is a critical factor influencing bioherbicide efficacy. In particular, new formulation techniques via multiple surfactants are required to improve the penetration and absorption of bioherbicidal compounds. Bioherbicidal compounds are challenging to utilize as a potential herbicide in the field because of their quick degradation and comparatively higher costs. To address these issues, the chemical industries can design effective formulations and generate relevant bioherbicidal compounds as it is more convenient than acquiring them naturally. Several chemical and pesticide companies worldwide are coming forward to develop and introduce bioherbicides globally. To date, companies such as Emery Oleochemicals (Malaysia), Verdesian Life Sciences (USA), Marrone Bio Innovations Inc. (USA), Bioherbicides Australia Pty Ltd. (Australia), Certis USA (USA), Premier Tech (Canada), Plant Advanced Technologist (France), Koppert (Netherlands), Philombios (Canada), Upjohn (USA), Mycogen (USA), Tifton Innovation Corporation (USA), and Abbott (USA) have launched several bioherbicides in the global market.

References

Abraham, E., John, J., & Pillai, P. S. (2016). Allelopathic effect of leaf loppings of homestead trees on ginger (*Zingiber officinale* Roscoe). *Journal of Tropical Agriculture, 54*, 60.

Anese, S., Jatoba, L. J., Grisi, P. U., Gualtieri, S. C. J., Santos, M. F. C., & Berlinck, R. G. D. S. (2015). Bioherbicidal activity of drimane sesquiterpenes from *Drimys brasiliensis* Miers roots. *Industrial Crops Production, 74*, 28–35.

Auld, B. A. (1993). Vegetable oil suspension emulsions reduce dew dependence of a mycoherbicide. *Crop Protection, 12*, 477–479.

Bailey, K. L. (2014). The bioherbicide approach to weed control using plant pathogens. *Integrated pest management* (pp. 245–266). Academic Press.

Bailey, K. L., Pitt, W. M., Leggett, F., Sheedy, C., & Derby, J. (2011). Determining the infection process of *Phoma macrostoma* that leads to bioherbicidal activity on broadleaved weeds. *Biological Control, 59*(2), 268–276.

Bashar, H. K., Juraimi, A. S., Ahmad-Hamdani, M. S., Uddin, M. K., Asib, N., Anwar, M. P., Karim, S. M. R., Rahaman, F., Haque, M. A., & Hossain, A. (2022). Documentation of phytotoxic compounds existing in *Parthenium hysterophorus* L. leaf and their phytotoxicity on *Eleusine indica* (L.) Gaertn. and *Digitaria sanguinalis* (L.) Scop. *Toxins, 14*, 561.

Batta, Y. A. (2016). Invert emulsion: Method of preparation and application as proper formulation of entomopathogenic fungi. *MethodsX, 3*, 119–127.

Bedi, J. S., Muller-Stover, D., & Sauerborn, J. (2002). Status of plant pathogens for the biological control of weeds. In O. Koul, G. S. Dhaliwal, S. S. Marwaha, & J. K. Arora (Eds.), *Biopesticides and pest management* (pp. 316–331). Campus Books International.

Boydston, R. A., Morra, M. J., Borek, V., Clayton, L., & Vaughn, S. F. (2011). Onion and weed response to mustard (*Sinapis alba*) seed meal. *Weed Science, 59*, 546–552.

Boyette, C. D., Hoagland, R. E., & Abbas, H. K. (2007). Evaluation of the Bioherbicide Myrothecium verrucaria for weed control in tomato (*Lycopersicon esculentum*). *Biocontrol Science Technology, 17*(2), 171–178.

Brun, T., Rabuske, J. E., Todero, I., Almeida, T. C., Junior, J. J. D., Ariotti, G., Confortin, T., Arnemann, J. A., Kuhn, R. C., Guedes, J. V. C., & Mazutti, M. A. (2016). Production of bioherbicide by *Phoma sp.* in a stirred-tank bioreactor. *Biotech, 6*, 230–238.

Charudattan, R. (2023). Use of plant viruses as bioherbicides: The first virus-based bioherbicide and future opportunities. Pest Management Science.

Chakraborty, A., & Ray, P. (2021). Mycoherbicides for the noxious meddlesome: Can colletotrichum be a budding candidate? *Frontiers in Microbiology, 12*, 754048.

Chehri, K., Salleh, B., & Zakaria, L. (2015). Morphological and phylogenetic analysis of *Fusarium solani* species complex in Malaysia. *Microbial Ecology, 69*, 457–471.

Chen, L., Li, J., Zhu, Y., Guo, L., Ji, R., Miao, Y., Guo, L., Du, H., & Liu, D. (2021). Caffeic acid, an allelochemical in *Artemisia argyi*, inhibits weed growth via suppression of mitogen-activated protein kinase signaling pathway and the biosynthesis of gibberellin and phytoalexin. *Frontiers in Plant Science, 12*, 1–15.

da Silva, T. L., Vidart, J. M. M., Silva, M. G. C., Gimenes, M. L., & Vieira, M. G. A. (2017). Alginate and sericin: Environmental and pharmaceutical applications. In *Biological activities and application of marine polysaccharides*. Published online by InTechOpen.

Dar, S. A., Wani, S. H., Showkat, M. S. H. A., Dolkar, T., & Dawa, T. (2021). Biopesticides: Mode of action, efficacy and scope in pest management. *Journal Advantage Research Biochemical Pharmacology, 4*, 1–8.

de Almeida, T. C., Spannemberg, S. S., Brun, T., Schmaltz, S., Escobar, O., Sanchotene, D. M., & Mazutti, M. A. (2020). Development of a solid bioherbicide formulation by spray drying technology. *Agriculture, 10*(6), 215.

El-Baky, N. A., & Amara, A. A. A. F. (2021). Recent approaches towards control of fungal diseases in plants: An updated review. *Journal of Fungi, 7*(11), 900.

Fenibo, E. O., Ijoma, G. N., & Matambo, T. (2021). Biopesticides in sustainable agriculture: A critical sustainable development driver governed by green chemistry principles. *Front Sustainable Food System, 5*, 619058.

Ferrell, J., Charudattan, R., Elliott, M., & Hiebert, E. (2008). Effects of selected herbicides on the efficacy of tobacco mild green mosaic virus to control tropical soda apple (*Solanum viarum*). *Weed Science, 56*, 128–132.

Geoffrey, K., Mwangi, A. N., & Maru, S. M. (2019). Sunscreen products: Rationale for use, formulation development and regulatory considerations. *Saudi Pharmaceutical Journal, 27*(7), 1009–1018.

Harding, D. P., & Raizada, M. N. (2015). Controlling weeds with fungi, bacteria and viruses: A review. *Frontiers in Plant Science, 6*, 659–664.

Harsha, K. N., Nitish, B., & Rana, S. S. (2018). Bioherbicdes. www.researchgate.net/publication/322851277

Hasan, M., Ahmad-Hamdani, M. S., Rosli, A. M., & Hamdan, H. (2021a). Bioherbicides: An eco-friendly tool for sustainable weed management. *Plants, 10*, 1212.

Hasan, M., Mokhtar, A. S., Rosli, A. M., Hamdan, H., Motmainna, M., & Ahmad-Hamdani, M. S. (2021b). Weed control efficacy and crop-weed selectivity of a new bioherbicide weedLock. *Agronomy, 11*, 1488.

Hasan, M., Mokhtar, A. S., Mahmud, K., Berahim, Z., Rosli, A. M., Hamdan, H., et al. (2022). Physiological and biochemical responses of selected weed and crop species to the plant-based bioherbicide WeedLock. *Science and Reports, 12*(1), 19602.

Hasan, M., Mokhtar, A. S., Mahmud, K., Rosli, A. M., Hamdan, H., Motmainna, M., et al. (2023). Weed control efficacy and soil activity of a new promising bioherbicide "WeedLock." *Sains Malaysia, 58*(8), 2223–2233.

Hintz, W. (2007). Development of *Chondrostereum purpureum* as a mycoherbicide for deciduous brush control. In *Biological control: A global perspective* (pp. 284–290). CAB International

Hossen, K., Ozaki, K., Teruya, T., Kato-Noguchi, H. (2021). Three active phytotoxic compounds from the leaves of *Albizia richardiana* (Voigt.) king and prain for the development of bioherbicides to control weeds. *Cells, 10*, 2385.

Hubbard, M., Hynes, R. K., & Bailey, K. L. (2015). Impact of macrocidins, produced by *Phoma macrostoma*, on carotenoidprofiles of plants. *Biological Control, 89*, 11–22.

Jiang, S. J., Qiang, S., Zhu, Y. Z., & Dong, Y. F. (2008). Isolation and phytotoxicity of a metabolite from *Curvularia eragrostidis* and characterisation of its modes of action. *The Annals of Applied Biology, 152*, 103–111.

Kato-Noguchi, H., & Peters, R. J. (2013). The role of momilactones in rice allelopathy. *Journal of Chemical Ecology, 39*(2), 175–185.

Kim, E. Y., Choi, Y. H., Lee, J. I., Kim, I. H., & Nam, T. J. (2015). Antioxidant activity of oxygen evolving enhancer protein 1 purified from *Capsosiphon fulvescens*. *Journal of Food Science, 80*(6), H1412–H1417.

Klein, A. (2022). Plant-based hot foam kills weeds as effectively as chemical spray. NewScientist, 20 September 2022. https://www.newscientist.com/article/2338128

Kostina-Bednarz, M., Płonka, J., & Barchanska, H. (2023). Allelopathy as a source of bioherbicides: Challenges and prospects for sustainable agriculture. *Review Environmental Science*, 1–34.

Kremer, R. J. (2005). The role of bioherbicides in weed management. *Biopestic International, 1*, 127–141.

Kumar, V., Aggarwal, N. K., & Malik, A. (2018). Bioherbicidal concept: A novel strategy to control weeds. In J. Singh, et al. (Eds.), *Microbial bioprospecting for sustainable development.* Springer Nature Singapore Pte Ltd.

Lederer, B., Fujimori, T., Tsujino, Y., Wakabayashi, K., & Booger, P. (2004). Phytotoxic activity of middle-chain fatty acids II: Peroxidation and membrane effects. *Pesticide Biochemistry and Physiology, 80*, 151–156.

Lee, S. M., Radhakrishnan, R., Kang, S. M., Kim, J. H., Lee, I. Y., Moon, B. K., Yoon, B. W., & Lee, I. J. (2015). Phytotoxic mechanisms of bur cucumber seed extracts on lettuce with special reference to analysis of chloroplast proteins, phytohormones, and nutritional elements. *Ecotoxicology and Environmental Safety, 122*, 230–237.

Liu, D. L., & Christians, N. E. (1997). The use of hydrolyzed corn gluten meal as a natural preemergence weed control in turf. *International Turfgrass Social Resources Journal, 8*, 1043–1050.

Martelloni, L., Frasconi, C., Sportelli, M., Fontanelli, M., Raffaelli, M., & Peruzzi, A. (2019). The use of different hot foam doses for weeds control. *Agronomy, 9*, 490.

Moraes, C., Monteiro, A. C., & Mochi, D. A. (2011). Reaction of *Bipolaris euphorbiae* to environmental factors and its tolerance to solar and ultraviolet radiation. *World Journal of Microbiology and Biotechnology, 27*(6), 1449–1456.

Motmainna, M., Juraimi, A. S., Uddin, M. K., Asib, N. B., Islam, A. K. M. M., & Hasan, M. (2021a). Allelopathic potential of Malaysian invasive weed species on weedy rice (*Oryza sativa* f. *spontanea* Roshev). *Allelopathy of Journal, 53*, 53–68.

Motmainna, M., Juraimi, A. S., Uddin, M. K., Asib, N. B., Islam, A. K. M. M., & Hasan, M. (2021b). Assessment of allelopathic compounds to develop new natural herbicides: A review. *Allelopathy of Journal, 52*(1), 21–40.

Motmainna, M., Juraimi, A. S., Uddin, M. K., Asib, N. B., Islam, A. K. M. M., & Hasan, M. (2021c). Bioherbicidal properties of *Parthenium hysterophorus*, *Cleome rutidosperma* and *Borreria alata* extracts on selected crop and weed species. *Agronomy, 11*(4), 643.

Motmainna, M., Juraimi, A. S., Uddin, M. K., Asib, N. B., Islam, A. K. M. M., Ahmad Hamdani, M. S., Berahim, Z., & Hasan, M. (2021d). Physiological and biochemical responses of *Ageratum conyzoides*, *Oryza sativa* f. *spontanea* (weedy rice) and *Cyperus iria* to *Parthenium hysterophorus* methanol extract. *Plants, 10*(6), 1205.

Motmainna, M., Juraimi, A. S., Uddin, M. K., Asib, N. B., Islam, A. K. M. M., Ahmad Hamdani, M. S., & Hasan, M. (2021e). Phytochemical constituents and allelopathic potential of *Parthenium hysterophorus* L. in comparison to commercial herbicides to control weeds. *Plants, 10*(7), 1445.

Motmainna, M., Juraimi, A. S., Ahmad-Hamdani, M. S., Hasan, M., Yeasmin, S., Anwar, M. P., & Islam, A. K. M. M. (2023). Allelopathic potential of tropical plants—A review. *Agronomy, 13*(8), 2063.

Muñoz, M., Torres-Pagán, N., Peiró, R., Guijarro, R., Sánchez-Moreiras, A. M., & Verdeguer, M. (2020). Phytotoxic effects of three natural compounds: Pelargonic acid, carvacrol, and cinnamic aldehyde, against problematic weeds in Mediterranean crops. *Agronomy, 10*, 791.

Mupondwa, E., Li, X., Boyetchko, S., Hynes, R., & Geissler, J. (2015). Technoeconomic analysis of large-scale production of pre-emergent Pseudomonas fluorescens microbial bioherbicide in Canada. *Bioresource Technology, 175*, 517–528.

Netland, J., Dutton, L. C., Greaves, M. P., Baldwin, M., Vurro, M., Evidente, A., Einhorn, G., Scheepens, P. C., & French, L. W. (2001). Biological control of *Chenopodium album* L. in Europe. *Biological Control, 46*, 175–196.

Park, J. M., Radhakrishnan, R., Kang, S. M., & Lee, I. J. (2015). IAA producing Enterobacter sp. I-3 as a potent bio-herbicide candidate for weed control: A special reference with lettuce growth inhibition. *Indian Journal of Microbiology, 55*(2), 207–212.

Poluektova, E., Tokarev, Y., Sokornova, S., Chisty, L., Evidente, A., & Berestetskiy, A. (2018). Curvulin and phaeosphaeride a from paraphoma sp. VIZR 1.46 isolated from cirsium arvense as potential herbicides. *Molecules, 23*(11), 2795.

Radhakrishnan, R., Alqarawi, A. A., & Abd-Allah, E. F. (2018). Bioherbicides: Current knowledge on weed control mechanism. *Ecotoxicology Environmental Safety, 158*, 131–138.

Radhakrishnan, R., & Lee, I. J. (2014). Effect of low dose of spermidine on physiological changes in salt stressed cucumber plants. *Russian Journal of Plant Physiology, 61*, 90–96.

Radhakrishnan, R., Khan, A. L., & Lee, I. J. (2013). Endophytic fungal pre-treatments to seeds alleviate salinity stress effects in soybean plants. *Journal of Microbiology, 51*(6), 850–857.

Rovetto, E. I., Luz, C., La Spada, F., Meca, G., Riolo, M., & Cacciola, S. O. (2023). Diversity of mycotoxins and other secondary metabolites recovered from blood oranges infected by *Colletotrichum*, *Alternaria*, and *Penicillium* Species. *Toxins, 15*(7), 407.

Šćepanović, M., Košćak, L., Šoštarčić, V., Pismarović, L., Milanović-Litre, A., & Kljak, K. (2022). Selected phenolic acids inhibit the initial growth of *Ambrosia artemisiifolia* L. *Biology, 11*(4), 482.

Snyder, A., Morra, M. J., Johnson-Maynard, J., & Thill, D. C. (2009). Seed meals from brassicaceae oilseed crops as soil amendments: Influence on carrot growth, microbial biomass nitrogen, and nitrogen mineralization. *Horticultural Science, 44*, 354–361.

Sousa, C. P. D., Farias, M. E. D., Shock, A. A., & Bacarin, M. A. (2014). Photosynthesis of soybean under the action of a photosystem II-inhibiting herbicide. *Acta Plant Physiology, 36*, 3051–3062.

Tessmann, D. J., Charudattan, R., & Preston, J. F. (2008). Variability in aggressiveness, cultural characteristics, cercosporin production and fatty acid profile of *Cercospora piaropi*, a biocontrol agent of water hyacinth. *Plant Pathology, 57*, 957–966.

Tigre, R. C., Silva, N. H., Santos, M. G., Honda, N. K., Falcao, E. P. S., & Pereira, E. C. (2012). Allelopathic and bioherbicidal potential of *Cladonia verticillaris* on the germination and growth of *Lactuca sativa*. *Ecotoxicology and Environmental Safety, 84*, 125–132.

Travlos, I., Rapti, E., Gazoulis, I., Kanatas, P., Tataridas, A., Kakabouki, I., & Papastylianou, P. (2020). The herbicidal potential of different pelargonic acid products and essential oils against several important weed species. *Agronomy, 10*, 1687.

Verdeguer, M., Sánchez-Moreiras, A. M., & Araniti, F. (2020). Phytotoxic effects and mechanism of action of essential oils and terpenoids. *Plants, 9*, 1571.

Weng, X., Li, H., Ren, C., Zhou, Y., Zhu, W., Zhang, S., et al. (2022). Calcium regulates growth and nutrient absorption in poplar seedlings. *Frontiers in Plant Science, 13*, 887098.

Zhang, W., & Watson, A. K. (1997). Efficacy of *Exserohilum monoceras* for the control of *Echinochloa* species in rice (*Oryza sativa*). *Weed Science, 45*, 144–150.

Zidack, N. K., & Quimby, P. C. (1999). Formulation and application of plant pathogens for biological weed control. In F. R. Hall, & J. J. Menn (Eds.), *Methods in biotechnology* (Vol. 5, pp. 371–381). Biopesticides: Use and Delivery, Humana Press, Totowa.

Epilogue

As we end this comprehensive journey through the realm of "Advances in Tropical Crop Protection," we reflect on the transformative paces made in safeguarding the lifelines of tropical agriculture. This scientific text has endeavoured to capture the essence of the dynamic field of crop protection, exploring innovative methodologies, emerging technologies, and sustainable practices that promise to reshape the landscape of tropical crop management.

In the chapters preceding this epilogue, we have delved into the intricacies of pest and disease management, scrutinized the interplay of environmental factors, and celebrated the triumphs of cutting-edge research and technological breakthroughs. Our exploration has not only illuminated the challenges faced by tropical crops but also illuminated the promising pathways toward resilient and sustainable agriculture in the tropics.

The collective efforts of researchers, scientists, practitioners, and policymakers have borne fruit in the form of novel approaches to crop protection. Furthermore, sustainable practices such as agroecology, integrated pest management (IPM), and organic farming have gained prominence, aligning our endeavours with the broader goals of environmental conservation and food security. As we adapt to the challenges posed by climate change, these sustainable approaches will play a pivotal role in ensuring the resilience and adaptability of tropical crops.

The significance of interdisciplinary collaboration cannot be overstated in the context of tropical crop protection. This epilogue serves as a reminder that our journey does not conclude with the final page of this text. The pursuit of knowledge is an ongoing process, and the scientific community must remain dynamic, adaptive, and collaborative in the face of evolving challenges.

M. Wong (ed.), *Advances in Tropical Crop Protection*,
https://doi.org/10.1007/978-3-031-59268-3

In closing, “Advances in Tropical Crop Protection” is more than a compilation of facts and theories; it is a testament to the collective intelligence and dedication of the global scientific community. As we turn the page from this epilogue, let it be a catalyst for continued exploration, innovation, and collaboration in the pursuit of a sustainable and secure future for tropical agriculture.

Nusaibah Syd Ali
Serdang, Malaysia

The manufacturer's authorised representative in the EU is Springer Nature Customer Service Centre GmbH, Europaplatz 3, 69115 Heidelberg, Germany. If you have any concerns regarding our products, please contact ProductSafety@springernature.com

Printed and bound by CPI Group (UK) Ltd, Croydon, CR0 4YY
01/10/2026
02225862-0009